PRAISE FOR

THE INFINITY PUZZLE

"A fascinating book. . . . Where Close really shines is in exposing the fraught process of recognition in science. . . . Close's history of the field is engaging and gives insight into how great theories are created." —*Nature*

"A great read." —*Discovery News*

"Written with pellucid prose, a keen eye for salient details, a talent for the illuminating metaphor, a passion for the topic, and a novelist's gift for portraiture, narrative, and suspense." —*Barnes & Noble Review*

"Eminently readable . . . a very manageable introduction to quantum physics for those who are interested in, but possibly intimidated by, understanding the inner workings of the fabric of our Universe." —*Ars Technica*

"[An] intriguing tale." —*American Scientist*

"Serious physics-history buffs . . . will find *The Infinity Puzzle* invaluable." —*Physics World*

"A masterpiece. . . . This book is essential reading—I never normally give five stars, but for this I'll make an exception." —John Gribbin, *BBC Focus*

"Brims with charming anecdotes about particle physics between the 1950s and 1980s, when breakthroughs came almost too fast to be comprehended and every scientist seemed to be maneuvering (and occasionally begging) for Nobel prizes." —*The American Scholar*

"Superb." —*Mathematical Association of America Reviews*

"An engrossing history." —*Publishers Weekly*

"A wonderfully written book." —*CHOICE*

"Lively." —*Kirkus Reviews*

"Absolutely fascinating. . . . It's a very rich and very human story, clearly and engagingly told. . . . If you're interested in what we *know* to be true about the universe and how it works, and how we put that knowledge together, I highly recommend this book." —Chad Orzel, Uncertain Principles

"Gripping. . . . Close has succeeded in humanising a dramatic era of physics in what is my science book of the year. . . . 'Hold Infinity in the palm of your hand,' William Blake wrote in the *Auguries of Innocence.* Frank Close does a fabulous job of reconstructing how physicists like Feynman and 't Hooft managed to do exactly that." —Manjit Kumar, author of *Quantum*

"A fascinating new book . . . go out and get a copy soon." —Peter Woit, *Not Even Wrong*

"It is a pleasure to read a book on recent advances in our understanding of the structure of matter by an author who not only understands the subject but also takes care to investigate conflicting accounts of how these advances came about." —Peter Higgs, Emeritus Professor of Physics, The University of Edinburgh

"Close belongs among the very first rank of scientist-authors. I strongly recommend *The Infinity Puzzle.*" —Steve Nadis, coauthor of *The Shape of Inner Space*

"Anyone who wants to understand why we built the LHC and what we hope to learn from it should read this book." —Dan Hooper, Fermi National Accelerator Laboratory and University of Chicago; author of *Dark Cosmos* and *Nature's Blueprint*

THE INFINITY PUZZLE

ALSO BY FRANK CLOSE

Neutrino

Antimatter

The Void

The New Cosmic Onion: Quarks and the Nature of the Universe

Particle Physics: A Very Short Introduction

Lucifer's Legacy: The Meaning of Asymmetry

THE INFINITY PUZZLE

QUANTUM FIELD THEORY AND THE HUNT FOR AN ORDERLY UNIVERSE

Frank Close

BASIC BOOKS
A MEMBER OF THE PERSEUS BOOKS GROUP
New York

Hardcover first published in 2011 by Basic Books,
A Member of the Perseus Books Group
Paperback first published in 2013 by Basic Books

Designed by Timm Bryson

The Library of Congress has cataloged the hardcover edition as follows:
Close, F. E.
The Infinity Puzzle : quantum field theory and the hunt for an orderly universe / Frank Close.
p. cm.
Includes bibliographical references and index.
ISBN 978-0-465-02144-4 (hardback) — ISBN 978-0-465-02803-0 (e-book)
1. Quantum field theory. 2. Higgs bosons. 3. Infinite. I. Title.
QC174.45.C56 2011
530.14'3—dc23
2011022966

ISBN 978-0-465-06382-6 (paperback)

10 9 8 7 6 5 4 3 2 1

FOR MAX AND JACK

Whose appearance out of The Void
Is an Infinite Puzzle

Old men forget. . . . But he'll remember with advantages,
what feats he did that day.

SHAKESPEARE, *HENRY V*

CONTENTS

Acknowledgments, xi

Prologue: Amsterdam, 1971 1

PART 1
GENESIS

1 The Point of Infinity 17
2 Shelter Island and QED 33
3 Feynman, Schwinger, . . . and Tomonaga (and Dyson) 55

INTERMISSION: *1950* *65*

4 Abdus Salam: A Strong Beginning 67
5 Yang, Mills, . . . and Shaw 77
6 The Identity of John Ward 93
7 The Marriage of Weak and Electromagnetic Forces—to 1964 107

INTERMISSION: *1960* *125*

8 Broken Symmetries 127

9 "The Boson That Has Been Named After Me," a.k.a. the Higgs Boson 151

INTERMISSION: *Mid-1960s* *183*

10 1967: From Kibble to Salam and Weinberg 185

11 "And Now I Introduce Mr. 't Hooft" 203

INTERMISSION: *Early 1970s* *229*

PART 2

REVELATION

12 B. J. and the Cosmic Quarks 233

13 A Comedy of Errors 257

INTERMISSION: *1975* *281*

14 Heavy Light 283

15 Warmly Admired, Richly Deserved 295

16 The Big Machine 313

INTERMISSION: *End of the Twentieth Century* 333

17 To Infinity and Beyond 335

Epilogue: The Bonfire of the Infinities 353

Postscript 371

Glossary, 373

Notes, 379

Bibliography, 429

Index, 431

ACKNOWLEDGMENTS

I am indebted to many people who gave their time for interviews, answered queries by correspondence, read versions of the script, or provided access to archive material.

Anne Barrett, Imperial College London Archives; John Hillsdon and Bodleian Library staff, Oxford University; Louise Johnson, for abstracts from Abdus Salam's diaries; and Lucio Visintin, ICTP Library, and the ICTP Directorate, gave invaluable help to me in obtaining original records, especially concerning Abdus Salam. Shelley Erwin at the Cal Tech Archives found Feynman's original notes from 1968, and Tini Veltman produced a copy of Steven Weinberg's manuscript that even Weinberg himself thought had been lost. Peter Higgs produced copies of his diaries, and Erick Weinberg made his Ph.D. thesis available, both of which helped to establish some chronology of events from forty years ago, in the midst of often confusing memories.

In addition I have interviewed or had correspondence with: Henry Abarbanel, Ian Aitchison, Tom Appelquist, David Bailin, Michael Birse, James "bj" Bjorken, Elliott Bloom, David Boulware, Stan Brodsky, Robert Brout, Hugh Burkhardt, Nicola Cabibbo, John Cardy, John Charap, Geoffrey Chew, Bill Colglazier, Bob Delbourgo, Norman Dombey, Hans-Peter Dürr, Gosta Ekspong, John Ellis, François Englert, Graham Farmelo, Tom Ferbel, Michael Fisher, Gordon Fraser, Jerome Friedman, Mary K. Gaillard, Fred Gilman, Shelly Glashow, Nigel Glover, Terry Goldman, Jeffrey Goldstone, Gerald Guralnik, Dick Hagen, Andre Hassende, Tony Hey,

Peter Higgs, Gerard 't Hooft, Chris Isham, Roman Jackiw, J. David Jackson, Cecilia Jarlskog, Bob Johnson, Louise Johnson, Marek Karliner, Tom Kibble, Andy Kirk, Peter Knight, Chris Korthals-Altes, Chris Llewellyn Smith, Giuseppe Mussardo, Lev Okun, Giorgio Parisi, Manny Paschos, Ken Peach, Don Perkins, David Politzer, John Polkinghorne, Sacha Polyakov, Chris Quigg, Mike Riordan, Dick Roberts, Graham Ross, Ian Sample, Ron Shaw, Andrew Steane, John Strathdee, Ray Streater, John C. Taylor, Richard (Dick) Taylor, Tini Veltman, Alan Walker, Erick Weinberg, Steven Weinberg, Frank Wilczek, Tony Zee, Nino Zichichi, Bruno Zumino, and George Zweig.

Ian Aitchison and Michael Marten read preliminary versions of the whole manuscript and made so many comments that my first draft bore their stamp no less than my own. I am indebted to my editor, T. J. Kelleher, whose transatlantic phone calls helped me to forge the final version, and discussions with Latha Menon and Emma Marchant at OUP, which involved a 5 minute walk from my office. Annette Wenda, Melissa Veronesi, and Paul Beverley went through the final manuscript very thoroughly and made several editorial suggestions. The energy and enthusiasm of my agent, Patrick Walsh, has helped this project throughout. I am deeply grateful for the support and help of my family during the years of research and writing. Finally, I invite all readers to examine the postscript and, if possible, to help add to the accuracy of this history.

Prologue

AMSTERDAM, 1971

And now I introduce Mr. 't Hooft, who has a theory that is at least as elegant as anything we have heard before.
—TINI VELTMAN, "AMSTERDAM INTERNATIONAL CONFERENCE ON ELEMENTARY PARTICLES," 1971

Tini Veltman is a contrarian: a forthright man who has never shied away from controversy. His single-mindedness has brought him success where others either gave up or didn't even dare to try. It is the characteristic that set him on course to a Nobel Prize for Physics. Part of the reason for his triumph was the fortune to have a student whose genius was in constructing a masterpiece by using tools that Veltman had forged.

Veltman and his protégé, Gerard 't Hooft, are like chalk and cheese. Veltman is a big man, with a fulsome beard, often found with a cigar stuck in the corner of his mouth or waved between his fingers as he holds court. His near-perfect English resonates with Dutch vowels as he dismisses some rival's work as "baloney" or "crap." This blunt approach can mislead, obscuring a sensitive and thoughtful personality, with deeply held convictions about the way science should be conducted. His nickname,

"Tini"—an abbreviation of Martinus—is ironic given his stature, in all senses of the word.

't Hooft, by contrast, slight in build, with thinning hair, dressed smartly in jacket and tie, and with a small mustache, could easily be mistaken for an English country doctor or an accountant. During discussions, I am often possessed by a sense that he already knows what he is being told and is politely waiting to hear something novel. When he speaks, there is no doubt that he is correct: His soft voice carries real force, aided by a dry sense of humor.

Forty years ago, their meeting would change the world of physics. However, today, Veltman—the teacher whose ideas enabled his star pupil to produce his magnum opus—and 't Hooft have drifted apart.[1] In Veltman's own book about particle physics, 't Hooft's appearance is limited to a photograph and a few lines of text. He describes 't Hooft's breakthrough as "a splendid piece of work," which, enigmatically, he was very happy with "at the time."[2] That is how it was in 1971, when Veltman "proudly introduced" his young maestro to the world.

THE INFINITY PUZZLE

A half century or so ago, and more than two thousand years after the philosophers of ancient Greece had first conceived of atoms, these basic pieces of matter had been revealed to consist of smaller particles, of lightweight electrons remotely encircling a bulky central nucleus.[3]

In the aftermath of Hiroshima, where the nuclear atom's explosive power had been revealed, understanding the nature of the atomic nucleus and the mysterious forces that control it was what defined the new frontier. That the nucleus of an atom has a labyrinthine structure of its own was already apparent; the surprise was that the closer that scientists looked at it, the more complicated things appeared to be. And to cap it all, strange particles—similar to those found on Earth, yet behaving in other ways—were discovered to be pouring down from the heavens, as the result of cosmic rays from outer space smashing into the atmosphere above our heads. Exotic forms of matter, whose existence had not been dreamed of by scientists in their earthbound laboratories, were changing our whole perception of nature. Any theory of the universe had to explain them.

This was a time when the pursuit of breakthroughs had become the physics world's equivalent of the Klondike gold rush.[4] Some theoretical high-energy physicists staked their claims with half-baked theories, which they published in obscure journals. The logic seemed to be that if your idea turned out to be wrong, few would notice and the paper would be quietly forgotten. However, if it turned out that a discovery proved your idea to have been correct, you could then refer the world back to your paper and claim priority.

Throughout this febrile period, one problem stood out, resisting all attempts at a solution. This was what I call the "Infinity Puzzle." Three great theories—Maxwell's theory of electromagnetism of the nineteenth century, Einstein's theory of special relativity of 1905, and Quantum Mechanics, developed in the 1920s—individually made profound predictions that turned out to be completely accurate: for example, the description of light as electromagnetic waves with a constant speed; the conversion of mass into energy via $E=mc^2$, where c is the speed of light; and the explanation of the stability of atoms, with a quantitative description of their beautiful spectra. In the 1930s the union of these theories gave birth to a complete theory of electromagnetic force and how light interacts with atoms, known as Quantum Electrodynamics, or QED. Initially, it appeared beautifully seductive, but what at first had appeared to be a Cinderella soon threatened to become an Ugly Sister. When the equations of QED were applied beyond the simplest approximations, they seemingly kept predicting that the chance of some things occurring was "infinite percent." Why is this a problem? The answer is that infinity is transcendent, beyond measure, signifying a failure of understanding rather than a real answer.

To put this into context, the probability of chance can range from zero (that I will never win the lottery, for instance, as I never buy a ticket) to an absolute certainty at 100 percent (death and taxes). "Infinity," by contrast, is boundless and immeasurable; it has no quantifiable meaning. In the context of the questions that the scientists were posing, the answer was nonsense, analogous to your computer giving you an error message: "computer violation" or "overflow." When this happens it is usually a hint that you have made some catastrophic error—such as instructing the machine to divide by zero. Or it may be a sign that there is a glitch in your computer, perhaps even that the machine itself has been assembled incorrectly.[5] Without doubt "overflow"—or in our example, infinity—is telling you that something is wrong; the problem is: What to do about it?

Nor was this a nonsense confined to some arcane piece of atomic science, for this enigma touched upon our ability to understand the principles underlying some of the most basic and far-reaching phenomena. Plants grow as their atoms absorb energy from light, for instance; radio waves result when electric charges are disturbed by electric or magnetic forces; and much of modern electronic technology involves the interactions between electromagnetic radiation and electrons. Each of these—whole industries and indeed many forms of life itself—depends on a simple underlying mechanism: an electron absorbing or emitting a photon, which is the basic particle of light. Yet QED seemed unable to agree with even this most rudimentary of processes. If, as QED seemingly implied, the chance of a photon being absorbed by an atom was infinite, then photosynthesis and indeed many chemical reactions would happen instantaneously. Life would have burned itself out long ago, if indeed it had ever begun.

For physicists, *infinity* is a code word for disaster, the proof that you are trying to apply a theory beyond its realm of applicability. In the case of QED, if you can't calculate something as basic as a photon being absorbed by an electron, you haven't got a theory—it's as fundamental as that.

One particular example of this catastrophe is the magnitude of an electron's magnetism, which experiments could measure relative to some standard scale. By using the standard theory, that is, QED, physicists expected to be able to compute this number. All that is required is to solve the algebraic equation describing an electron absorbing a single photon.

This is standard fare in undergraduate physics, and I can well recall the joy I felt when, back in 1967, I first carried out the calculation myself. I thought that at last I had qualified as a theorist. Unfortunately, I then learned that this was just the first of a whole series of calculations that would be needed in order to arrive at the true answer; furthermore, my tutor had glossed over the fact that if I were somehow able to do this momentous task, and then to add up the total, the answer would turn out to be infinity. Unknown to me at that time, a few hundred miles away, in Holland, I had a contemporary named Gerard 't Hooft, who was also being exposed to the mysteries of infinity and within five years would gain scientific immortality by solving them.

The reason that there was so much more to do lies with the fact that, according to QED, the electron in question is not alone in the void: A vacuum is not empty but seethes with transient particles of matter and antimatter, which bubble in and out of existence. Although these will-o'-the-wisps are invisible to our normal senses, they disturb the photon and electron in the moment of their union and contribute to the number that the experiment measures.

QED contains the means of calculating the effect of each of these disturbances, one by one. There is an infinity of them, the contributions of all but a few being so trifling that they can be ignored—so long as you are prepared to accept some limit to the precision of what you are computing. The trick is to start with the most important (which is what my student calculation had done, naively thinking it to be the lot), then add in the next, and then to continue by including the effects of smaller and smaller contributions, the sum total approaching the "true" answer ever more accurately.

This can be difficult to do, but there is nothing necessarily wrong here, as an infinite sum can have a finite answer (such as $1 + \frac{1}{2} + \frac{1}{4} + \frac{1}{8} + \ldots = 2$). After the first two terms you are already within 25 percent of the answer; add in the next couple, and your inaccuracy is less than 10 percent. It is merely a pragmatic question of how precise an answer you need as to how many terms, and how much work, you have to do.

Or so physicists thought in their early explorations of the implications of quantum mechanics and QED. However, by contrast to the previous sum, which gave the desired answer of 2, what they found instead was a series that was more like $1 + \frac{1}{2} + \frac{1}{3} + \frac{1}{4} + \ldots$. At first sight this looks good too—after just three terms the sum is already within 10 percent of 2. But add in the next one, ¼, and you will find that the running total has overshot: 2.08. Add in further terms and it continues to get worse: infinitely worse. The sum 1 + ½ + ⅓ + ¼ + . . . = infinity.

In their search for precision, physicists had utterly lost accuracy. Attempt to calculate an electron's electrical properties, such as the size of its charge or magnetism, and your answer would turn out to be infinity; if you wanted to know what would happen when a photon hit an electron, and listed the odds of this or that possibility, each one would turn out to have the chance "infinite percent."

While QED describes how light interacts with matter, it alone cannot confront the stability of matter itself. There are two other forces acting in and around the atomic nucleus, known as the weak and strong nuclear forces, their names alluding to their strengths relative to that of the electromagnetic force when acting on atoms here on Earth. The strong force is the binding force that holds atomic nuclei together; the weak force, by contrast, destabilizes nuclei, causing a form of radioactivity that plays an essential role in the way that the sun produces its energy (see Figure at right). The theories of these forces also ran into problems.

The theory of the weak force gave a series of diminishing terms, similar to QED, which led to infinity also. The strong force was an even greater enigma, as in its case the infinite sum explodes; instead of a gentle approach to infinity, like 1 + ½ + ⅓ + ¼ . . . , there was an unnerving sum like 1 + 4 + 9 + 16 + . . . , where each successive term is bigger than all that went before it. This was so daunting a result that physicists decided some other way of explaining the strong force was needed.

For the particular case of QED, a way of abstracting useful numbers from the morass was found in 1948, as we shall see in Chapter 2. The basic trick, which works but has never made everyone, including those who created it, totally satisfied, is as follows.[6]

There are many properties of atoms and their constituent particles that you may compute in QED, each of which gives the answer infinity, but the key discovery was that whatever you calculated, the way that infinity emerged from the mathematics was the same from one process to the next. For example, when physicists calculated one quantity, they found a horrible infinite thing multiplied by, say, the number 1. Then they calculated some other quantity and found the very same "horrible infinite

THE FORCES OF NATURE

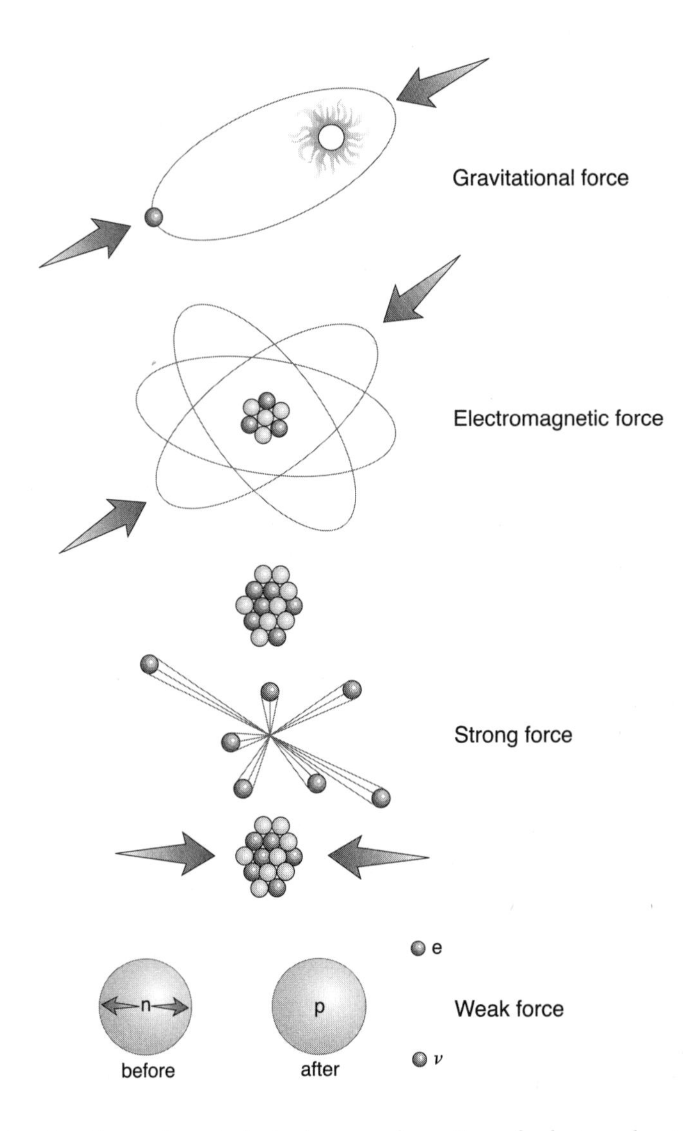

Gravity is attractive and controls the large-scale motion of galaxies, planets, and falling apples. Electric and magnetic forces hold electrons in the outer reaches of atoms. They can be attractive or repulsive and tend to counterbalance in bulk matter, leaving gravity dominant at large distances. The electrical repulsion between protons in an atomic nucleus would prevent the existence of stable nuclei but for the existence of the strong force, which gives a powerful attraction between protons or neutrons when they touch. The weak force can change one form of particle into another. This can cause transmutation of the elements, such as turning hydrogen into helium in the heart of the sun. (Credit: Reproduced with permission of Oxford University Press)

thing," but this time multiplied by, let's suppose, 2. So this second quantity was predicted to be twice the size of the first. If an experiment had already measured the true (finite!) value for the first quantity, QED could then confidently predict the magnitude of the second as being twice as great, and experiment confirmed this to be true. So the horrible "infinity" could be subsumed, hidden from view as if it didn't exist, leaving an apparently pristine theory on display. As I said, no one was entirely happy, yet it worked.

This is how. The values of an electron's electric charge and its mass have been measured. The miracle is that these two known quantities are sufficient to provide benchmarks for anything else that we may wish to compute in QED. We cannot use QED to calculate the electron's charge or mass from theory—for were we to do so, we would get infinity—but we can use QED to calculate everything else relative to these experimentally determined quantities. The marvel is that instead of infinity, all the answers now turn out to be finite, and, even better, the values are correct. Today, some quantities have been calculated this way that agree with experiment to an accuracy of one part in a trillion, which is an order of magnitude much like the diameter of a hair when compared to the width of the Atlantic Ocean.

Although inelegant, the recipe for extracting sensible answers for QED worked. However, the explosive numbers that erupted in the case of the strong force convinced people that some other route was needed there (we shall see in Chapters 12 and 13 how that breakthrough came about). For three decades, both the weak and the strong nuclear forces appeared to be beyond quantitative description. However, in the case of the weak force there was a tantalizing hope that something akin to the miracle of QED might be replicated. Unfortunately, here, too, everyone who tried soon hit a seemingly impenetrable barrier.

The Infinity Puzzle for the weak force resisted the physics world's greatest minds for a quarter century. Some tried to solve the problem but failed; most ignored it and hoped that it would go away. The nature of this impasse, how it was defeated, and the arguments over priority for Nobel Prizes that it has spawned are the themes of this book.

The saga is a paradigm of how science happens in the real world, as opposed to the steady heroic progression portrayed in some winners' accounts. Instead of a direct line linking theoretical idea and experimental discovery, there are numerous wrong turns, partial answers, and mislaid arguments. The picture of science as a sequence of great discoveries and Nobel Prizes, which is presented in some narrative histories, and which forms many people's idea of the field, is really an attempt to make easy narrative sense of the whole saga, with hindsight. In practice, scientific research is a series of twists and turns; scientists experience the same emotions, pressures, and temptations as any other group of people and respond in as many ways.

You may experience the euphoria of making a great discovery, only to find out that someone else has beaten you to it. Or you may have been first, but not been ready, or brave enough, to go out on a limb and publish—perhaps wanting more time in order to be certain, or even not realizing at the time the significance of what you had achieved. As we shall see, even at the top level people often don't know if their idea is world changing or a mere fancy until later events determine which. This is like Paul McCartney, years later, admitting that at the time of writing his songs he didn't know which would sell millions and which would fail.

For composers of music, or literature, there is no limit to the number of possible creations—it is infinite. If you don't make your composition public, it is unlikely that anyone else will create the very same symphony. For theoretical physics, on the other hand, nature already has the solution, and it is we who are trying to reveal it for ourselves. So there is a sense of uniqueness, a right or wrong, which experiment or further advances in theoretical understanding may ultimately reveal. Discover what it is, publish first, and the credit will be yours. However, if you do not, and someone else independently, later, publishes what you might have done, how do you react when the world takes notice? History records the winners' names in the pantheons of science; the names of "Nobel Prize runners-up" are as memorable as those of the losing semifinalists in Grand Slams or World Series.

Such are the realities of science, where scientists' emotional responses to these pressures may be far from the dispassionate ideal of popular belief.

Our story, spanning more than a half century, has examples of all of these, and more.

AMSTERDAM, 1971

Among those who thought that they knew how to solve the Infinity Puzzle for the weak force were Abdus Salam and Tsung-Dao (T.-D.) Lee. The person who actually did solve it, however, was Gerard 't Hooft, Veltman's brilliant student. It was 1971 when 't Hooft convinced his teacher that he had indeed succeeded. Veltman decided to launch his apprentice, already a master craftsman, in dramatic fashion.

In June that year a major international physics conference was scheduled to take place in Amsterdam. Veltman, a senior physicist at the University of Utrecht, had been asked to organize a series of presentations in theoretical physics. He invited Lee and Salam to present their ideas on how to solve the Infinity Puzzle.

Chinese American theoretical physicist T.-D. Lee had already won a Nobel Prize, shared with his colleague Chen-Ning Yang, for showing that the world behind the mirror is essentially different from the real world. Whatever it is that is responsible for the radioactive decays of atoms seems in the real world to be controlled by a mysterious subatomic left-handed screw. Viewed in a mirror this would appear to be right-handed. Had Alice known of radioactivity, she would have been able to tell whether she was in the world behind the "Looking Glass" or in the real one. The discovery in 1956 that nature is left-handed was a huge cultural shock and guaranteed Lee and Yang scientific immortality. By 1971, Lee had decided that the Infinity Puzzle was the one to crack, and he thought he knew how to solve it. However, thanks to 't Hooft, Veltman knew better.

In 1971 Abdus Salam had not yet won a Nobel Prize but was ambitious to do so. Later, Veltman would not be slow to remind people of this, hinting at the lengths Salam would go to lobby the committees. Salam was a visionary, head in the clouds, ideas flowing as if he were in a perpetual brainstorm, ready to publish anything and hope for the best. His style didn't gel with Veltman, who held Salam's oeuvre in less high esteem than some. Salam thought that he knew how to solve the Infinity Puzzle for the weak force, and from some ambiguous remarks in talks may even have convinced himself that he had the answer. However, he never quite convinced others, certainly not Veltman.

That summer, Salam believed that the key to finding the solution was to incorporate gravity into the mix. Here too, thanks to 't Hooft, Veltman knew better. He invited Salam to open the proceedings.

The venue was a small room off the main hall in the Amsterdam Congress Centre. More than 2,000 scientists attended the conference, but of them only a few dozen were present at what at first appeared to be a sideshow to the main proceedings.

Salam spoke first, saying that he was convinced that gravitation was the key. Veltman let Salam talk about his "baloney" before calling on T.-D. Lee, who then described his own attempts to solve the puzzle by inventing hitherto unknown particles with weird properties.[7] Lee completed his presentation, answered questions, and returned to his seat in the auditorium. The moment had at last arrived: "And now I introduce Mr. 't Hooft," Veltman announced, "who has a theory that is at least as elegant as anything we have heard before."

't Hooft's talk lasted just ten minutes, and to those in the audience, unaware of the significance of what they were witnessing, the occasion appeared to be simply a means for Veltman to push a promising student to wider attention. The "before" in Veltman's introduction was assumed by members of the audience to mean "in this session,"[8] and as few regarded Salam's or Lee's ideas with much enthusiasm, this introduction did not seem unreasonable, nor did it heighten expectations. However, what Veltman meant by "before" was "in the past thirty years," for 't Hooft had found the philosopher's stone.

Most did not understand his talk, let alone realize that they were present at a singular moment in the history of science. Salam certainly did not. In the written version of his own talk, which he revised after the conference, he added a note "welcoming G 't Hooft's theory," also advertising that "the same theory" had been proposed in 1964 by himself and a colleague, J. C. Ward, and then included this afterthought: "Gravity . . . is likely [to be needed] to give the right numerical values."[9] However, as the passage of time has shown, incorporating gravity would not be necessary.

Salam's postscript shows how even an expert failed to appreciate the full import of what he had just heard.

't Hooft, by contrast, did not write up his talk. He was still completing his thesis and wanted all the arguments to be presented there, carefully, where they could be spelled out like a legal document for experts to examine the logic of the proof until convinced that it was watertight. As one colleague present recalled years afterward, some in the audience had caught a flavor of what had happened, and the delegates were asking one another, "Veltman's student—'t Hooft—is he really claiming to have solved the Infinity Puzzle?"[10] Discussions in the corridors afterward convinced them that indeed he was.

When the news began to spread, the reactions of two Nobel laureates were typical. Steven Weinberg remarked, "I had never heard of him so my first reaction was: this can't be right."[11] Sheldon Glashow retorted, "Either the guy's a total idiot [to be making such an outrageous claim] or he's the biggest genius to hit physics in years."[12]

"Genius" was correct. 't Hooft and Veltman would share the Nobel Prize for Physics in 1999 for this achievement. Given their initial reactions, there is irony too that Glashow, Weinberg, and Salam (but not Ward) would themselves share a Nobel Prize in 1979 for their own work, which 't Hooft's breakthrough was about to bring to center stage, for 't Hooft's entrance was a pivotal moment in the development of understanding during the second half of the twentieth century.

In simplified accounts, Veltman's role was much like that of John the Baptist, preparing the way with the tools, the blueprints, and the machinery to fit everything together; 't Hooft was the true Messiah, the genius that physics had awaited for years who built the theory, and the structure, that would lead to a golden age. Forty years later, their legacy includes the largest and most ambitious experiments in physics that have ever been attempted: the simulation at the Large Hadron Collider (LHC) at CERN in Geneva of the first moments in the universe after the Big Bang.

For more than two thousand years, until 't Hooft, a central aim of philosophy and science had been to identify the fundamental pieces of matter, the "atoms," and, latterly, the elementary particles. Following that breakthrough, the focus has changed: Our conceit today is that we may be able to reveal how matter itself was created and how our universe of shape and form came to be.

The first half of this narrative describes how 't Hooft, and others, made the crucial breakthroughs that culminated in the triumph of 1971. The remarkable developments that have come to pass since that seminal moment will be the theme of the later chapters. There I shall trace the path from a sideshow of a talk in Amsterdam to a multibillion-dollar worldwide scientific collaboration that hopes to answer such questions at the Large Hadron Collider.

PART 1

GENESIS

Chapter 1

THE POINT OF INFINITY

Abdus Salam arrives in Cambridge from India in 1946 and becomes a theoretical physicist by chance. Paul Matthews tells him that the textbooks on atomic physics are out-of-date. A beautiful theory of atoms and light—Quantum Electrodynamics—is in crisis, as its equations give nonsense, "infinity," as the answer for quantities that are known to be finite.

One evening, around 1960,[1] I first came across the remarkable physicist Abdus Salam. In those days, the BBC had a mission to inform, and their weekly science broadcast featured leading figures speaking about the latest discoveries in a popular but serious style. On a winter's night I listened to our old analog radio as Salam was describing some breakthrough that promised to revolutionize our understanding. I hadn't much idea what it was about, but his presentation was captivating. Whatever was going on in physics, it sounded truly exciting. I knew that I wanted to be a part of it.

Years later, when he won the Nobel Prize, I wrote a brief congratulatory note and told him the effect that his words had had on me. In return I received the most charming and humble letter; the fact that his broadcast had so inspired another into the wonders of science seemed to be more important to him than the prize that he was about to collect. Those who knew him will have their own memories of this complex man, the first Muslim Nobel scientist and an international statesman and champion for science in the developing world, whose visions for science and society were equally singular and who raised high emotions among his colleagues, described by some as a genius with deep intuition, by others as an opportunist.

Salam was an iconic figure in what was known in the 1960s as the Third World. Full of energy, charisma, and politically adept, he was the inspiration behind UNESCO's funding of the International Centre for Theoretical Physics. For intellectuals from the Third World, the "ICTP" became a scientific Mecca. Salam's spirit is still there, as are his collected papers.[2]

"Why," I asked him, around the time of his Nobel Prize in 1979, "did you choose to create a Centre for Theoretical Physics, and particle physics no less, instead of, say, a Centre for Engineering, or Medicine, either of which would have far more practical use for their homelands?" His response astonished me. "It really doesn't matter," he replied. "In many of those countries, intellectuals spend most of their time trying to preserve their heads on their shoulders." Here, as in his science, Salam saw problems in a different way than most others.

Abdus Salam was born on January 29, 1926, in Jhang, Punjab, in what was then India. His father was a schoolteacher who trained his son's memory by asking him to repeat the tales that his parents had read to him.

Salam became a theoretical physicist by chance. The Indian High Commission had offered a scholarship at St. John's College, Cambridge, to a postgraduate student of English literature. However, in August 1946, this student withdrew, and the college said that in his place they would prefer an undergraduate. Abdus Salam already had a master's degree in mathematics from Lahore and so could have entered graduate school directly. However, his family had another agenda: Their sights were set on his entering the Civil Service, and the vice chancellor of Punjab University advised that the cachet of St. John's College, Cambridge, would clear the way. An undergraduate mathematics course therefore would provide the means. This was an opportunity too good to miss; if repeating undergraduate mathematics was the price for entering Cambridge, let it be.

He received the telegram with the offer on September 3 and eight days later was en route aboard the ocean liner *Franconia*, bound for Liverpool, completing the journey by train and taxi. His entire possessions, clothing and books, accompanied him in a large cabin trunk. This odyssey brought him to the main gate of St. John's College, at which point he was confronted with the problem of how to get his belongings to his room. Upon asking the porter for help, Salam was offered a wheelbarrow.

What he found on arrival was utterly different from anything he had previously experienced. England in 1946 was recovering from the privations of the war. It was the coldest winter for decades, and coal to warm his room was limited to three small buckets a week.[3] Food was scarce; meat, eggs, and sugar along with many other items required payment both in cash and in coupons from a ration book. Obtaining this document was Salam's most urgent need, and as soon as he had one, the college commandeered it: Students' ration books enabled the college to obtain the necessary food for communal dining. Other than these meals, Salam would later recall, he had lived on apples, which were the only things that could be bought without the precious coupons, or "kosher sausages and other delights that defied the rationing allocations" at the house of his friend Paul Matthews.[4]

In what was still a colonial, insular, and almost entirely Caucasian society, prejudice was in those days patronizing rather than openly malicious.[5] Within academia, Abdus Salam was less aware of this. Life in college in 1946 was hugely privileged. College servants were like personal butlers—waking you in the morning, making your bed, cleaning your room, polishing your shoes. The dining hall, hundreds of years old, high-ceilinged with stained-glass windows and portraits of grandees, consisted of the High Table, where the master and fellows of College dined, and three rows of oak tables, which stretched the length of the room and catered for the undergraduates. But for the monotonous menus, which reflected the postwar austerity, evening dinners were like medieval banquets.

Having already completed a master's degree in mathematics at Lahore, Salam duly won First Class Honours at the end of his first year at

Cambridge. He decided to proceed straight to the advanced course known as Part 3, which in the academic year 1947–1948 included a series of lectures on quantum mechanics given by Paul Dirac—Nobel laureate and one of the greatest theoretical physicists of the century. This experience changed the course of Salam's life.

Until that moment, he had a vision of success at Cambridge providing the entrée for his return home and a career in the Civil Service. But Dirac's lectures revealed to Salam a profound beauty—the mysterious ability of mathematics to resonate with the workings of the physical universe. To the devout young Muslim, this seemed to be a calling "to carve mathematical monuments to Allah's work."[6] By 1949 he was ready to start on his chosen career—research in theoretical physics.

MATTHEWS AND KEMMER

Now is the moment to introduce two people who entered Salam's life at this point and determined its course for the next two decades.

One is Paul "P. T." Matthews, who, like Salam, came from India, though he had spent all but the first seven years of his life in England. He would become Salam's research supervisor—who at the end of each term generously gave Salam the "8 guineas supervision fee, which the University of Cambridge had paid him"—and eventually his best friend, who piloted him toward a Nobel Prize.[7]

Matthews, however, was not Salam's original choice for supervisor. Instead, he had sought out Nick Kemmer, the second in our duo. Kemmer will turn out to have a central role in the later development of this story—conducting the orchestra if not actually being one of the virtuoso soloists.

Nick Kemmer had been born in St. Petersburg, Russia, in 1911, and arrived in London just before the 1917 October Revolution, never to return. Moving to Germany, and then to Switzerland, before returning to England in 1936 to a post at Imperial College, Kemmer absorbed languages as easily as physics. It was years later when I first met him. I remember him as a wonderful man, educated during the days when quantum mechanics was in its infancy, and always ready to tell stories about the great physicists with whom he had worked.

Multilingual, his English was so good that he enjoyed cryptic crosswords, once greeting me by asking if I knew the compiler of the *Guardian* crossword, as he had seen my name in a clue. I knew nothing of this and asked him to explain. The clue had been "Breakfast for Frank Close? [4,6]"; the "[4,6]" indicating that the answer consisted of two words, the first with four letters and the second with six. He then revealed that the answer was Corn Flakes, and with a chuckle asked if I had ever realized that my name was an anagram of that cereal. I had to confess that this was news to me, and seeing my embarrassment he added, "Well, always remember that on the packet under the name it says 'The original and best.'"[8]

In the 1930s Kemmer had been a research student of Wolfgang Pauli, an overpowering Austrian theorist who had given him such a tough problem to investigate that Kemmer had nearly given up theoretical physics there and then. This experience had scared him so much that, to protect students from a similar fate, he was reluctant to suggest any problems to them at all. Instead, he recommended that they read the literature and find something for themselves.

Although Salam's undergraduate record showed him to be a potential star, Kemmer was cautious and, being overloaded with students, he suggested that Salam go to Birmingham to work under nuclear theorist Rudolf Peierls. However, Salam, having at last settled into Cambridge, had no wish to leave and pleaded with Kemmer to supervise him "if only peripherally." As a compromise, Kemmer suggested that Salam talk to Matthews: "He is finishing his PhD this year. Ask him if he has any problems left."[9]

Thus, Matthews met the young Salam for the first time. Upon discovering that he was studying a textbook on quantum mechanics, Matthews pointed out that a lot of exciting things were happening, such as the discovery of how to get rid of infinity in Quantum Electrodynamics, and the textbooks were already becoming out-of-date. He explained to Salam that the new essential pieces of reading were papers by the likes of luminaries named Feynman and Schwinger, to be found not in the bookstore but on the shelves of the physics library, cataloged under the heading "The Physical Review."

Realizing that textbooks contain what is established but none of the news from the frontier is one of the rites of passage from undergraduate to research student. The key is to have advice from someone in the know, able to point the way. In Matthews, Salam had found his mentor.

THE ATOMIC BAR CODE

The textbooks may have been out-of-date, but that doesn't mean they were useless. As knowledge advances, new ideas subsume the old but rarely prove them to be "wrong." Thus, Einstein's theory of Special Relativity contains Newton's Classical Mechanics as a special case, while the discovery of quantum mechanics extended the application of these theories to the world of very small things, such as atoms and their constituent parts. The textbooks on atomic physics in the bookshops of Cambridge, which Salam read in 1949, told much that was true then, and remains so today.

In popular imagination, an atom is often pictured as a miniature solar system. In this naive analogy, the nucleus plays the role of the sun, and electrons are like the remote planets. Whereas the force of gravity controls the motion of the latter, it is the electrical attraction of opposite charges—positively charged nucleus and negatively charged electrons—that holds atoms together.

Analogies can be dangerous if stretched too far, and the case of the planetary electrons is a cautionary example: Atoms built like that could not survive for a moment if they obeyed Isaac Newton's laws of mechanics. The same force of gravity that governs the motion of the planets is degrading their orbits over the eons. The solar system is vast and gravity relatively feeble; as a result, the erosion of the orbits is so gradual even the most sensitive measurements cannot detect it.[10] Atoms, by contrast, are very tiny, and the electrical forces are much more powerful than gravity. The resulting changes in the orbits of the electrons would occur faster—much faster. Had electrons in atoms encircled the central nucleus like planets orbiting the sun and obeyed Newton's laws, they would have spiraled into the nucleus within a mere fraction of a second. An atom, once formed, would self-destruct in a flash of light almost immediately; matter, including you and me, would not exist.

The fact that we are here shows that very small things, such as atoms and their constituent parts, follow different laws from those of Newton, which explain the behavior of objects that are large enough to see. Today we know these laws. Instead of an electron being able to go where it pleases in an atom, it is limited, like someone on a ladder who can step only on individual rungs. Electrons in atoms follow a fundamental regularity, each rung corresponding to a state where the electron has a unique amount of energy. Danish physicist Niels Bohr discovered this in the summer of 1912, following a remarkable observation, in 1885, by a Swiss schoolteacher, Johann Balmer.

Light, that rainbow or spectrum of colors, consists of electromagnetic waves whose electric and magnetic fields oscillate hundreds of trillions of times each second; what we perceive as color is the brain's response to the different frequencies of these oscillations. Albert Einstein—most famous for his theory of relativity—won his Nobel Prize for showing that light rays, rather than being a continuous stream, consist of a staccato burst of particles—photons. A photon has no mass, but traveling at a speed of about 300,000 kilometers every second, it has energy. The energy of a photon is proportional to the frequency: Thus, a photon at the high-frequency violet end of the rainbow has roughly twice the energy of one from the low-frequency red end.[11]

A hot sodium or mercury vapor lamp glows with a characteristic yellow or turquoise hue. Lamps shine because heat is shaking photons loose from the atoms' electromagnetic fields; they have characteristic colors because the photons emerge with energies, or frequencies, unique to the parent atoms of each element.

These colors identify the pattern of energy levels available to the electrons within those atoms. When an electron drops from a rung with high energy to one that is lower down, the excess energy is carried away by a photon of light. Conversely, if an atom is hit by a photon whose energy exactly matches the gap between two rungs, the atom will absorb that photon, in the process lifting the electron up the ladder.

The hot sun, like all stars, emits electromagnetic radiation across the entire spectrum. There is a lot of gas in its outer atmosphere, containing a smorgasbord of elements. In sunlight, those photons whose energies happen to match the gaps between rungs in the atomic ladders are absorbed by the atoms of these elements and never reach Earth. These "missing" photons show up as dark lines in what, at first sight, appears to be a continuous spectrum of colors. By viewing starlight through a diffraction grating—a piece of glass that has been scratched with many close packed grooves—it is possible to split the light into its component colors, such that sharp, bright lines become visible.[12]

The lines are like some fundamental bar code identifying the elements present in the sun, or in other stars. Being able to know the makeup of the heavens is Promethean. In the nineteenth century, the beauty of spectra, and their powerful application to stellar physics, inspired both experiments to determine the patterns and also a question: What causes these specific lines? Why does each atomic element have such a unique character?

Today we know that an atom of hydrogen consists of a single electron encircling a nucleus consisting of a single proton. It is this basic simplicity that made hydrogen the Rosetta Stone for deciphering the atomic bar codes.

In 1885, Johann Balmer—a schoolteacher in Basel—discovered a remarkable feature about hydrogen's spectrum: The frequencies of its lines fitted a simple formula, each being proportional to a common quantity multiplied by the difference of two numbers, which themselves followed a simple rule. These two numbers were ¼—written as $(\frac{1}{2})^2$—and $1/n^2$, where $n = 3, 4, 5$, and so on. The fit was so perfect that Balmer waxed eloquently that these frequencies are due to the "vibrations of a material" and form the "overtones of one specific keynote."[13] He was so confident in his formula that he predicted the existence of lines for any values for m and n in the formula[14] $1/m^2 - 1/n^2$.

Balmer's simple formula described the spectrum of the hydrogen atom perfectly, but not its cause. By luck or judgment, Balmer had stumbled upon a great truth. The question was: Why does his magical rule work?

In 1912 Niels Bohr found the explanation, courtesy of quantum theory. In quantum theory any particle can take on a wavelike character. What is familiar for photons and electromagnetic waves occurs for electrons also.[15] We can visualize the waves for electrons in atoms as if they were wobbles on a length of rope. When coiled in a circle, like a lasso, for a wave to fit perfectly into its circumference, the number of wavelengths in the circuit has to be an integer. Imagine this circle like a clock face. If the wave peaks at twelve o'clock, with a dip at six o'clock, the next peak will occur perfectly at twelve: The wave "fits" into the circle. However, a peak at twelve followed by a dip at five o'clock would have its next peak at ten and be out of time with the beat of the wave—out of "phase" in the jargon of physics: The wave will not "fit."

Electrons circulating in atoms cannot go anywhere they please but can go only on those paths where their waves fit perfectly on the lasso. The numbers n and m that Balmer had identified turn out to be the numbers of wavelengths in a single circuit. A single wave ($n = 1$) corresponds to the electron being on the lowest rung of the ladder; two waves will find it on the second rung, three on the third, and so on. The energies of the rungs on Bohr's ladder miraculously explained Balmer's discovery: When an electron drops from a high-energy rung to a lower one, the difference in energies is radiated as light in accord with Balmer's formula.[16]

Bohr's model took no account of relativity. In 1928 Paul Dirac completed the picture of the electron, with his celebrated equation marrying quantum mechanics with Special Relativity. In the meantime, the electron had revealed a strange duality,[17] where it acts like a miniature magnet with a north and south magnetic pole. Dirac's equation incorporated this in an elegantly natural way. His equation also predicted that subtle deviations from Balmer's formula would show up if very precise measurements were made, and this indeed turned out to be the case.[18]

Dirac was reluctant to work out the implications, however, fearing that they might not agree with nature and thereby ruin what Nobel Laureate Frank Wilczek has referred to as an "achingly beautiful" creation, its symmetry and balance "almost sensual."[19] Dirac need not have worried: His equation is today recognized as the seed of everything that underpins chemistry, biology, and in principle life itself. Instead, Dirac made a great synthesis, a relativistic quantum theory of both electron and light known as "Quantum Electrodynamics," or QED for short.[20]

ALPHA

The beauty of QED is that it unites the nature of light and matter. In QED, as light waves act like particles, so also do particles of matter, such as the electron, behave like a wave spread over some region of space: The essence of both Einstein's and Bohr's pictures emerges naturally. I, and most physicists, find it easier to visualize miniature billiard balls than diffuse waves—hence the name: *particle* physics. Think of electrons and photons as little particles playing subatomic billiards according to the laws of quantum mechanics, and you will have the essence of QED.

Whereas real billiard balls exist all the time, some subatomic particles can come and go. You are seeing this page because its atoms are pregnant with particles of light—photons. Not just atoms but electrically charged particles have this ability to radiate or absorb photons. For example, an electron at one point in space may emit a photon, which carries away energy and momentum. When this photon hits another charged particle, it sets that particle in motion. Thus, in QED, the electromagnetic force is transmitted by the action of photons,[21] which bump into other particles and jostle them.

As photons and electrons come together, merge, and separate in cosmic terpsichore, QED encodes the likelihood of them interacting in a number, known as "*alpha*."[22] *Alpha* sets the scale of nature—the size of atoms and all things made of them, the intensity and colors of light, the strength of magnetism, and the metabolic rate of life itself. It controls everything that we see.

Experiments have determined its value to be 0.007297, which seems unremarkable until you notice that 1 divided by this number is almost exactly an integer: 137. Almost immediately following that discovery, this number took on a sense of mystery, which has fascinated physicists ever since. In *137*, apparently, science had found nature's PIN code.

In Cambridge, England, in the 1930s, astronomer Sir Arthur Eddington, seduced by this numerology, inspired a Pythagorean cult.[23] There have been spoofs connecting 137 to the biblical book of Revelation;[24] one of the fathers of Quantum Electrodynamics—Julian Schwinger—had 137 as the vanity license plate on his sports car;[25] and eighty years on, many of us continue to receive unsolicited papers from people who believe that they have found the true path to enlightenment with an explanation of this number. Physicist Wolfgang Pauli collaborated with Carl Jung, the psychologist, in a fruitless attempt to find deep significance in its value.[26]

Richard Feynman himself described it as "one of the greatest damn mysteries of physics: a magic number that comes to us with no understanding by man," adding that if the "hand of God" wrote that number, "we don't know how He pushed his pencil."[27]

This glimpse of the "hand of God" has tantalized physicists, and mystics, for eighty years. Recently we have discovered where to find an explanation. Experiments at the LHC may reveal the answer, as we shall see later.

INFINITY APPEARS

Even though the reason for the size of *alpha* was a mystery, the fact that it is empirically small is a godsend for theoreticians. In QED the equations include *alpha*, sometimes once, or repeated—"*alpha-squared*," "*alpha-cubed*," and more. When you square a small number, what you get is even smaller. For example ½ times ½ is ¼. *Alpha* is 1⁄137; *alpha-squared* is less than 1⁄10,000. This inspired a way of simplifying the sums: As a first approximation, ignore contributions that include *alpha-squared*, *alpha-cubed*, or higher powers, relative to those with simply *alpha*. Every extra presence of *alpha* merely perturbs slightly the previous estimate.

By means of this technique, at the end of the 1920s Quantum Electrodynamics successfully explained many phenomena: what happens when a photon bounces off an electron, how electrically charged particles scatter from one another, how an electron behaves in a magnetic field, and so forth. These innovative calculations used what is called the "first approximation" —they ignored *all* equations where *alpha* appeared more than once. Seduced by this initial success, theorists began to calculate what should have been small "corrections" to these numbers, by including contributions that they had neglected in the first approximation.

They did so in the hope of achieving more precise predictions. Instead, what happened horrified them. The first "correction" turned out to involve the small number—*alpha*—multiplying a quantity, which, as we are about to see, the theory predicted to be infinite. The next level of correction was, at the outset, the even smaller number—*alpha-squared*—but, once more, multiplied by infinity. Calculations of the mass of the electron, its charge, and the probabilities for any electromagnetic process that you could think of all turned out to be infinitely large. By the mid-1930s, serious doubts had begun to emerge as to whether the theory is at best an approximation; valid only for photons, electrons, and positrons in a limited set of circumstances; or at worst fatally flawed. This is how the disaster unfolded.

The spectral lines of hydrogen, as encoded in the Balmer formula, had been beautifully explained by Dirac's equation. The problem was that in its original form, Dirac's equation encodes the properties of an isolated electron in otherwise empty space. This is the simplest situation imaginable and a seductively natural starting point. However, this seemingly straightforward circumstance never occurs in practice, because empty space, though easy to imagine, is impossible in reality.[28]

A massive body, such as the Sun or Earth, sends out gravitational tentacles in all directions uniformly, creating a kind of tension—known as the gravitational field—in otherwise "empty" space. The field manifests itself by producing forces on objects that happen to be in the vicinity. An analogous set of remarks can be made for electric fields emanating from an electrically charged particle.[29]

The concept of field gives clues to the mystery of how a force can occur between two apparently disconnected bodies. Precisely what stuff the field consists of is a question for philosophers; describing its effects is the purview of physics. However, when the implications of such ideas are worked through, in some cases they lead to apparent nonsense. To see how, let's start with something as familiar as a battery, such as you might use in a flashlight or to power a radio.

Such a battery provides a few volts, and with its positively and negatively charged plates separated by the order of a millimeter, the resulting electric field between them will be up to a thousand volts per meter. In high-energy particle accelerators, electric fields of tens of millions of volts per meter may occur. This technology gives far greater electric fields than in a simple battery, but in turn is trifling compared to what occurs within atoms. Inside an atom of hydrogen, some ten volts is the gap between the electron and proton, which are separated on the average by only a tenth of a billionth of a meter. The resulting electric field is more than a thousand times greater than we can achieve in macroscopic technology, though this vast magnitude is restricted to atomic dimensions.[30]

Now we meet the enigma. In Dirac's equation, the electron appears as a fundamental indivisible point of electric charge, and in the immediate vicinity of an electron with no physical extent, the field becomes infinitely strong. This could perhaps have been dismissed as a mathematical curiosity but for the physical implication. The electron interacts with its own field, like a snake biting its own tail, and gains energy known as "self-energy." For the electron described by Dirac's theory, the self-energy is infinite. Einstein's theory of special relativity tells us that an amount of energy E represents an amount of mass E/c^2 where c is the speed of light. The paradoxical result is that by interacting with its own electromagnetic field, an electron gains an infinite amount of inertia, or mass. As the mass of an electron has been measured, this infinite theoretical result is manifestly nonsense.[31]

As if this were not enough to worry about, in QED the vacuum is also seething with transient particles of matter and antimatter. These ephemera are invisible to our direct senses, yet according to QED they affect the motion and properties of particles, such as the electron's mass, and of the forces that act on them.

Wolfgang Pauli realized that all of these contributions would have to be taken into account. In 1929 he gave this task to a new research assistant who had arrived at Zurich to work with him. The theorist in question was J. Robert Oppenheimer, later famous for his role in the Manhattan Project developing the atomic bomb. Quick thinking, impatient, he was either utterly charming or exceedingly annoying to those around him. He thought fast, and calculated even faster, making mistakes constantly. He was full of confidence, and Pauli proposed that he calculate the spectrum of hydrogen, not by using Dirac's equation, for that had already been done and worked beautifully, but via QED.

Oppenheimer's calculation had to take account of the fact that, in QED, an electron can emit a single photon, and then reabsorb it. In classical physics this is impossible, because energy is conserved and it would require a source of energy to make the transient or "virtual" intermediate step. However, in quantum mechanics energy need not be conserved, at least for exceedingly short time spans.[32] The intermediate state consisting of the electron accompanied by a single photon can occur, and moreover can have any amount of energy, ranging all the way to infinity. The total effect therefore involves a sum over all of these possibilities.[33] This was a sum that Oppenheimer and Pauli hoped would be finite—like 1+ ½ + ¼ + ... = 2. However, Oppenheimer found an unsettling answer: The sum is infinite.[34]

His arithmetic was correct. Far from an electron emitting a characteristic color of light as it moves from one rung to another, QED implied a nonsense: Infinite amounts of energy are emitted; atomic spectra do not exist. The atomic bar codes, so perfectly described in the most naive of approximations, first by Balmer and Bohr, and then by Dirac's equation, dissolved into uniform gray according to the more sophisticated version of QED.

As a description of physical reality, the entire enterprise was beginning to look ridiculous. Pauli was a great critic. He wrote to Dirac expressing his opinion that QED was useless and became so depressed at the false promise of the theory that he even considered quitting physics to write novels.[35]

THE TEXTBOOKS ARE OUT-OF-DATE

That was the state of knowledge, or of ignorance, in the books available to Salam in 1949: QED, the wonder theory, whose acronym had seemed so apt, had been undermined by the plague of infinity. But Matthews then told Salam what the textbooks did not yet have: Everything had dramatically changed; in the previous twelve months, Quantum Electrodynamics had been transformed; infinity had been banished and the electromagnetic force understood. QED could now explain the behavior of electrons in atoms so precisely that in every case the results of experiment and theory agreed. Matthews explained that Salam would need to learn all about this, and in Chapter 2 we shall see for ourselves. In the meantime, the Infinity Puzzle for QED was yesterday's story; Matthews had already moved on to the new frontier—the mysterious nucleus at the heart of the atom, where further forces are at work.

In the nuclei of elements such as iron, gold, and lead, large numbers of protons are packed close together. Each of these is positively charged. Yet the golden rule of electric forces is that like charges repel, which creates a paradox for the existence of atomic nuclei. That they survive the electric disruption is because there is a powerful attractive force between "nucleons" (protons and neutrons), the so-called strong force.

As the electromagnetic force is mediated by particles—photons—why not a similar story for the strong force? In 1935 Hideki Yukawa, a Japanese theorist, had proposed that this was so and predicted the existence of a particle—the "pion"—as the carrier of that force. The pion had been discovered experimentally in 1947, and Matthews's thesis was the first attempt to incorporate this particle in a theory of the strong nuclear force, analogous to what QED had achieved for the electromagnetic force.

It was Matthews's breakthrough that had inspired Kemmer to introduce Salam to him. Kemmer had captured Salam's interest by telling him, "All theoretical problems in QED have been solved. Paul Matthews has done nearly the same for [pion] theories." Hence the significance of Kemmer's further advice: "See if he has any problems left."

And indeed he did, for Matthews was not yet convinced that the Infinity Puzzle in QED had been completely solved. Having spoken with one of the main architects, Matthews had become suspicious. His reasons, and what he proposed they do about it, he would tell Salam all in good time. First, Salam needed to learn how the puzzle had been solved and to meet the new ideas, which the textbooks did not have.

Chapter 2

SHELTER ISLAND AND QED

Infinity is banished; QED works thanks to the discovery of "renormalization." Feynman and Schwinger, two youthful giants of physics, compete for attention. First Schwinger astounds science with a bravura performance, but after embarrassing defeats, Feynman finally wins the day.

If there is a single moment marking the start of the modern conceit that we have an outline for the final theory of particles and forces, it is when the American Willis Lamb took the floor at the legendary conference on Shelter Island, New York, in June 1947.

The Shelter Island meeting was the first postwar gathering of scientists aimed at assessing the status and prospects for development in physics. It is hard today to realize how cataclysmic an impact the Hiroshima and Nagasaki bombs, of uranium and plutonium, had had on the international scientific community. These were explosions as much of the mind as of physics, bringing one arena of the war to an end but also showing all the world's physicists the very real effects of unleashing the power contained within an atomic nucleus. There was a worldwide sense of awe at how theory had turned into reality. Indeed, when the delegates were en route to the conference in a bus, they were pulled over by a policeman on a motorcycle who asked, "Are you the scientists?" On receiving confirmation that they were, he proceeded to escort them with sirens clearing the way. It turned out that this was not a security measure but a "thank-you," from a policeman who had been a marine in the Pacific and was grateful for the scientists' development of the atomic bomb, but for which, recalled one of the participants, "he might not have been there to thank us."[1]

Not every physicist had been working on the atomic bomb, however. Lamb had spent World War II working with microwaves for radar, and when peace was restored he realized that with his experience he would be able to measure the energies of electrons in atoms far more precisely than ever before. This he did, and beyond a shadow of doubt he succeeded in establishing that there is a subtle shift in the energies of electrons in hydrogen atoms relative to what Dirac's equation predicted. It was this news that he brought to the gathering at Shelter Island.

Quantum mechanics explains the spacing of the rungs on the atomic ladder and predicts the frequencies of radiation that are emitted or absorbed when an electron switches from one to another. According to Dirac's equation, which was the state of the art in 1947, in hydrogen two of these rungs have identical energy.[2] However, Lamb's measurements showed that they differ by about one part in a million. This tiny but significant difference was at odds with Dirac's description of the hydrogen atom.

Lamb's experiment was so precise that it had revealed the subtle effects of quantum physics on the atom's electromagnetic field, which can momentarily convert into matter and antimatter—an electron and its doppelganger, a positron. These particles disappear almost instantaneously, but in their brief mayfly moment of existence they alter the shape of the atom's electromagnetic field slightly. This in turn affects the motion of the electron and leads to the subtle shift in energy that Lamb had measured.

Lamb was thus the first person to observe experimentally that the vacuum is not empty, in contrast to what Dirac had implicitly assumed, but is instead seething with ephemeral electrons and positrons. This is what QED had been designed to account for. But, as we have seen, QED had predicted that the magnitude of this disturbance, far from being trifling, should be infinite. At Shelter Island, Lamb was announcing that he had measured the value. At a tiny one part in a million it was not zero, as Dirac's original equation would have required, but nor was it infinite, as QED seemed to imply. Lamb's discovery led physicists to rethink the basic concepts behind the application of quantum theory to electromagnetism.

It had been 1929 when J. Robert Oppenheimer exposed the Infinity Puzzle in QED. Eighteen years and a world war later, what irony that it should be Oppenheimer who was chairing the meeting as Lamb addressed it. Oppenheimer's report of the discussions recorded that Lamb's presentation on the very first morning showed that "a new chapter in physics is upon us." The challenge was to compute the value that Lamb had found; "infinity" would not be accepted as the answer.

The baton was being passed to a new generation of theorists. The gurus would be the two youngest members of the audience at Shelter Island: Still under thirty, and already veterans of the scientific war just ended, they were Julian Schwinger and Richard Feynman. Only later would it become known that in Japan, completely independently, Sin-Itiro Tomonaga had already solved the puzzle.

SCHWINGER AND FEYNMAN

Julian Schwinger and Richard Feynman were exact contemporaries. Born in 1918, in New York City, both were brilliant theorists, but there the comparisons end.

Schwinger, from Upper Manhattan, was a small, heavy man, a natty dresser who spoke eloquently without notes, filling the board with equations written deftly with both hands. He aspired to elegance in all things, and achieved it. Feynman, by contrast, was akin to the street-smart kid, a prankster, antiauthoritarian to a degree that became obsessive. His memoir, *Surely You're Joking, Mr. Feynman*, is full of examples of his irreverence and unsophisticated behavior where his natural cleverness nonetheless ensures that in all adventures he comes out on top. A colleague once described him as "half genius; half buffoon."[3]

Julian Schwinger was the veritable wunderkind. He attended Townsend Harris High School—in those days New York's leading academy—and in

1932, at just fourteen, he heard a lecture by Dirac, which inspired his interest in quantum field theory. He matriculated as an undergraduate at New York's City College at fifteen, and was soon writing research papers with his instructors—themselves already doctoral research students at Columbia University in New York. His fame spread rapidly, Hans Bethe[4] writing of the seventeen-year-old Schwinger in glowing terms: "His knowledge of Quantum Electrodynamics is certainly equal to my own, and I can hardly understand how he could acquire that knowledge in less than two years and almost all by himself,"[5] adding that "Schwinger already knows 90% of physics; the remaining 10% should only take a few days."[6]

With such a glowing testimonial, Schwinger transferred to Columbia. By 1937, still only nineteen years old, he had published seven research papers, enough to qualify for a Ph.D., even though he had not yet taken the bachelor examinations. Even at this stage, while still an undergraduate, physicists of immense stature, such as Enrico Fermi and Wolfgang Pauli, would meet with him to discuss issues at the frontiers of research. Holding court with Pauli—infamous for his low opinions of others, and caustic in dismissing sloppy thinkers—shows how Schwinger was revered.

He had to wait until 1939 to receive his Ph.D.; university bureaucracy required time to pass and regulations to be satisfied. Schwinger by now had outgrown Columbia and transferred to Berkeley to work with Oppenheimer.[7] This exposed him head-on to the enigmas of infinity in QED. The Second World War, however, was about to interrupt his blossoming career.

The youthful Feynman had also proved to be a remarkable mathematician, but had been living a more normal teenage life—"hanging out," in modern terminology, and playing practical jokes. His perfect grades in science and mathematics were not matched by his performance in other subjects. While Schwinger was mesmerizing the faculty at Columbia, Feynman's application to enter was rejected: In the 1930s, U.S. colleges had limited admission quotas for Jews. He went to MIT. Even though Albert Einstein had had a ticker-tape parade, theoretical physics was not a major pursuit in American universities before the Second World War.

By 1941 Schwinger was available for hire, but a long tradition of anti-Semitism may have been a reason for his lack of job offers.[8] He accepted a lowly position at Purdue University, on the condition that his physics course would not start before noon. Purdue agreed.

By the time the United States entered the world war, both Feynman and Schwinger were being heralded as stars. By this stage Schwinger had presumably learned "the remaining 10%" of physics; as for Feynman, Bethe rated him in the new generation of world physicists as "second only to Schwinger." Feynman was seconded to Los Alamos to work on the atomic bomb; Schwinger felt uncomfortable with that, and instead he helped develop microwave radar at the MIT Radiation Laboratory.

Once the war ended, the stature of physics—which had produced the atomic bomb—changed utterly. The U.S. government poured money into research; physicists had become heroes; Einstein—previously described as a mathematician—was now reinvented as a physicist. Awareness of the Holocaust, and the role that Jewish scientists had played in winning the war, meant that universities were queuing up to hire the brightest stars, without the historical prejudice. In February 1946 Schwinger accepted a professorship from Harvard; Feynman took a similar post at Cornell.

With Schwinger's and Feynman's reputations already established, it is no surprise that they became the chosen representatives of the new generation who, when the academic year ended, joined with the select few at Shelter Island.

Even though only a few months had elapsed since the war's end, they had each already homed in on the Infinity Puzzle. Their contrasting approaches typified their individual characters. Schwinger, deeply mathematical and erudite, his arguments complex and crafted like a lawyer's brief, was if anything too precise; if he had a fault it was that his performance seemed more designed to show what he was capable of than to enable others to repeat the feat. Feynman, by contrast, was driven by physical intuition, and would not be satisfied until he had worked out things in his own unique way. Quantum mechanics was a perfect example of this. He was, in effect, redesigning quantum mechanics from the bottom up, led by intuition as much as by formal mathematics.

FEYNMAN'S ACTION

The challenge of classical mechanics, such as determining the motion of planets, is that if you know where some objects are now, where will they be at some future moment? In the seventeen century Isaac Newton stated the laws of motion: If no forces act, bodies move at a constant velocity, whereas a force gives them acceleration. This inspired the concept of energy, such as the energy associated with motion—"kinetic energy"—and latent or "potential" energy, where the situation of a body gives it the potential to gain kinetic energy, the sum of the potential and kinetic energies being constant. It is with such principles that most of us first meet mechanics. We learn the Newtonian methods, exploit the principle of the conservation of energy, and work out how things move.

This applies to large objects, but on the atomic scale Newton's classical mechanics gives way to quantum mechanics. The original constructions of quantum mechanics had imitated Newton's approach. However, there is another technique for solving classical mechanics, invented in the eighteen century by French Italian mathematician Joseph-Louis Lagrange. In 1942 Feynman reconstructed quantum mechanics by using Lagrange's methods and by building on ideas that Cambridge mathematician Paul Dirac had pioneered ten years before.

Instead of focusing on the (conserved) sum of kinetic and potential energy, Lagrange considered their *difference*. The magnitude of this difference at any point on an object's trajectory is called the Lagrangian. Then all you have to do is to add up the values of the Lagrangian along the path, from beginning to end. This sum, or "integral," is known as the action.[9] The remarkable feature is that the path taken by an object to get from one point to another in a specified amount of time is the one with the least action.[10]

The principle of least action leads to Lagrange's equations of motion, with which students can solve problems easily in classical mechanics that would be exceedingly complicated using Newton's techniques. In every case, the results are the same.[11]

We tend to regard the behavior of large objects as obvious, whereas that of the quantum world is mysterious. Thus, whereas billiard balls bounce from one another in a determined way—indeed, by minimizing the action—beams of atoms scatter in some directions more than others. The atoms end up spread in areas of intensity or scarcity, like the peaks and troughs of water waves that have diffracted through an opening. As young children we experience the macroscopic world and build our intuition accordingly; wavelike atoms are not part of the scenery. However, the concept of action reveals unexpected mystery in what at first seems familiar and makes sense of what otherwise appears mysterious. Focusing on the action makes the quantum world the one that appears relatively natural and reveals that the classical laws emerge from the underlying fundamental quantum mechanics.

The purposeful aspect of the action in classical mechanics is actually rather eerie. Does a body really follow a uniquely prescribed trajectory by having first sampled all possible routes, calculated their actions, and decided on the magic solution? The idea that inanimate bodies somehow send out explorers on forays, like colonies of ants, seems unreal. Yet it is as if the system knows beforehand how to get to where it wants to be. The natural tendency of a body, free of external forces, to travel in a straight line, rather than on infinite possible zigzags or curves, is actually quite mysterious when thought of in this way. Feynman's genius was to realize that here was a case where the quantum world made more sense than the large-scale one and to use his insight to develop a novel approach to quantum mechanics.[12]

The unfamiliar weirdness of the quantum world arises because particles seem able to go anywhere—it's all a matter of chance. Feynman took this as a starting point. He assumed that all paths *are* possible, not just those with the least action: The ants spread everywhere. Feynman imagined time sliced into pieces and asked, if a particle is at some point at time zero, what is the probability of its being at some other place at a specific later time? In his formulation, the probability is the square of a complex number known as the probability amplitude, which is simply related to the action.[13]

The idea here is first to calculate the value of the action for each path, including trajectories that are absurd in normal experience. In Feynman's picture, which incorporated relativity, these even included paths where a particle could move backward and forward in time. In effect, in quantum mechanics an individual particle has an infinite number of possible paths. However, when a group of particles is gathered together so as to form a large object, such as a molecule, their individual amplitudes mutually cancel out for all paths except those that are very near to the classical one.

For a truly macroscopic body, such as a planet, only the unique trajectory of classical mechanics survives.

These ideas may seem strange, but they are actually rather familiar: They parallel how the ordered geometry of light rays emerges from spreading undulating waves of electric and magnetic fields, which radiate from a source in all directions.[14] The golden rule was discovered by Fermat in the seventeenth century: Out of all possible paths that light might take between one point and another, the actual path is that where the light takes the least time.[15] Waves set out in all directions, and if they hit a mirror, for example, they are reflected in all directions also. The different waves mingle, adding in some directions, canceling in others. In the case of bouncing off a mirror, all the overlapping waves cancel to nothing except along a direct line to the mirror, which is reflected at the same angle. Along this route they appear as simple "rays."[16]

Feynman's insight was that an analogous model could be made for the quantum mechanics of electrons. In Feynman's vision, nature is utterly democratic, placing no constraint on where an electron goes. An electron, in his theory, could sample all possible paths in both space and time. The waves would mutually self-destruct everywhere but for the shortest "optical" path, thereby giving an appearance of traveling in rays, as particles do. He then focused on these trajectories and built his theory around them. His pictorial representation of the particle rays, or trajectories, that ensued would eventually win the day in his competition with Schwinger, but only after he had first suffered several defeats.

PICTURES OR PROUST?

Stories in children's comics in the 1950s were often published in more than one format. You could read the full literary version, as in a conventional book, or instead you could skim a cartoon strip, where the text was minimal, restricted to essential speech bubbles and atmospheric illustrations. The latter were more popular. Years later, even students who studied English literature in college admit that they started with the comic-strip version.

If Schwinger's development of QED was akin to reading Proust, then Feynman's was the cartoon version. Today students meet Feynman's pictorial approach and from these diagrams deduce the mathematical expressions appropriate to the situation; Schwinger's methods have by and large been consigned to history.

Feynman represented the path of an electron by a solid line and the effect of the electromagnetic force by a wobbly line, the latter symbolizing the transfer of a single photon between the electrically charged particles involved. These cartoons gave a pleasing visualization of how particles are born, spend their lives, and die. However, the diagrams are more than this: The hieroglyphs are code for mathematical equations, some simple, others complicated. Using his pictograms, Feynman could calculate and solve problems in a few minutes that Schwinger's cumbersome mathematics took pages to work through.

These "Feynman diagrams" are basic tools today; however, in 1948, although Feynman knew what he was doing, no one else could understand why. He had spent months constructing a new set of equipment for the quantum mechanic's tool kit. He knew how to select the requisite pieces, knock the construction together, and set it working; he could drive the machine, but no one understood what was going on under the hood.

By the time of Shelter Island, both Feynman and Schwinger were already well on the way to completing their new formulations of Quantum Electrodynamics. During the formal presentations, Schwinger was rather quiet, listening to the news from Lamb, and also of the measurement of the magnetic strength—the "magnetic moment"—of the electron. Feynman was also enthralled by the news and during informal discussions in the evening after dinner would display his wares, using them to make calculations at lightning speed. Abraham Pais, a distinguished physicist who was another of the relative youngsters at Shelter Island, and later became a respected historian of science, recalled that "whatever he was doing had to be important—but I did not understand it."[17]

RENORMALIZATION

Dirac's equation assumed that an electron is no more than a piece of electric charge at a point in a spatial void. QED implied that as you voyage toward that elusive entity, you do so in the presence of an unseen swarm of ghostly spectators. These include the electromagnetic fields surrounding the electron, plus virtual electrons and positrons bubbling in and out of the vacuum. The "physical" electron is not the same as the ideal of Dirac's equation. Instead, what experimentalists interpret as the electron's mass is the result of Dirac's naked electron interacting with its own electromagnetic field and also with the "vacuum polarization" that fills the void. A "real" electron is a much more sophisticated thing than Dirac's equation describes.[18]

The infinities in the QED calculations of an electron's mass or electric charge hint at what you would find if you were able to measure with infinitely perfect resolution—revealing Dirac's ideal point. An example of the illusory quest to reach such a goal can be found on the arid Nullarbor Plain in South Australia, which stretches as flat as the eye can see in every direction. Cutting across this wasteland is the longest stretch of straight railroad track in the world. For three hundred miles, parallel lines of iron head off toward the horizon, the perspective of distance appearing to bring them together in the distant heat haze. However, as you travel along the tracks, the horizon and the apparent point of convergence recede as fast as you move toward them.

As the meeting point of two parallel lines is in practice illusory, so it seems is the concept of a point. While a sphere has three dimensions, a circle two, and a straight line one, a point has none. This is a place of true nothingness, of nonexistence. Shrinking a circle ever smaller until it becomes a point is like trying to reach that far-off place of convergence on the Nullarbor railroad: Like the end of the rainbow, it is forever out of reach, nonexistent in the real world.

The values of an electron's charge and mass that you actually measure are what matter in practice. Even the most powerful microscopes do not probe to truly infinitesimal distances. In general, the properties of an electron depend on how closely you look. For QED to be pragmatic, these subtleties of measurement had somehow to be taken into account. The solution became known as renormalization: a change of scale from the infinitesimal to that actually relevant in an experiment.

The philosophy of renormalization is one of the most difficult, and controversial, in all of particle physics. Yet it underpins modern theory. Analogies are dangerous, as I have said earlier: The following gives a feeling for the ideas, but their implementation is more profound.

An experiment to determine what a hydrogen atom consists of, or even as straightforward as measuring the mass or electric charge of a free electron, is analogous to viewing something on a computer screen, where the image is made of pixels. At low resolution you see some level of detail, but if you increase the density of pixels, the picture becomes sharper. For example, if an image with a few pixels shows a river, representations built from larger numbers of pixels will reveal finer details of its swirling waters. Likewise, at low resolution, an atom of hydrogen can just be made out to contain a powerful electric field, which grips a single electron in the atom's outer regions. The electron itself can be discerned as a fuzzy lump of electricity, filling a single pixel and seemingly just beyond the limit of resolution.

In order to see what that haze of charge is really like, we must take a closer look. Suppose we increased the pixel density by a hundred, so that for every individual large pixel that was present in the previous case, we now have a hundred fine-grain ones. Where before the electric charge was located within a single large pixel, we now find that it is situated inside the single minipixel at the center, the ninety-nine surrounding minipixels containing negatives and positives: the whirlpools of virtual electrons and positrons in the vacuum.

Now increase the density by another factor of a hundred. The charge of the electron, which previously filled a single minipixel, is now found to be concentrated into a central micropixel, which is surrounded by further vacuum whirlpools. And each of the whirlpools that had previously shown up in the image made of minipixels is revealed to contain yet finer eddies.

And so it continues. The image of an electron is like a fractal, repeating over and over at finer detail, forever. The electric charge of the electron is focused into a single pixel of an ever-decreasing size, surrounded by a vacuum of unimaginable electrical detail.

I cannot imagine it, nor, I suspect, can any other physicist. We are content to follow what the mathematics and experimental data reveal. Anyway, it is not the surrounding vacuum that concerns us here; our quarry is the electric charge of the original electron.

As we zoom in toward the kernel, and the smeared lump of that electron's charge is concentrated into a smaller and smaller pixel, the density of the electric charge increases. The way that this number changes with the scale of the resolution is called the "beta function." Going uphill, in the sense that the density of charge is increasing, the slope is positive, while downhill, decreasing, it is negative.[19]

In QED the slope of beta is positive, meaning that as your microscope zooms in, the charge concentrates more and more. According to QED, if the pixel size were to become infinitesimal, we would find the charge to be located totally within that infinitely minute point. Consequently, it would be infinitely dense.

Calculations in QED, which gave infinity as the answer, were in some cases doing so because they had summed up the contributions of all pixel sizes—from large scale all the way down to the infinitesimal extreme. However, this is not what a real experiment measures. What you see depends on the resolving power of your microscope.

In our voyage into inner space, current technology enables us to resolve matter at a scale as fine as one-billionth the size of a single atom of hydrogen. This is very small, to be sure, but it is not truly infinitesimal. In practice, what we want to be able to calculate are the values of measurements made at this particular scale of resolution. Adding up the contributions of all pixel sizes is like a travel agent charging you for trips to outer space, and including in the bill exotic destinations that are as yet impossible to reach. For example, you may have traveled on the International Space Station, the limit of what is currently on offer, and expressed an interest in Lunar exploration and even a trip to Mars when these become practical. The travel agent includes these in the bill, along with twenty-first century trips to deep space that are presently science fiction. With the potential trips extrapolated into the indefinite future, the bill comes to infinity.

By any rationale, this is absurd. What you really want to know is the cost of a trip to the moon or to Mars. To do so, we may agree with the accountant that the price for a trip to the space station (plus an option on possible future trips to infinitely far other destinations) is some amount: X. Then we might expect to get a sensible answer, in terms of multiples of X, for the cost of a trip to the moon or Mars. In fact, it would be determined by the finite difference in the distances between those destinations.

Analogously, infinity emerges from a typical sum in QED because the calculation has included the effects of scales of length—pixel sizes, if you prefer—that are infinitesimally small. With hindsight, this too is absurd. And as in the case with the travel bill, we need some way in QED to do the accounts for what we can measure and not include in the sums the wonders of an inner space that lie beyond vision. If we have some point of reference, where we know the correct value, the equivalent of the known cost of going to the space station, then we may be able to compute another quantity relative to the first one. That was at the core of what Feynman and Schwinger were developing.

This technique of using known values, such as the charge of the electron (the cost of going to the space station in our analogy) to calculate other values, such as its magnetism (the cost to the moon or Mars), became known as renormalization.[20]

ROUND 1: SCHWINGER

Four days after the end of the conference, Schwinger was married. He spent the honeymoon ruminating on the experimental result for the magnetism of the electron, which had been measured with a precision that showed disagreement with the value predicted by Dirac's theory. Admittedly, this was only a small discrepancy—about one part in a thousand—but it was a real difference nonetheless. What appealed to Schwinger was that explaining this would be an ideal test of his new relativistic approach to QED.

Feynman, meanwhile, tried calculating the value of the Lamb shift with his own techniques. By the end of the year Feynman was confident in his theory. Where before equations for the mass and charge of an electron had given the answer "infinity," now Feynman knew how to rewrite them in terms of the physical mass and charge. Upon inserting these known values into his equations, he computed the answer for the Lamb shift without any tinkering. At last, the result was finite.

How do you do multiplication using roman numerals? It is difficult, and slow, but you can get there in the end. Thankfully, we have the decimal system whose notation is more efficient. Schwinger's approach to quantum field theory may be compared to the roman numerals.

This is not to imply that Schwinger was somehow clumsy—far from it: Schwinger was one of the deepest and most brilliant theoretical physicists of the twentieth century and a hero for those who knew him. Feynman's charismatic personality certainly has played a role in elevating his profile in public perception, but for physicists, this competition is more an example of how notation can be the key to success.

In this metaphor, Feynman created the analog of the decimal system. Had Feynman not existed, progress in QED and much else in theoretical physics would have been much slower, or even in some cases impossible. Today, Feynman's diagrams and techniques are the staple education of all students and the practical route to solving problems by professionals. I learned the Feynman rules as a student and have spent four decades as a theoretician using them; I confess that I had never read Schwinger's papers until I was researching this book. I have asked my colleagues for their experiences, and the story is similar. There was no need. There is a mathematical Darwinism at work, where the fittest survive: Feynman rules.

However, it was not always thus. Schwinger was first to build the edifice, establish the procedures, and present to the public his revolutionary results, which agreed with the experimental measurements for the Lamb shift, the anomalous magnetism of an electron, and more besides.

Ironically, what Schwinger was doing is most easily understood by means of Feynman's cartoons. Feynman represented an electron by a solid line and a photon by a wobbly one, which you can imagine representing their motion through space and time, from left to right. The simplest approximation to an electron producing its electromagnetic field is if it radiates a single photon (Figure 2.1a). An electron responding to an electromagnetic field is represented by a wobbly line—photon—being absorbed by the electron (Figure 2.1b). In Feynman's picture, an electron interacting with its own electromagnetic field is illustrated by the electron radiating a transitory photon, which it later absorbs, like Frisbee players catching their own disks rather than exchanging them (Figure 2.1c).

The calculation of the electron's magnetism using Dirac's equation in effect corresponds to computing only the simplest diagram—Figure 2.1b.

FIGURE 2.1

ELECTRONS AND PHOTONS IN FEYNMAN DIAGRAMS

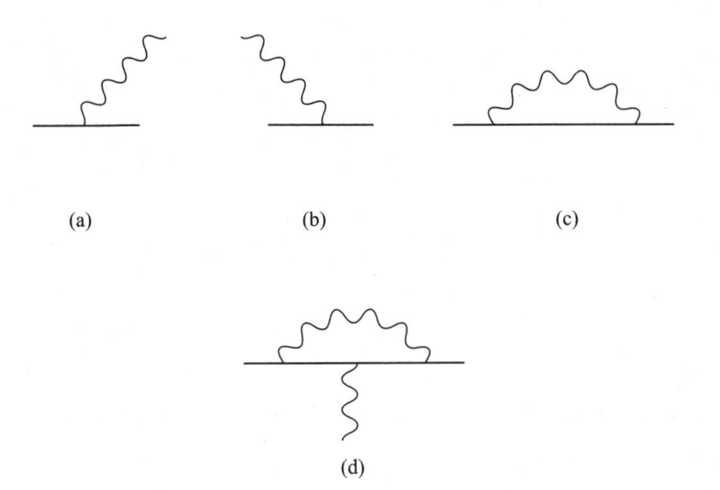

An electron (solid line) emits (a) or absorbs (b) a photon (wiggly line). In (c), the electron emits and then absorbs the same photon; this contributes to the electron's self-energy. In (d), this self-energy interaction occurs while the electron is interacting with a magnetic field, represented by a single photon that is the wiggly line entering from below. Imagine time and motion running from left to right. Where lines separate and merge after curving back, as in (c) and (d), this is for ease of illustration and not to be taken literally.

That is what I had done as a student years ago, naively thinking that I had computed the exact answer. In fact, the answer is within better than a percent of the experiment even with this first approximation.

Schwinger was interested in calculating the interaction between an electron and a magnet, like Figure 2.1b—together with its own electromagnetic field, like Figure 2.1c. Putting these two together, the relevant Feynman diagram is that in Figure 2.1d.

Whenever you have a diagram where lines form one or more closed loops, the infinity disease erupts. Look at Figure 2.1c and you can see how the problem arises. In this case, the energy and momentum of the electron is shared between the electron and photon. There is an infinite number of ways this can be done: All of the energy could be taken up by one particle, or they could each share it 50:50, or in some other ratio. Each and every one of these possibilities has to be included in the eventual accounting. This is the source of the infinite sum where the problems of the Infinity Puzzle were born.

Adding up all the contributions to the magnetism, which arise from the infinite possible ways of sharing motion between the photon and electron in the transitory state, was in effect the challenge that Schwinger took up in his own formalism.[21] The way that infinity emerged in Schwinger's calculation, Figure 2.1d, was similar to the infinite expression in Figure 2.1c. However, the latter contributes to the self-energy, or mass, of the electron, which is experimentally known. So Schwinger could "renormalize" his calculation by replacing the infinite expressions with this known, finite, quantity.

With his honeymoon and setting up home having occupied the summer, Schwinger was able to start serious calculations only in September. By the end of November he had the results.

Feynman heard on the grapevine what Schwinger had done. He also heard that Schwinger's approach was very difficult and time-consuming. Feynman had no interest in playing catch-up by learning Schwinger's language, and instead used his own techniques, which verified the answer easily. Having done it his own way convinced him that finite results could indeed emerge, as long as you take care when interpreting the measured values of quantities such as an electron's mass and charge.

Have you measured a value that is true under all circumstances, or does your result depend on the spatial resolution of the microscope with which you made the measurement? Feynman, and Schwinger too, realized that the latter is the case. Feynman had built his equations to take this into account. That they did so in a direct way confirmed the power of his techniques. Nonetheless, only Feynman understood what he was doing; as of late 1947, Schwinger was getting all the attention.

In January 1948, Schwinger's theory made its sensational debut one Saturday morning at the meeting of the American Physical Society, in New York. The hall was full to overflowing, and he was asked to give an encore that very afternoon. This again played to a rapt audience, which included several who had already heard him in the morning and were hoping to understand more the second time around.

Schwinger was using traditional tools but in novel ways. His equations included Greek symbols inside parentheses with script capitals alongside and had strange curlicues and summation signs in unfamiliar places. To the uninitiated the collection looked like Mayan script and as unintelligible. A collection of symbols on the left of the board and another set on the right were linked by =, which would be recognized by anyone as a sign of equality, though what precisely equaled what took some decoding.

He explained how the magnetism of the electron was affected by interaction with its own electromagnetic field. The anomalous magnetism of the electron—which experiment had found to be about one part in a thousand—came out in Schwinger's calculation as a beautiful combination of fundamental quantities: The universal strength of the electromagnetic interaction, *alpha* ($1/137$), divided by twice *pi*. This was in remarkable agreement with the data.[22] The *New York Times* was soon reporting: "Theorists regarded [Schwinger] as the heir apparent to Einstein's mantle and his work on the interaction of energy and matter as the most important development in the last 20 years."[23]

Feynman was in the audience, hearing for the first time Schwinger explaining his theory. At the end of the talk, Feynman announced that he too had obtained the same results. He immediately felt foolish, like a small child who has cried out, "I did it too, Daddy."[24] Feynman desperately needed to display his own wares. The chance came on March 31 at the sequel to the Shelter Island conference. However, for all the good it did him, it might have been more appropriate if his talk had taken place twenty-four hours later, on April Fools' Day.

SHELTER ISLAND 2: POCONO MOUNTAINS

Within a year after the end of the Shelter Island conference, Oppenheimer organized a reunion, which lasted from March 30 to April 2, 1948. This meeting, to update progress, took place at the Pocono Manor Inn, a large hotel in the Pocono Mountains of Pennsylvania, which offered the same undisturbed ambience as had the first conference at Shelter Island. The original group was joined by two of the founding fathers of the quantum revolution: Niels Bohr and Paul Dirac.

The talks during the first day concentrated on the discoveries of "strange particles," which were showing up in cosmic rays. There was also news of the first results at the particle accelerator in Berkeley.[25] Few realized that they were witnessing the birth of a new science: high-energy particle physics, of which more later. The main agenda began the following day,[26] with Julian Schwinger's presentation of his new relativistic theory of QED.

Schwinger described every step of his argument, like a prosecutor presenting the evidence with no stone left unturned. Members of the audience would cut in, like defense counsel making objections, or a judge seeking clarification of obscure points. This went on throughout the entire morning until the lunch break. However, Schwinger was still far from finished and after lunch continued from where he had left off. He began by showing the foundations of his method and then calculated the self-energy of the electron. He demonstrated that the sum diverged—like $1 + \frac{1}{2} + \frac{1}{3} + \frac{1}{4} \ldots$ —and that the result was independent of the motion of the electron. This was a consequence of his use of relativity, and itself a major new development.

He then turned to the case of an electron that is moving in an external magnetic field, which brought him to the Lamb shift. This contained the same divergent expression that he had found in his calculation of the self-energy. By interpreting the latter in terms of the experimentally known mass of the electron, he finally displayed his formula for the Lamb shift. After all the buildup, Schwinger came to the climax: Upon inserting numerical values into his formula, the sum agreed with the measured value that Lamb had reported at Shelter Island.

Schwinger's lecture lasted until late in the afternoon, leaving his audience both exhausted and hugely impressed. Feynman, who was scheduled to speak next, also had endured this masterly marathon performance. Like a batsman at cricket, waiting for his turn at the wicket while watching the openers dominating the opposition for much of the day, Feynman at last had his turn. In such circumstances, after hours of an unreleased buildup of tension, it is not uncommon for the newcomer to fail. Not surprisingly,

Feynman himself was nervous: He was about to present his ideas, for the first time, to an audience containing two dozen of the greatest physicists in the world, who had just been serenaded by Schwinger.

Science itself is a pure idea, but its practitioners are driven by the same emotions as anyone else. Like many of the highly successful, Feynman was competitive and felt that Schwinger had gained all the attention, first with his bravura performance at the American Physical Society in January, where he had repeated his talk to overflow audiences, and here again at Pocono. Feynman now had his chance to show what he had achieved.

By the time of the Pocono conference, Feynman had wondrous gems to display, having succeeded in reworking the theory of QED into his own language and in a form with which he could calculate the answers very rapidly. Unfortunately, his presentation made it appear less a language than a cartoon, as he had abolished the electromagnetic field in a fundamental sense and subsumed it in pictograms. In his favor was that he had obtained a lot of results—a relativistic formulation of QED, renormalization of the electron's mass, rules for calculating probabilities for the various outcomes when electrons interact with photons—and from all this could derive the anomalous magnetic moment of the electron and the magnitude of the Lamb shift. However, he had not yet published anything of it, as he did not have the mathematical proofs of all of his methods.

So Feynman had to demonstrate not only his calculations but also the elaborate new construction of quantum mechanics upon which it was founded. Adding to the pressure, his audience contained Dirac and Bohr, who had helped to build quantum mechanics in the first place. Schwinger also felt the competition and was eager to hear for the first time what his rival had done and how it compared to his own results.

As Feynman himself would later admit, his whole thinking was physical, doing calculations by "cut and try methods" that he himself had invented.[27] He had at that stage discovered a mathematical expression from which all of his diagrams, rules, and formulas emerged, but the only way he could be certain that his formulas actually worked was when he got the right results by using them. He recalled that members of the audience kept asking, "Where does that formula come from?" to which he could only answer that it didn't matter, insisting, "It's the right formula." "But how do you know that?" Answer: "Because it gives the right results." This then led to the question: "How do you know the results are right?" to which he would respond: "I'll show you by using it to do one problem after another."

It was like a match between chess champions, whose opening moves are automatic thrust and parry, until the pieces are positioned in some novel configuration, at which point the real game begins. Feynman, however, never managed to get that far. His ideas were so novel and radical that the grand masters either did not understand them or objected at the basic foundations, which Feynman was taking for granted. In their view of the world, past and future are intuitively distinct and each played a well-defined role in quantum mechanics; Feynman, however, treated them on an equal footing, with electrons free to travel either backward or forward in time. Dirac asked a question, which was meaningful in the standard view but which Feynman could not answer, as it had no analog within his new vision. Bohr then objected to some technical issues, and when Feynman attempted to respond, Bohr took over the blackboard and began a discourse of his own for several minutes.[28]

At this point, Feynman realized that he had lost: His basic ideas were simply too novel. His own memory was that each member of the audience, for different reasons, thought that his presentation contained "too many gimmicks." His presentation had left his audience confused and, in comparison to the impression left by Schwinger's talk, underwhelmed.

Schwinger too had felt that Feynman was making intuitive guesses and lacked mathematical rigor. However, they compared notes in discussions away from the main agenda, over lunch and in the evenings, and although neither of them could understand the other's approach, they found with delight that they obtained the same answers. There was no doubt that Feynman and Schwinger had each found a way to calculate in QED, which worked. Quite why, or how, no one yet knew.

Enrico Fermi, realizing that he was witnessing a historic event, took voluminous notes of Schwinger's lecture. When he returned to Chicago, several doctoral students spent all of April and May working through these notes, digesting them, gradually understanding what Schwinger had done. Among these was a brilliant Chinese American, C. N. "Frank" Yang, for whom Schwinger's talk was an epiphany that would frame his own entry into this saga—as we shall see in Chapter 5.

After six weeks working through Fermi's notes of Schwinger's lecture, one of the students recalled that someone had mentioned that Feynman had also spoken. They asked Fermi what Feynman had said; neither Fermi nor two other colleagues could remember.[29] All that they could recall was that Feynman had used an idiosyncratic notation. That memory was fairly typical of the lasting impression of Feynman's contribution to QED, as of 1948.

Few, if indeed anyone at all, foresaw that Feynman's techniques would eventually become the staple diet of physics. Nor did anyone at the conference know that Feynman and Schwinger were not alone in their creation. The unusual circumstance in which the "third man" enters this story, and the irony behind his existence becoming known, forms our next thread.

Chapter 3

FEYNMAN, SCHWINGER, . . . AND TOMONAGA (AND DYSON)

Feynman and Schwinger are not alone: In Japan Sin-Itiro Tomonaga has also solved the puzzle. Freeman Dyson proves that all three theories are fundamentally the same. Feynman, Schwinger, and Tomonaga win the Nobel Prize for QED; Dyson misses out.

In Japan, Sin-Itiro Tomonaga had kept theoretical physics alive throughout the war. In 1943, intellectually isolated and driven purely by theory, Tomonaga had created a mathematical framework that confronted the infinities of QED and published it in Japanese. Lamb's shift and the anomalous magnetic moment of the electron were still for the future; indeed, Schwinger and Feynman's colleagues in the American war effort were only then developing the experimental techniques that would lead to Lamb's measurement.

The first Tomonaga knew of the dramatic experiments that had been presented at Shelter Island was not from any formal paper in the scientific literature, but an item in a Japanese newspaper. He confronted the results with his theory, discovered that they agreed, and promptly sent a letter to Oppenheimer.

What irony. Tomonaga had been creating the theory in Japan, while Oppenheimer had been overseeing Feynman, and indirectly Schwinger also, in work that would bring Japan to its knees. Only once this had been done would they in turn come up with their versions of the theory. Now Oppenheimer was the conduit that would bring the fruits of Japanese theoretical physics to the leaders of U.S. science.

English translations of papers by Tomonaga and his students soon followed. The message was becoming clear: There were now three theories on how to make infinity finite where only months before there had been none. At this point the young English mathematician Freeman Dyson entered the scene.

FREEMAN DYSON

I met Dyson in 1986, when along with scientist and author Paul Davies we were guests of the University of Adelaide, at some celebrations[1] where each of us was due to give a talk. The venue was a huge hall, which held up to a thousand people. I was pleased to see that it was at least half full for my own talk, but when Dyson's turn arrived, it was standing room only. It turned out that this was not entirely due to his reputation but due to a typing error in the publicity for his talk.

His presentation was to be based on his memoirs, titled *Disturbing the Universe.* These included his reminiscences of times with Feynman and Schwinger, which brought Dyson to his own great contributions in the unraveling of QED. However, at a time of student unrest, his talk appeared even more exciting, due to a typo in the widely distributed advertising, which promised that he would talk on the theme of "Disturbing the University."

Once it had become clear that his talk was to be about physics rather than anarchy, the numbers fell somewhat. Consequently, the social revolutionaries missed a treat. Dyson told us many stories from his long career in science and in private elaborated on his memories of the early days of QED.

During 1947 and '48, Feynman would talk about his ideas to anyone prepared to listen. Dyson was at Cornell for nine months, from September 1947 to June 1948, and listened. Early in the summer, Feynman traveled to New Mexico, and Dyson joined him on a completely unplanned trip by car for three days between Cleveland and Albuquerque. During this time he had Feynman for twenty-four hours at a stretch[2] and began to understand Feynman's philosophy.

After this experience, Dyson went to the University of Michigan in Ann Arbor, where between July 19 and August 7 Schwinger was lecturing about the Lamb shift and his approach to QED, in summer school. The lectures were in the mornings, enabling Dyson to take notes, spend the rest of each day working through them, and then press Schwinger on obscure points in the evenings.

He recalled the lecture content as a "bewildering morass" but that discussions in the evenings were very transparent. He found Schwinger's public and private personae quite different. He spent five weeks in all, talking with Schwinger about his ideas, after the lecture series had ended, until he finally understood Schwinger's techniques in depth.

By this stage he had managed to demonstrate that Tomonaga's and Schwinger's approaches were very similar, whereas Feynman's was utterly different. Yet all three gave the same results. It was clear that they were the same, but it was not at all obvious why.

At the end of August, Dyson went to California for a vacation, mulling over these ideas all the while. It was on the bus trip back from there to Chicago that he at last understood what Feynman's theory really was. Dyson could see that in fact Feynman and Schwinger had really composed the same symphony, though in different keys and with different notations. Having been so much in contact with Feynman, he realized that he could translate Feynman's ideas into a form that others would be able to understand and to provide a foundation for Feynman's "cut and try." This led to Dyson's celebrated paper[3] showing that Feynman, Schwinger, and Tomonaga had all, in their different ways, discovered the same fundamental truths. The implications were far-reaching and profound.

Recall that in the 1930s, QED gave qualitative success until pushed beyond the first approximation, at which point it gave meaningless infinite answers. Many physicists in those prewar days had been ready to give up the theory and, in the absence of any alternative, even to quit physics. The opinion was that the appearance of "infinity" indicated some fundamental flaw in the whole construct.

Dyson made great strides in proving that in Feynman's, Schwinger's, and Tomonaga's theory, the infinities could be tamed not just for the Lamb shift and the anomalous magnetic moment of the electron, but for any process involving light and electrons that you might care to compute. What had been achieved, in Dyson's opinion,[4] was the "rescue [of QED] without making any radical innovations . . . [which] was a victory for conservatism." By this Dyson means that Feynman, Schwinger, and Tomonaga had retained Dirac's basic theory and merely changed the "mathematical superstructure."

Today QED has stood the test of time for more than sixty years. Dyson described the theory as the "middle ground" of all physics. If you disregard gravity on one wing and nuclear forces on the other, QED describes the laws of atomic and molecular structure; the creation, transmission, and absorption of electromagnetic radiation; and all structures that are held together or based on the electromagnetic force. These include the physics of solids, liquids, and gases; lasers; electronics; as well as chemistry and, in principle, biochemistry, biology, and the genetic code. The failure of QED in practice to derive the properties of amino acids from first principles is not believed to be a fault of its equations but of the inability to solve them in all but relatively simple cases. For example, an engineer doesn't need QED, but the empirical rules of stress and strain ultimately result from its fundamental laws.

FEYNMAN'S DIAGRAMS

By 1949 Freeman Dyson had proved that Feynman's, Schwinger's, and Tomonaga's works are equivalent to one another and made inroads into the eventual proof that the theory was free of infinity. In the United States, Schwinger's encyclopedic formalism had attracted all the plaudits, though

Dyson alone seems to have plumbed its full depths. Feynman's pictogram scheme was much easier to apply and would become the staple diet of students forevermore.

It was at the American Physical Society meeting in January 1949 when Feynman discovered that he had "gotten ahead of the world" with something truly remarkable.[5] A physicist named Murray Slotnick had given a talk, only to be crushed by Oppenheimer, who announced to the hall that Slotnick must be wrong because his work "violated Case's theorem." No one knew what this was, not least as Kenneth Case, a postdoc at Oppenheimer's institution, had not yet published his "theorem." Slotnick was devastated, unable to reply. Oppenheimer announced that Case would give a talk about his theorem the next day.

Feynman was "angry at Slotnick's embarrassment."[6] That evening he made the calculation himself and verified that Slotnick was right. The next morning he buttonholed Slotnick and told him. Slotnick was astounded: He had spent two years on the problem, including six months of lengthy calculations, and Feynman's techniques had completed the whole project in a single evening.[7]

Slotnick and Feynman joined the throng to hear Case's talk, at the end of which Feynman stood up and announced that he had confirmed Slotnick's result. No more was heard of Case's "theorem."

Feynman, however, knew that he had created something special. For him that was the moment when he felt the "fire,"[8] which exceeded even the pleasure of his Nobel Prize. With newly found confidence, it was Feynman who carried the day at the third meeting in the quest to solve QED, held at Oldstone on the Hudson, forty miles north of New York City, on April 11–14, 1949.

By this time the renormalization problem was solved. Earlier we met the analogy of pixels, travel agents, and the need for known reference values. In QED the measurements of electric charge and of mass are two examples of such values. It turns out that everything involving the interaction between light and matter can be related to one or other of these. That the electric charge and mass are sufficient to enable QED to be renormalized means that, in our analogy of the travel accounts, there are just two agents at work: one doing the sums for the electric charge

FIGURE 3.1
FEYNMAN DIAGRAM WITH A SINGLE VERTEX

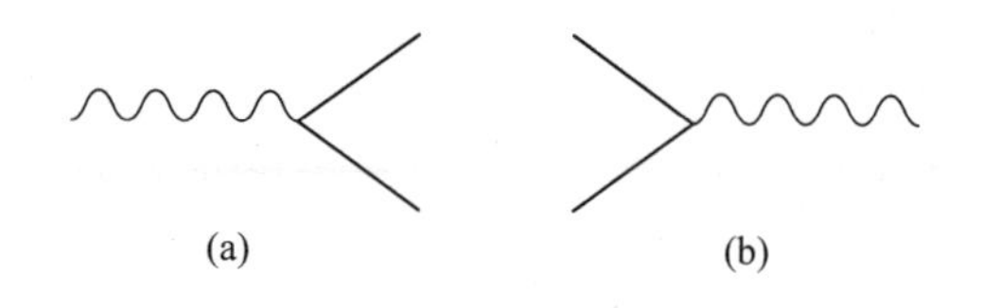

(a) One photon turns into an electron and positron, (b) or is produced by the annihilation of electron and positron.

FIGURE 3.2
FEYNMAN DIAGRAM WITH TWO VERTICES

Combine the individual templates from Figure 3.1 to give (a), a photon turns into a virtual electron and positron, which convert back to a photon. (b) An electron emits a photon and then absorbs it.

and another for the mass. As the experimental magnitude of each of these has been measured, you have enough information to compute the true value of any other quantity by comparison. That is the miracle of QED—it is possible to rescale or "renormalize" the accounts to determine the genuine, finite, amount.[9]

At last, all the calculations gave finite answers. From then on it was just a matter of being prepared to do more work, attaining higher accuracy by computing more of the "Feynman diagrams." Feynman was aware that there is actually an infinite number of diagrams, all of which would have to be computed if an exact answer is to be found. However, his diagrams formed a hierarchy, whereby some were more important than others. The strategy was to calculate the most important ones and ignore those that were insignificant. This is how it was done.

Feynman had incorporated physics into his cartoons. A photon's energy can materialize as an electron and its antiparticle, the positron. Hence, the wobbly line (photon) splitting into two—the electron and positron—in Figure 3.1a. Electric charge is conserved, so an electron lives forever unless it meets and annihilates a positron—its antiparticle—their energy being given over to the photons that emerge. The simplest diagram involves a single photon (Figure 3.1b). A photon that momentarily converts into an electron and a positron, which then annihilate, leaving a photon once more is then illustrated by the diagram in Figure 3.2a.

According to Feynman's rules, every time a further photon is linked to an electron line, the relative probability reduces by a factor of *alpha*—$\frac{1}{137}$. Consequently, the more photon-electron links there are, the less important the diagram's contribution is likely to be.

The rule of thumb with Feynman's diagrams is to draw those with the fewest such linkages, compute them, and regard the result as an approximation to the total answer. Then draw diagrams with the next fewest occurrences of *alpha*; the resulting answer will be a better approximation than before. The process is known as "perturbation" theory—you compute a number from a handful of diagrams and assume that all the more complicated diagrams at worst "perturb" the answer. This is the basic idea that had started the endeavor in the 1930s, which we met in Chapter 1, but Feynman's diagrams exposed the physical ideas explicitly.

When an electron interacts with an electromagnetic field, the simplest diagram is merely a photon sprouting from the line that represents the electron. We saw this in Figure 2.1b: It corresponds to the "first approximation," which had proved so successful for QED in its infancy. The "corrections," proportional to *alpha-cubed*, involve three junctions.[10] There are three possibilities, which are illustrated in Figures 3.3a, b, c.

Each of these diagrams contains three places where a photon joins with an electron line. They each contribute to the total sum of diagrams at "order *alpha-cubed*," which means that diagrams with four or more junctions are neglected, but those with one, two, or three are included. There is one further possibility for this example of an electron interacting with an electromagnetic field, which contains three such junctions. It is where the photon, before being absorbed by the electron, has momentarily

FIGURE 3.3
FEYNMAN DIAGRAM WITH THREE VERTICES

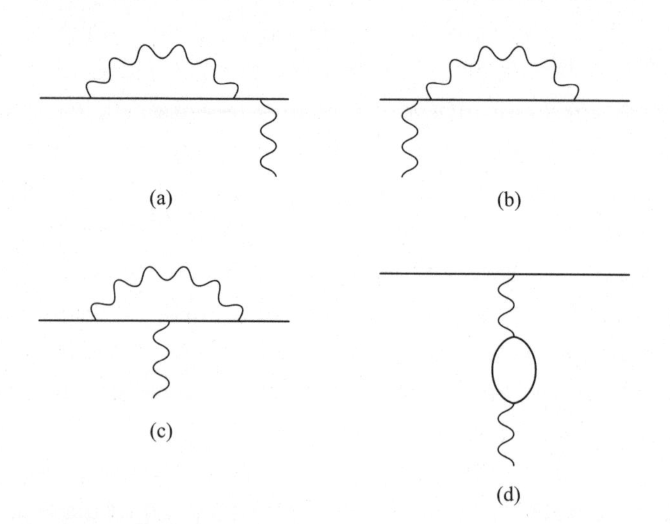

(a) An electron first emits a photon, then absorbs it, finally interacting with a magnetic field. (b) The electron interacts with the magnetic field before emitting and absorbing a virtual photon. (c) The interaction with the magnetic field occurs in the midst of the photon emission and absorption. (d) The magnetic field itself converts into a virtual electron and positron before interacting with the electron.

transformed into an electron and a positron (Figure 3.3d). Each of these diagrams contains a closed loop, which causes the mathematical expressions for each individual diagram to be plagued with the problem of infinity. It is only after all of them are taken into account that a finite answer emerges.

This accounting procedure makes the unwanted infinity disappear; in addition, the finite numbers that emerge like a rabbit from the magician's hat turn out to be precisely right—hence, the analogy in the Prologue of this being like measuring the breadth of the Atlantic Ocean to a precision of the width of a single human hair.

To calculate to this precision in QED requires evaluating diagrams with many loops. The mathematics is complicated, and the accounting involves the use of computers. I was at a seminar at Stanford in 1972 where a visiting speaker was describing his calculation of just one diagram. This had taken many hours of computation, and the seminar threatened to last not much less. Feynman was in the audience. His frustrations grew as the speaker started talking about renormalization in order to get the final result. Feynman had had enough: "What's all this renormalization about?" boomed a voice in a pronounced Brooklyn accent. The speaker, not realizing that this was Feynman himself making the interjection, started to explain renormalization as if to a rather dim student. Feynman cut in: "When I invented all this twenty-five years ago . . . ," and with the speaker turning pale as realization dawned, Feynman continued to outline his frustration that no better way had been found to solve QED other than to use perturbation theory, compute more and more diagrams, and then have to invoke renormalization to get a sensible answer.

Feynman, in common with many physicists then and since, was never completely happy with renormalization.[11] Nonetheless, it works. At the end of a long computation, the number according to theory agrees precisely with the data: Quantum Electrodynamics truly merits its acronym.

The theory had been established in the two years between the Shelter Island and Pocono conferences and has paved the way for building fundamental theories of particles and forces ever since. Oppenheimer declared that the purpose of the meetings had been achieved and closed the proceedings.

Nearly twenty years elapsed before Feynman, Schwinger, and Tomonaga shared the Nobel Prize in Physics in 1965. Perhaps the delay was because the committee had difficulty deciding how to apply the "maximum of three" rule to four such deserving scientists. As we shall see, this will not be the last time that a fourth worthy candidate got left out. In the case of QED, at least, the Nobel committee must have felt that Feynman, Schwinger, and Tomonaga had written the symphony, whereas Dyson was more the maestro who interpreted it. Others, including some highly respected physicists, feel that a truer analogy is that the trio had written individual melodies, from which it was Dyson who created the symphony. The decision not to have recognized Dyson with an award during six decades is a sad failure.

Intermission: 1950

We've reached 1950.

Infinity has been banished from QED, which successfully describes the interactions of light and matter and phenomena controlled by the electromagnetic force.

Abdus Salam is now fully briefed and understands how QED works.

Matthews suspects that there is, however, a loophole in the proofs of its viability, which Salam resolves.

Salam begins his research at the new frontier—seeking theories of the strong and weak nuclear forces, which lie beyond the reach of QED.

Chapter 4

ABDUS SALAM: A STRONG BEGINNING

Matthews reveals his suspicions; Salam meets Freeman Dyson; Salam and Matthews solve the strong force—and then nature pulls a surprise.

The news that QED was solved, with the magic ingredient being renormalization, had brought Abdus Salam up-to-date. Matthews told him that he himself was now attempting to prove the same to be true for a theory of the strong nuclear force. This would be the culmination of a program that Nick Kemmer had started before the war.

By 1935 atomic nuclei were known to consist of protons and neutrons: twins, whose only distinguishing feature was that one, the proton, carried electric charge, while the other, the neutral neutron, did not. In 1936 Kemmer—then at Imperial College in London—realized that for Yukawa's idea of the pion as carrier of the strong nuclear force to work, there would have to be a triplet of them, distinguished solely by their electric charges—one positive, one negative, and one neutral. The discovery of the electrically charged pions in 1947, together with the successful rebirth of QED soon after, gave Kemmer one of his rare moments of confidence in suggesting a research problem to a student.[1]

Following the war Kemmer had moved from Imperial College to Cambridge and became research supervisor to Paul Matthews. Thus, it was to Matthews that he made the challenge: Create a theory of the strong nuclear force involving pions and "nucleons" (a collective name for the proton and neutron) and find out if the infinities can be tamed; is it renormalizable?

Matthews had succeeded in the first step of this endeavor, but there were still some loopholes to be filled in proving its renormalizability. He suggested that Salam look at the papers by Freeman Dyson and see if the way that Dyson had managed to complete the proof of renormalization for QED could be taken over to the theory of the strong force.

THE CASE OF THE OVERLAPPING INFINITIES

Salam read Dyson's papers and a few days later burst into Matthews's office announcing that he had solved the problem. Matthews listened politely and then gently pointed out that Salam had verified what was already known, but there remained the crucial and unsolved problem of the "overlapping infinities."

Infinity can occur in an infinite number of ways. The simplest way in Quantum Electrodynamics is when an electron emits a photon and then absorbs it back again (Figure 4.1). As we have seen, in quantum theory the conservation of energy can be put on hold for a brief time; the more that the energy account is overdrawn, the shorter the time that can elapse between the emission and reabsorption of the photon. The total accounting for such a process involves the sum over all possible time scales: large energy mismatches lasting for trifling time spans, all the way down to small mismatches lasting for longer times. This infinite sum gave an infinite contribution, but Dyson and others had shown how to remove this from the accounts—the renormalization of QED. Their prescription also worked if successive examples of emission followed by absorption occurred (Figure 4.1b).

However, there was another possibility, where the successive emissions and absorptions overlap (Figure 4.1c). Instead of borrowing and then paying

FIGURE 4.1. FEYNMAN DIAGRAM WITH LOOPS

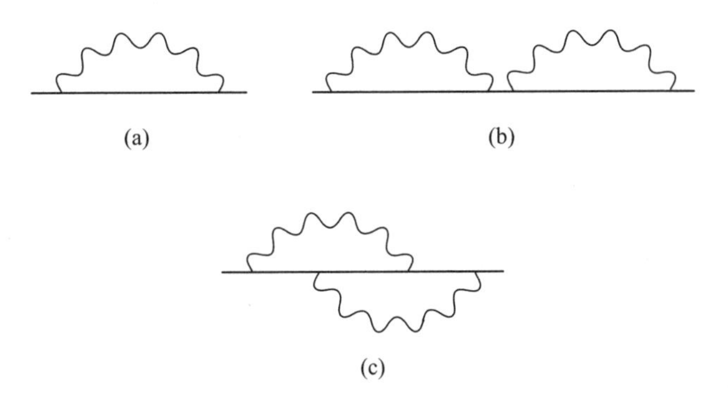

An electron emits and absorbs a single virtual photon. In (b), two of these occur in succession. In (c), emissions and absorptions overlap.

back sequentially, as in the previous examples, the overlapping case was akin to making a second withdrawal before the first was paid off. The mathematics here is trickier; each individual bubble contributing infinity, with the separate mathematical sums, each of which is infinite, intermingling en route.

If such a concept makes you blanch, then you are not alone. This was as technically complex as its name—overlapping infinities—suggests. Dyson had supposedly solved it in the case of QED, and now Matthews revealed his hand.

Matthews told Salam that Dyson had recently been the examiner of his Ph.D. thesis and during the course of the viva had asked if he had considered the case of the overlapping infinities. Matthews was surprised, as he thought that Dyson had assured everyone that this threatening tangle of mathematical complexity was actually not an issue. Matthews had taken this for granted in his own work, citing Dyson's publication as justification. How strange, Matthews thought, that Dyson should raise such a question when, apparently, everything had already been solved.

Dyson asked no more questions about it; the thesis examination ended, and Matthews passed. Nonetheless, he was uneasy. In QED, which Dyson had examined, these beasts were rare, but in the equations describing the strong force that Matthews was investigating, they came up all over the place. As he was about to go on vacation before taking up a post at

Princeton, Matthews suggested that Salam look into Dyson's overlapping-infinities solution for QED during the summer and see what progress he could make, and then when Matthews returned to Cambridge briefly before departing to the United States, they could review the state of play.

Dr. Johnson famously said that there are two types of knowledge: knowing a fact or knowing where to find it. As Dyson was normally based in the United States but was by chance at that time visiting Birmingham, Salam decided to consult the oracle. So he phoned Dyson, whom he had never met, to ask about the overlapping infinities.

Picture Dyson, preparing to depart for America, receiving this call from a complete stranger. Dyson dealt with the insistent Salam by saying that as he was leaving the next day, Salam had better come at once if he wanted to discuss the matter.

Salam had been offered a chance that was too good to miss. He set off immediately by train to Birmingham.

The trip proved invaluable because when Salam asked Dyson to explain the solution to the problem of the overlapping infinities, he was astonished by the response, "I have no solution. I only made a conjecture."[2]

A conjecture is not a guess; it is a confident belief, based on the experience gained in completing other proofs, that the conclusion will indeed be as one expects, even though the technical proof has not yet been found. Why was Dyson so sure? Unfortunately, Dyson now announced that he really had to leave for the airport. This required a two-hour train journey to London, and so Salam decided to join him.

In those days, British trains had individual compartments, where you could sit opposite one another in some comfort, as steam from the engine billowed gracefully past the windows. In the privacy of this pinnacle of nineteenth-century technology, Dyson explained to Salam his ideas about the frontiers of twentieth-century quantum physics, including the reasoning behind his conjecture. After they had parted, Salam completed his round-trip, courtesy of a further one-and-a-half-hour train journey, from London to Cambridge. During this final leg of his rail tour, he began to see how he might be able to complete the proof that Dyson had glossed over. By the time that Matthews returned from vacation, Salam had solved the problem and by September 1950 had submitted his first major paper to *Physical Review*, where it was published in April 1951.[3]

A STRONG SOLUTION—TEMPORARILY

Salam had been discussing his work with Dyson in several letters exchanged during the autumn of 1950,[4] one of which, on November 8, 1950, turned out to be special. Dyson was based at the Institute for Advanced Studies, in Princeton, New Jersey. The director of the IAS was Oppenheimer, and Dyson had spoken to him about Salam.

Dyson told Oppenheimer that the triumph of QED had been joined by Salam and Matthews's scheme for the strong force. He added that this could soon be "developed to the point where it can be compared with experiment." This led to Salam's being formally invited to Princeton, Dyson urging by writing, "Certainly you must [come] as it would be absurd to refuse such an opportunity."[5] He added a critique: that Matthews and he agreed that Salam's draft manuscript of a paper on QED was "badly written." Dyson noted that it was an "extremely difficult piece of work to write up," and for Salam, still a research student, writing research papers was a skill yet to be mastered. Dyson suggested that Salam rewrite it completely, advising him that he need not hurry, as no one else would do it ahead of him. Dyson advised further that it was crucial for Salam's reputation as a scientist, that he should "finish [the research] completely and take trouble to write it up so that people can understand it." Salam clearly took all of this to heart: Nearly two years elapsed before the paper in question appeared,[6] and his articles later in his career became notable for their fluent style.[7] Finally, Dyson offered the student Salam a further piece of advice: "Always give other people more credit than they deserve; [you will] never regret it."

Salam duly moved to Princeton early in 1951. By June 1951 he and Matthews had completed their theory of the strong force and, crucially, could prove that it was renormalizable. Salam and Matthews, it appeared, had provided the means to make precise descriptions of the atomic nucleus comparable to what QED had just achieved for atomic electrons.

For a year following their proof, they were euphoric. They believed that they had *the* theory to explain the strong force, that in the space of five years, the infinities had been expunged both here and in QED, paving the way for the "theory of everything" in atomic and nuclear physics.

However, it didn't turn out that way, though through no fault of Matthews or Salam. What they had done was mathematically correct, but the physical universe turned out to be richer than anyone had imagined. Where the duo thought that they had explained the whole strong force, discoveries in experimental physics soon showed that they had mapped but a mere corner of a vast land.

During the 1940s clear signs were pouring from the skies of things that had been undreamed of in earthbound laboratories. Cosmic rays revealed hitherto unknown varieties of particle, such as the so-called strange particles, or the muon—seemingly a heavy version of the electron. "Who ordered that?" exclaimed Isador Rabi, the physicist in whose laboratory some of the most important precision measurements of the electron had been made, which in turn had stimulated the revolution in QED. (Decades later, people are still looking for the answer to his question about the muon.) Particle accelerators were also beginning to simulate the effects of cosmic rays hitting Earth's atmosphere, spawning new particles in the process.

Matthews and Salam's theory of the strong force involved pions and nucleons, and nothing else. In the Institute for Advanced Studies at Princeton, they were divorced from experiment, isolated in a theoretical bubble. Already, the discovery of "strange" particles in cosmic rays hinted at a richer universe than that of their theory. The death knell came in 1953 when an experiment in Chicago discovered that a pion and nucleon can fuse to form a short-lived new particle, known as the "Delta Resonance." The Delta had no place in the Matthews-Salam theory. Some beautiful mathematics had been spoiled by the fact that nature sings to a different tune.

THE SMILE OF THE CHESHIRE CAT

Although Matthews and Salam's theory of the strong force explained but a small piece of what was turning into a more complicated vista, QED remained robust. No one knew why the electron had a heavier sibling, the muon, but QED worked perfectly for the muon no less than the electron.

Not everyone was so convinced that the files on QED were closed. Even Dirac was never fully happy, declaring in 1951, "A final judgment on [QED]'s merits can only come if we can define its limits." The problem was renormalization, which worked yet seemed to him to be a rule of thumb rather than a true part of the foundations. When Dyson, one of the principal architects of renormalization, asked Dirac for his opinion, Dirac replied that he would have accepted it "if it hadn't been so ugly." In 1981 Dirac told Ed Witten—widely regarded as his successor today—that the most important challenge in physics was to "get rid of infinity."[8] Feynman also felt unsatisfied with renormalization, though his famous reference to it as "brushing infinity under the rug," made when he received his Nobel Prize in 1965 for his part in establishing QED, was in response to the persistent media obsession with sound bites, such as requests to describe his Nobel Prize work "in less than 30 seconds." Feynman, in despair, finally responded, "If I could describe it in 30 seconds, it wouldn't be worth a Nobel Prize."[9]

As we shall see, today renormalization is seen not as a threat but as an opportunity. So finely balanced are the infinities that extracting finite numbers by canceling them is like walking a tightrope over Niagara Falls. The slightest deviation in the construction of QED would lead to disaster, no less than for a would-be Blondin.[10] That QED successfully works is such a marvel, it has become the paradigm in the search for the ultimate theory of all of nature.[11]

However, a serious disease still remained within the bowels of the theory—a hint that having assuaged the scourge of infinity, that very success threatened to destroy the logical foundations of the entire enterprise. Dirac was all too aware of this, but it was Lev Landau, one of the Soviet Union's most influential theoreticians, who put the problem in its sharpest form in the mid-1950s. He realized that if you tried to avoid the meaningless notion of an infinitely dense charge by supposing, instead, that the density in the central pixel was finite, its influence would be nullified by the surrounding vacuum whirlpools. What this would mean physically is that the electron's charge would be completely screened, its electrical influence made ineffective, so that it would be rendered invisible to other charged particles (which themselves would be similarly neutralized). In an atom of hydrogen, for example, between the central atomic nucleus and the remote electron, the effect of the virtual particles of matter and antimatter bubbling in the vacuum would blanket and effectively extinguish the positive charge of the central proton and the negative charge of the remote electron. Since it is the electromagnetic pull between electrons and protons that holds atoms together, the conclusion would be that atoms, and matter, could not exist.

Landau had shown that the success of Quantum Electrodynamics was like the smile on the face of the Cheshire cat. While QED's ability to compute real numbers enigmatically remained, its logical basis had completely melted away, taking atoms, you and me, and Cheshire cats with it.

Landau believed that with this argument he had destroyed quantum field theory as a way of reconciling quantum mechanics and special relativity. Under the influence of Landau, a generation of Russian physicists ignored quantum field theory.

What no one had foreseen was that there was a loophole in Landau's argument. In every mathematical example that he had examined, the sign of the beta slope (Chapter 2) was positive—as in QED, where the electric charge concentrates ever more into the innermost pixel; he never considered theories where the slope of beta is *negative*. For them, as you zoom deeper into inner space, the amount of charge in the innermost pixels becomes increasingly dilute. No paradoxes of infinite density, or of vanishing forces, arise in such theories. Landau, however, was unaware of this, not least because in 1950 no such example was known, let alone that this would turn out to be what nature uses.

A theory that would solve Landau's paradox was about to be discovered by a Cambridge research student—Ronald Shaw. His supervisor was Nick Kemmer, but in 1953 Kemmer left Cambridge to become a professor at Edinburgh. He managed to get Salam, now established as a bright star, appointed as his successor. And so, along with the job, Salam inherited Shaw.

Shaw was about to make a discovery of immense importance—a theory that solved Landau's paradox and today is the foundation for understanding particles and forces. However, it was only later that they realized its significance, after others had independently done the same. Too late would Salam realize the import of what his student had achieved.

If Salam had an Achilles' heel, it was in assessing the true importance of ideas. The case of Ron Shaw haunted him for the rest of his career. The circumstances may help also explain some of Salam's actions two decades later, when he was himself in line for a Nobel Prize. But first, here is the story of Ronald Shaw.

Chapter 5

YANG, MILLS, . . . AND SHAW

Ron Shaw makes a great discovery but doesn't publish it. Yang and Mills independently have the same insight a month later, publish, and are credited with the invention of "Yang-Mills theories"—the key to modern ideas about the fundamental forces.

Ron Shaw was born in 1929 in Stoke-on-Trent, the center of the British pottery industry. Shaw was perhaps the last person one would have expected to enter academia, as there was not a single book in his parents' home.[1] The terraced house in Pittshill, which shared an outside lavatory with three other houses, typified life for workers in the Potteries in those days. The well-off lived on the west side of town, in the hills where their mansions rose like islands above the smog emanating from the pottery kilns; the workers—such as the Shaw family—lived in the smoke, near the polluted canal and the factories.

Pictures by artist Reginald Haggar show the Potteries as they were a century ago, in the days when Shaw's parents grew up. They had left school at twelve, his father soon faking his age to enter the army in the First World War, which he survived to become an insurance agent. Ron

Shaw's mother, who was a bright girl, had also left school at the same age, much against her wishes, having been forced to end her education, Ron told me, "because she was a girl."[2]

Although by modern standards Ron Shaw might be thought to have had a deprived childhood, he recalls that it was happy, helped by being successful at school. His mother often took him to Tunstall Park, where at age three he became fascinated by watching people playing bowls, and observing the behavior of the moving bodies, colliding on the sloping greens. His natural aptitude for symmetry showed early. At the age of six he was put in charge of his school's milk register. He noticed that the Saturday morning page was completely blank, and so he "filled it in with mythical entries to complete the pattern," for which he was "severely rebuked"—English schools in those days were geared more toward rote learning than encouraging individual genius, even supposing that they were able to recognize it.

In 1940 he entered Hanley High School and the present of a chemistry set two years later sparked his interest in science, or at least in verifying whether the information about explosives, in the school textbooks, was correct. A physics teacher, who had noticed his remarkable natural talent, enrolled him for the entrance scholarship exam at Trinity College, Cambridge. The choice of Trinity appears to have been entirely because the teacher admired Isaac Newton, the college's most distinguished alumnus.

Shaw arrived there in 1949 and moved into room K9, overlooking Jesus Lane. There is nothing particularly special about this room other than the coincidence that its previous occupant was Freeman Dyson. Trinity has always hosted a galaxy of stars, and that year was no exception. Shaw's contemporaries included Michael Atiyah, the distinguished mathematician; John Polkinghorne, future professor at Cambridge and winner of the Templeton Prize for his work on science and religion; and Roger Phillips, who was later head of theory at Britain's Rutherford Laboratory, and my boss for the first twenty years of my own research career.

As a student Shaw became an excellent bridge player, which he did "most of the time," even playing in national tournaments. He became fascinated too by "fairy chess," where the moves of the chess pieces can continue indefinitely, reflecting from the sides of the board, together with other fiendish possibilities. These included knights allowed to double their jumps, making them "knightriders." Shaw "spent hours after midnight trying to decide such things as: can two knightriders force mate against a lone king?" Sixty years later, Shaw admitted to me, "I got the strong impression that mate cannot be forced, but cannot say anything authoritative." This typifies the sort of puzzle that drove him. John Polkinghorne remembers Shaw as brilliant and unworldly, with no obvious personal ambition, driven totally by curiosity and in love with ideas.

After completing his undergraduate degree he enrolled for research in 1952. His supervisor was Nick Kemmer.

Kemmer's reluctance to suggest research problems fitted nicely into Shaw's style of independent curiosity, but after a year with nothing to show for it, Shaw thought that Kemmer might propose something. However, at that point Kemmer left for Edinburgh. Early in 1954, Shaw became Salam's student.

Shaw remembers Salam as the opposite of Kemmer. He told me that Salam "had a tendency to suggest something quite different to research each time I visited." He decided to steer clear of Salam as much as possible and carried on along his own path.

One day in January 1954 he was in the library and, by chance, saw a paper by Schwinger that someone had "left lying around." It was a preprint—a copy of a paper sent out to selected institutions prior to publication. Shaw recalls its being typical of the style common in the 1950s: cyclostyled in purplish ink on rather low-quality paper.[3] In this, Schwinger had shown how a special property of the equations describing an electron requires the existence of the electromagnetic field and the ensuing force.[4] This was not new, but what struck Shaw was that the mathematics contained two strands, reflecting the north-south dual magnetism of the electron. It looked strangely familiar yet subtly different from something that he had learned from Kemmer.

Kemmer's prediction in 1936 that there are three varieties of the pion distinguished solely by their electric charges had used the mathematics of group theory known as SU2—"ess-you-two"—where numbers were replaced by arrays, that is, matrices, containing two rows and two columns. The details aren't important here (we will revisit them later), but it was this "twoness" that gave Shaw his big idea. He told me that it came to him "in a flash" when he saw Schwinger's paper. He could take Schwinger's equations, and replace their twofold structure with the mathematics of matrices: SU2.[5]

He worked out the consequences and showed them to Salam in early 1954. He recalled that he did so "in a rather disparaging way" because he had realized that his equations implied that there should be analogs of the photon, which would have no mass but carry electric charge. The problem for Shaw was that there are no such particles in nature. I asked his contemporaries if they had any memories of that episode; John Polkinghorne told me that he thought that Salam had commented that the mathematics was very interesting,[6] but with its physical implications seemingly useless, he made no push for Shaw to publish it.

Unfortunately, at the time no one realized the profound consequences of what Shaw had done. Shaw himself certainly did not, and as some years passed before Salam began to advertise this achievement, it is clear that Salam had not seen its implications, either. Shaw's mathematical game play had unwittingly stumbled on the key to the explanation of not just the strong but of all of the forces. He had discovered the equations that form the basis of what today are known as "Yang-Mills theories." The irony is that Yang and Mills themselves had not yet done so.[7]

YANG-MILLS THEORIES

Chen-Ning "Frank" Yang was born in China in 1922 but left during the war to study with Enrico Fermi at the University of Chicago. He adopted the name "Frank," after Benjamin Franklin.[8] His father had been a mathematician, with a consuming passion for that subject. Frank Yang took after his father: An awareness of the mathematical beauty of physical theories shines throughout his oeuvre.

In 1948, when the excitement of the Shelter Island and Pocono conferences was intense, Yang had been one of the graduate students, whom we met on page 53, working carefully through Fermi's notes of Schwinger's presentation. One of the many intriguing features of Schwinger's construction was his insistence that the results in QED do not depend on the particular accountancy scheme that one uses—a property known as "gauge invariance."

For example, the message in this book is the same whether you read it in English or Chinese or some other translation: It is "language invariant." However, there is not a simple replacement word by word. Concepts in Chinese, for example, differ from those in English; the nuances are not the same; some analogies may be easier in one language than another. So it is also with gauge invariance. The final result must be identical, whichever scheme you use to do the calculation, but along the way, some aspects are easier to follow in one gauge, others in another. If you want artistic beauty, then Chinese symbols have much to offer, but if you are composing text at a keyboard, using an alphabet is easier.

Gauge invariance is profound and can place severe restrictions on what is possible. In his talk at Shelter Island, Schwinger showed that gauge invariance is the principle that both causes the electromagnetic force to exist and ensures that the photon has zero mass.[9] This link between gauge invariance, the existence of a force, and the vanishing mass of a photon was a profound result, which in the course of time would have far-reaching implications.

Yang was fascinated by Schwinger's use of gauge invariance. Yang's initiation into physics was thus immersed in this ethos, and gradually he began to realize that gauge invariance might run deeper than most physicists then realized.

So, first let's take a detour to see what this central feature is.

GAUGE INVARIANCE

Midday in New York is five o'clock in London, but it takes the same number of hours to fly the Atlantic whether your watch is set to Greenwich mean or eastern standard time. The choice is up to you. The journey cannot depend on how you choose to count.

The accounting scheme or measure (GMT or EST in this example) is called the gauge; independence of the phenomenon (flying time across the Atlantic) on the choice of gauge is known as "gauge invariance." This concept now underpins our theories of nature's fundamental forces.

For gravity, when an object falls from a table, the speed that it hits the floor is the same whether that table is on the ground floor or in a room at the top of a high-rise building. It is the change in the gravitational potential—the height from the tabletop to the floor—that matters, not their individual absolute elevation. Similarly, for electric current to flow, you need a change in the voltage. Whether the difference is 240 and 0, or 1240 and 1,000, your power plugs will work equally well.[10]

For either of these examples, a single number was all that is needed—the height or the number of volts. We refer to such potentials, where a single number suffices, as "scalar." The next level of complexity involves vectors—quantities that have both size and direction. A familiar analog here is air pressure and wind speed. Air pressure gives the "potential" for wind. The wind's speed depends on how the air pressure varies from place to place: Tightly packed isobars hint at high-speed gales; widely spaced isobars are associated with calmer weather. But direction is now also important: North winds feel different from those coming from the south.

A map of the wind often portrays it as a field of arrows, the size of arrows denoting the wind speed, and they point in the wind's direction. These arrows are an array of what we call "vectors."

Air pressure is the analog of voltage: Variation in voltage leads to a "wind" of electric charge–electric current. Here, too, direction matters: An electric field—a map of the electric wind's speed and direction—is an array of vectors. The same is true for a magnetic field.

Whereas for air pressure you need just one number at each point to define the weather map, for the electric and magnetic fields you need four: three directions of space and one of time. We say that the potential is a "four-vector." The accounting is more complicated, but the essential ideas are the same. Nonetheless, the possibilities for making mistakes are larger.

If you are feeling intellectually stretched, you are not alone. Assimilating such concepts takes time for students of physics, and we are grateful for gauge invariance as a means of checking that we have not strayed from the correct path.

As if this were not enough, adding to the complexity is the fact that in quantum mechanics, atoms and their constituent particles have a wavelike character. The mathematics that describes the interactions between atomic particles must therefore encode the rise and fall of individual waves, in order to compute the result of their intermingling—determining the location of peaks and troughs of intensity, for example. Each individual wave rises to a peak and then falls away again, only to repeat itself. The accounting thus has to track how far along any individual wave the rise-and-fall has progressed.

In quantum field theory, we need two numbers at each point in space to do this. We can represent such a pair on a sheet of paper by the first number corresponding to a distance along the horizontal axis and the other representing a distance up the vertical. Alternatively, if you prefer, you could keep the accounts by representing the two numbers in terms of the radial distance from the zero point and the angle relative to the horizontal axis.

The radial distance represents the magnitude, showing whether the peak of the wave is a gentle swell or a surge, and the angle is the "phase," which shows how far one is between one peak and the next. If the range of angles from 0 to 360° were represented by the face of a clock, then relative to 12:00, 90° corresponds to 3:00, 180° to 6:00, and so on. Then, as on page 25, a wave cycle can be represented by a peak at 12:00, falling to a dip at 6:00, returning to a peak at 12:00, over and over. These possibilities are illustrated in Figure 5.1.

Drawing the axes horizontally and vertically in the figure is a choice of measure—of "gauge." You could instead choose the axes to be oriented at some other angle relative to these; so long as they are perpendicular to one another, their direction is irrelevant. The invariance of electromagnetic effects to the choice of the phase is the root of electromagnetic "gauge invariance"; basically, this symmetry says that you are free to orient the axes wherever you like—the rules are the same if you rotate the page, or your head, while looking at the figure.

The difference in phase[11] is something everyone has to agree on, but the actual value is a choice: six hours' difference on the clock face—180° in angle—always corresponds to two waves being "out of phase," whereas coincidence or a difference of twelve hours corresponds to being exactly "in phase." Two waves that mingle "in phase" give a big splash; two that are maximally out of phase will cancel.

For QED one has to keep track of all this—both the four-vector potentials and the quantum phases. There are ample opportunities for making

FIGURE 5.1
PHASE

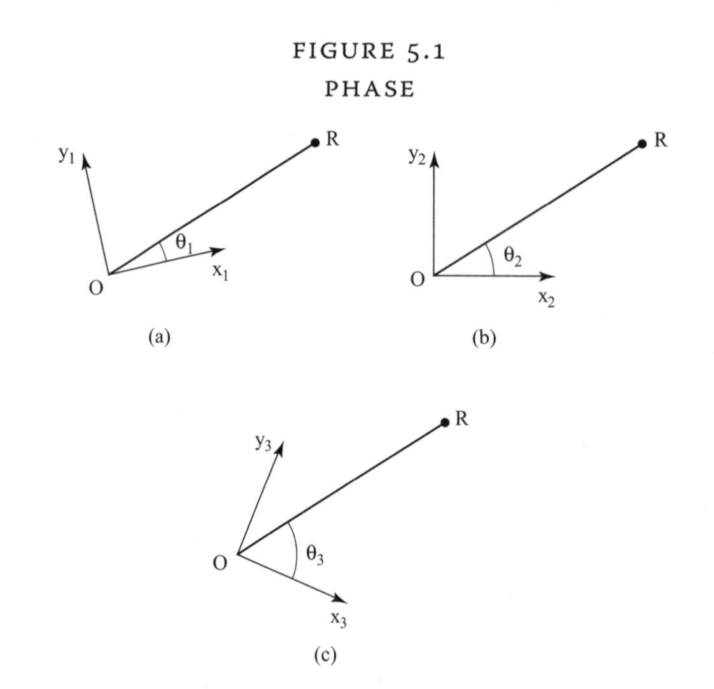

The line OR is the same in each case. The "horizontal" and "vertical" axes are identified by x and *y* in each case. Their orientation is arbitrary; all that is required is that they are orthogonal—at 90° to one another. The angles $\theta_{1,2,3}$ are the "phase" angles relative to the reference axes, the latter being the choice of "gauge."

mistakes in calculations, and some means of checking them is helpful. This is where gauge invariance comes to the theorist's aid.

One way of verifying your result is to compute it in two different gauges and confirm that this gives the same answer, even though en route the individual steps may look very different. For example, in Figure. 3.2, the magnitude of any individual diagram depends on the choice of gauge; only the sum of all diagrams with the same number of appearances of "*alpha*" give a fixed "gauge-invariant" amount.

Some of the difficulty in comparing the work of Schwinger and Feynman had been their preference for different gauges. At that early stage in building the theory of QED, gauge invariance was primarily being used as a way of checking the calculations. However, Schwinger had realized that in addition it has a profound deeper significance: Gauge invariance implies the existence of the electromagnetic force.

If gauge invariance is true at every location in the universe, the laws of electromagnetism remain the same if we alter the gauge at one location while leaving it unchanged elsewhere. If only electrons exist, it turns out that such "local" gauge invariance would be impossible. To maintain the invariance for electric charge requires some means of connecting the two places, in effect, an ability to transmit information about the local change of gauge to electric charges elsewhere. Local gauge invariance requires the agent that is carrying this information from one electron to another to be a massless particle, which has a sense of direction—hence known as a "vector." The photon has all of these characteristics—in the jargon it is a massless "vector boson," more commonly referred to as a "gauge boson."[12]

It is not necessary to add the photon into the theory; gauge invariance automatically forces the equations to include it. Thus, the existence of the electromagnetic field and its progeny, electromagnetic radiation, is a consequence of gauge invariance. If we had never seen a photon but had discovered the principle of gauge invariance for electric charge, the equations would have predicted that photons must exist. Thus, it is ultimately the symmetry of gauge invariance that enables an electron in a distant star to emit the light that another identical electron, in your retina, receives. The photons themselves carry energy and momentum from one electric charge to another, transferring motion across space. In a nutshell: Gauge invariance leads to forces and dynamics.

It turns out that gauge invariance also underpins Einstein's theory of gravity—General Relativity. As this principle led inexorably to successful descriptions of the two most familiar of nature's four fundamental forces, Yang began to wonder[13] if gauge invariance could be used to produce theories of the other two forces, the weak and strong nuclear forces.

YANG MEETS MILLS

Yang's musing about gauge invariance as the key to creating theories of other forces was right, but in 1948 he was ill-prepared to develop it. Instead, he completed his Ph.D. thesis, and moved to the Institute for Advanced Study in Princeton, New Jersey.

His idea lay dormant for five years. It was only in 1953, when Yang was spending some time at Brookhaven National Laboratory, on Long Island, New York, that at last it bore fruit. During his visit he was sharing an office with Robert Mills, a Ph.D. student from Columbia University in New York. When one theoretical physicist meets another, their conversations quickly focus on what each is interested in. Yang told Mills about his unfulfilled idea, and they decided that this was the question to pursue.

Gauge invariance doesn't come free of charge. For a theory with this mathematical feature to be successful, it must describe phenomena where some physical entity is conserved. In the case of space and time, which is the fabric on which Einstein's theory of gravity is woven, energy and momentum are conserved. For QED, it is electric charge. Yang and Mills had realized that the strong interaction also contains a conserved quantity, and they used this as the foundation for their theory.

It is easiest to appreciate the subtlety of Yang and Mills's theory by first saying what is *not* conserved. Neither the number of protons nor that of neutrons is separately conserved. Kemmer, in 1936, had realized that a neutron or proton can turn one into the other by absorbing or emitting electrically charged pions, and so cannot in general be conserved. However, Kemmer's insight gave the necessary clue.

The neutron and proton have almost the same mass and appear identical in many respects, but for the fact that a proton has electric charge. The latter affects how a proton responds to electromagnetic forces but plays no role in its affinity for the overwhelmingly powerful strong force. The strong nuclear force between a pair of protons is identical to that between two neutrons, or to that between a neutron and a proton. As this strong force could not see the difference between a neutron and a proton, the theory treated the pair as two versions of a single particle, called the nucleon. The strong interaction conserves the number of nucleons.[14]

Yang and Mills built the mathematics of gauge invariance for the pair of nucleons, which has all the tools of QED but more besides. In QED the single electric charge of an electron is conserved; in the strong interaction, by contrast, there are four possible ways for conserving nucleons. A neutron may remain a neutron, or convert into a proton; conversely, a proton may stay as it is, or convert into a neutron. The nucleon is two-headed, like Janus, and the tools that were the basis of QED—four-vectors and phases—must be modified to take this into account. The way that these variations are encoded makes use of two-by-two arrays of numbers—the same SU2 matrices that Kemmer had employed in 1936. Yang and Mills were following the same direction as Shaw—generalizing the equations of QED by replacing numbers with SU2 matrices.

They found that it is possible to take the equations of electromagnetism and allow the four-vector potential that had been at the heart of QED to have three times as many terms (hence twelve in place of four). From here on, everything is similar to QED, but with an added richness arising from the threefold number of terms.

This triplication relative to QED has an immediate physical consequence: Instead of the electrically neutral massless photon, which emerges from the gauge invariance of electric charge, in Yang-Mills's theory of the strong force, there are three such massless "gauge bosons." One of the trio is electrically neutral, like a photon, whereas the other two are electrically charged: one positive and the other negative.

Their theory therefore had a clear prediction: that there had to be massless electrically charged bosons that transmit the strong force. Unfortunately, experiment shows that such things do not exist: There are no such things in the real world, notwithstanding the fact that they flowed out of Yang and Mills's equations. For most people, this has a clear message: You may have written a beautiful melody, but it's not one that nature uses. Or to put it more directly: Mathematics that does not describe nature is of no use to physics. Yang, however, found the music too beautiful to throw away. Sometimes the theory looks so elegant that one hopefully turns a blind eye to an isolated empirical problem in the hope that someday a loophole in the arguments will be found and the theory will reach successful maturity. There was no sign of what that might be in 1954—the solution would come ten years later, as we shall see in Chapter 8—but inspired by the mathematics, Yang gave a talk about the theory, at Princeton.

PAULI'S INTERVENTION

Unknown to Yang, Wolfgang Pauli, the acerbic theorist, had also had this idea some months earlier. He too had concluded that there had to be these massless particles; he also knew that no such things exist. So, although it was an interesting mathematical game, Pauli decided it was just that—a game, of no significance to the real world—and he dismissed it from his mind.

At least he did until February 23, 1954. That day Pauli was in the audience for the weekly seminar at Princeton. The speaker was Frank Yang—"so young and already so unknown," to paraphrase one of Pauli's cruel aphorisms. Yang recalled the occasion, with anguish, in his *Collected Papers with Commentary.*[15]

He had hardly begun to speak about his theory of the strong force when Pauli interrupted: "What is the mass of these vector bosons?" The mathematics implied that the answer was zero, but Yang was fully aware that nature had no such things as massless electrically charged particles, and so he said in reply that he did not know.

Pauli sulked for a minute or so and then repeated the question. This time Yang hedged, saying that he and Mills had thought about this, but that it was complicated, and they had not reached any definite conclusion. Ten years later the answer would be found, but in 1954 no one had reached that deeper level of insight. Pauli leaped in for the kill: "That is not sufficient excuse," he stormed.

Yang was shocked. After a few moments he decided that he could not continue with his talk and sat down. Amid general embarrassment, Oppenheimer, who was chairman of the seminar, said that they should let Yang continue. Yang did so, and Pauli said no more. The next day Yang received a note from Pauli saying, "I regret that you made it almost impossible for me to talk to you after the seminar. All good wishes. Sincerely yours, W. Pauli."

It is remarkable that Yang's career did not come to a premature end there and then, his confidence shattered. Pauli's reputation was such that a story is told that, when he died, he was given a special audience in which God would reveal the answer to one problem that had defeated Pauli in life. As we have seen, Pauli was interested in 137, and so asked for an interpretation. God explained, and Pauli listened for a while, only to make one of his classic interruptions: "No! You have made a mistake!"

As Pauli's reputation was such that he could be imagined to correct errors even in the Creation, how did Yang survive? The answer is that he loved mathematics, which held greater charm than the physics for him.

In Yang's judgment, "The beauty of the idea alone merited attention."[16] A month after Shaw had shown his own equation to Salam and decided not to publish, Yang and Mills went ahead, eventually publishing their work in October 1954, even though it did not match experiment.

YANG, MILLS, AND SHAW

In January 1954, Salam had appreciated the beauty in his student Ron Shaw's theory but had not realized its profound significance. Later that year he saw the paper by Yang and Mills and showed it to Shaw.

He later recalled in a letter to Shaw: "I still remember asking you to publish this, and you were very shy at that moment because Yang-Mills had published it already although you had done the work independently."[17] Shaw never did publish, other than in his thesis. This haunted Salam ever after; years later he always referred to "Yang-Mills-Shaw" theory.

Salam's attitude to the stumbling block—that there are no massless electrically charged particles in nature—seems to have altered following the appearance of Yang and Mills's paper. Inspired by the mathematical beauty, Salam decided to leave the enigma of the vector-boson masses as a problem for the future and carry on with the theory's patterns regardless. On such vague feelings, some physicist's sixth sense, Yang and Mills had taken the plunge to publish, Pauli's condemnation notwithstanding, and Shaw completed his thesis.

Events would prove Yang and Mills to be right, though their attempts to make a gauge theory of the strong nuclear force were doomed. In that quest they were barking up the wrong tree. While the electron is fundamental, and the source of the electromagnetic force is as expressed in QED, the basic seeds of the strong interaction are not protons and neutrons. Today we know that protons and neutrons are made of smaller pieces—the quarks—and it is at this layer of the cosmic onion that Yang, Mills, and Shaw's theory is realized in nature. But quarks were unknown, and undreamed of, in 1954. The glorious culmination of the Yang-Mills theory for the strong force would come eventually—but not for nearly two decades.

To claim ownership of an idea requires both articulating it and recognizing its significance. Yang, at least, promoted the idea from the start, even in the face of powerful objections, such as from Pauli. There is no evidence that Shaw and Salam saw great physical significance in what had been done, and it was only years later, when the profound implications of Yang-Mills theories had become accepted, that Salam habitually appended Shaw's name to theirs. Today these ideas are uniformly referred to as "Yang-Mills theories."

Shaw today is remarkably modest and sanguine about his role in a Eureka moment. The written record in his thesis states, "The work described in this chapter was completed . . . in January 1954, but was not published. In October 1954, Yang and Mills adopted independently the same postulate and derived similar consequences." His opinion is that although their publication date was in 1954, after he had completed his work, "Yang and Mills must have priority since it seems that their research was completed in 1953."[18]

Fourteen years later, in 1967, Salam himself would have a similar experience, where failure to realize significance and publish led to his being scooped. Perhaps the sense of déjà vu explains the very different outcome then, with Salam actively promoting his work, as we shall see in Chapter 12. After 1954, however, even though Salam always insisted on adding "and Shaw" whenever anyone mentioned Yang-Mills, Shaw was soon forgotten. For his part, Shaw himself has no apparent regrets, saying, "Actually I am quite glad of this! I like a quiet life, and would not have enjoyed being pestered throughout the decades by lots of queries from researchers expecting me to be up to date with the latest developments."

A MISSED OPPORTUNITY

The fundamental particle that was known in 1954 was the electron. QED is the gauge theory built upon its electric charge. Within two years, an electrically neutral partner of the electron was found. Known as the neutrino, its existence had been predicted in 1930—by Wolfgang Pauli, no less.

The electron and neutrino are a pair of fundamental particles known as leptons, a name given to particles of matter that do not feel the strong force. They do, however, feel the weak force. Along with the electromagnetic and strong forces, this is the third and final force that measurably affects individual particles (the gravitational force being trifling for all but bulk matter). The weak force changes one variety of matter into another, as in beta radioactivity, in which a neutron converts into a proton and an electron and neutrino also emerge.

The electron and neutrino respond to the weak force as siblings, analogous to the proton and neutron's perceived twinning by the strong force. With hindsight, the fact that the electron and neutrino could be perceived by the weak force as a symmetric pair, analogous to the way that the proton and neutron were sensed by the strong force, could have led Yang and Mills to build a theory of the weak force. But that is not what happened.

Here we have an example of a theme that will permeate our narrative. This history is not a straightforward victory march, where we set out on a quest and end up at the goal by a set of direct logical steps. Instead, we are following people and their mistakes no less than their brilliant insights. When Yang and Mills produced their original work, the pieces were in place that, with benefit of hindsight, could have led to a theory of the weak interaction in the 1950s. As we shall see, such a theory eventually emerged courtesy of Yang and Mills's construction, but it is not what they had set out to do. Their motivation had been to explain the strong interaction.

Reaction to their ideas in 1954 had been cool, at best. Massless particles with electric charge do not exist, which appeared to doom the Yang-Mills theory as an explanation of either the strong or the weak interactions. Yang became intrigued by other properties of the weak interaction—its weird ability to distinguish the real world from that behind the mirror, known as violation of parity symmetry—for which he shared a Nobel Prize with his fellow Chinese American T.-D. Lee. The Yang-Mills theory, at least for Yang, lay dormant. But it would turn out that this class of theory is the key to everything that we now understand about the fundamental forces of nature.

Chapter 6

THE IDENTITY OF JOHN WARD

John Ward—a titan of QED, and self-proclaimed father of the British H-bomb. Ward has an idea on how to construct a theory of the weak force and contacts Abdus Salam.

Now is the moment to introduce Salam's longtime collaborator, the paranoid and complex physicist John Clive "J. C." Ward, whose work on Quantum Electrodynamics led Andrei Sakharov to describe him as one of six "titans" of modern physics. A young postdoctoral fellow in 1964 recalled that Ward, by then a senior figure, was hugely impressive, telling me, "There's a lot of very smart people in physics, and it takes you some time to realize who they are, but with Ward it was obvious: He radiated intelligence." A former Oxford colleague gave me another memory of him in the 1950s, when Ward had just produced his great work on QED and graduated from Oxford University. My friend's impression was of a "charming but reserved highly intelligent upper-class amateur," typical of the fashion at that time of "achieving great things without appearing to work hard."[1]

A "titan" indeed. Three of the other "titans," Feynman, Schwinger, and Tomonaga, were to receive the Nobel Prize in 1965 for their contributions to QED. Ward never did, though he won the Dannie Heineman Prize, the premier U.S. distinction for mathematical physics, in 1982, as well as the Hughes Medal of the Royal Society in 1983.

Although Ward never won the Nobel Prize, he could already have been on the short list for his work on QED in the 1950s. He certainly should have been a contender for what he did next, in setting the foundations of the theory that unified the electromagnetic and weak forces. Ward is the fulcrum between these two great sagas of the latter half of twentieth-century theoretical physics.

Ward's career and achievements are enigmatic. Quite why he did not share in a Nobel Prize, for which he had done much of the work that led to glory for others, will remain hidden in the files of the Nobel committees until at least 2029. More secrets about Ward may be buried in Whitehall, as he contacted British prime minister Margaret Thatcher in the 1980s requesting recognition for his claim as the "father of the British H-bomb." While this is regarded as an exaggeration, there is no doubt that he did play a significant role, which brings with it an irony: Abdus Salam, who would become his longtime collaborator, friend, and nemesis, was also intimately involved with the development of atomic know-how in Pakistan.

But we are getting ahead of ourselves. To have any hope of explaining why Ward received what is widely regarded as a raw deal, we first need to understand his personality, as revealed in his unusual career. A half century ago, he seemed almost fated to go from one disaster to another, rescued at the last moment like the heroes in popular movies of those times.

Born in East Ham, London, on August 1, 1924, Ward said nothing until, at the age of four, he spoke a complete sentence. Here we already see evidence for someone who absorbs what is going on around them but has difficulty in socializing. His subsequent education hardly helped.

In 1938 he was sent to a boarding school, where the staple diet of cold baths and sports made this unsporty loner exceedingly unhappy. In his opinion the aim of the school was to keep the boys out of danger of developing bad habits, such as "thinking by oneself." As he had no talent in the humanities, he was put into the science stream, but as the teachers "seemed to know very little science," he taught himself in the library.[2] Ward was already a confirmed outsider, and this habit of solitary contemplation "became a life-long resource." Its first success came with the award of a scholarship to join Merton College, Oxford University.

He found Oxford to be both "repelling and fascinating," where seemingly everyone was out of step but him. This remained the case until he came in contact with Maurice Pryce, the newly appointed professor of theoretical physics, with whom he produced his first important research paper, solving a problem that had defeated Dirac himself.

Apart from this piece of work, it seems that Ward did little else during his five years in Oxford, by which time he felt that he had reached a dead end. He had not written up a doctoral thesis, and in "desperation" he applied for a lectureship in mathematics at the University of Sydney, in Australia.

The Australians must have been eager to have him, as "by return of post" he learned that he had been awarded the job. He duly escaped from Oxford, only to discover upon arrival in Australia that all was not as it had seemed: His job wasn't really in the university at all. Instead, he found out, he was a tutor at a college with only a tenuous university association.

Feeling cheated and somewhat disillusioned, he spent the time enjoying the Sydney beaches, giving some mathematics courses, and waited out the year before resigning. His employers, however, were not prepared to let him go so easily. They reminded him that they had paid for his passage from Britain to Australia for what they had expected would be a long-term association: If Ward was to renege on this, then his sponsors demanded that he repay the cost of the travel that had brought him from Britain. Apparently, this condition had escaped his attention. As he also now had to buy himself a ticket to return home, he was penniless.

So by 1948 Ward found himself back in the Physics Department at Oxford, where he still had to write and submit his doctoral thesis. As he had published the paper with Pryce, which had turned out to be highly regarded, he had no worries about what was required. Indeed, in the United States, John Wheeler—the supervisor of Feynman, no less—had independently solved the same problem and, what's more, had won a national prize for his efforts. Furthermore, in Ward's opinion his own solution was superior and more complete than Wheeler's.

Ward had spoken about his work at Cambridge, two years before, where people there had also tried and, at that stage, failed to solve the challenge that Ward had completed. His host that day had been Nick Kemmer. As Kemmer was fully aware of Ward's priority (much had moved on in the interim, with others adding bricks to the foundations that Ward and Pryce had laid down), and would be the examiner of Ward's thesis, nothing could go wrong.[3]

Or so it may have seemed, and if Ward had submitted a conventional thesis, possibly nothing would have. However, his thesis was folklore among students in Oxford for years thereafter for its singular nature.

A doctoral thesis at Oxford requires not just originality but also evidence of scholarship, establishing the work within the context of the field, and other such minutiae. Ward had no time for such things. His thesis consisted of little more than his paper, made no attempt to set the work in the context of the field, and, at least in student folk wisdom, extended to fewer than ten pages. This was not a good start, and things got worse: Kemmer was unable to make the trip, and so Rudolf Peierls, professor at Birmingham University, took his place. Peierls declared the thesis unsuitable for a doctorate.

According to Ward, once the examination was ended, Peierls took him aside and privately offered him a consolation prize of a temporary job at Birmingham. Ward "refused on principle." But, like the hero of the comic strips, Ward came out on top. The Oxford authorities decided that his thesis was acceptable, Peierls's report notwithstanding, and, in Ward's words, "Peierls retired hurt from the contest."

I was a student at Oxford some twenty years after these events, and Peierls was by then professor and head of the Theoretical Physics Department. He did not confirm, at least to me, that Ward's thesis was only a few pages long, but it was open knowledge that the oral examination had lasted much of an entire day. Whether this was factual or an attempt to ensure that students in the 1960s did not make the error that Ward had done, I do not know. Having seen Ward's thesis, it seems to me that his version of events is somewhat self-serving and over-egged.[4] What is certain is that, following his successful thesis, and with Pryce's support, Ward won a scholarship from the government. This enabled him to stay in Oxford for two more years, during which he did the first work that made his name.

WARD'S IDENTITIES

In 1949, Ward delved into the deep waters of the newly born Quantum Electrodynamics. Willis Lamb had announced his discovery of the Lamb shift at the Shelter Island conference; Feynman had unveiled his new techniques, and Dyson had shown how to relate them to existing methods. Along the way, Dyson had invented a way of keeping track of the mathematics in the form of graphs, a technique developed by Feynman into what today are known as "Feynman diagrams." The tools were all there, awaiting use.

Dyson had shown how all the infinities that had polluted QED could be conjured away by reference to the electric charge and mass of the electron, which Ward saw as "an astonishing result." In addition, as we saw earlier, Dyson had conjectured, but not proved, that the infinity that emerged from certain graphs—the "overlapping infinities"—would cancel.

Ward now entered the fray. In a paper that was as short as its implications were huge, Ward announced the discovery of what have been known ever since as "Ward's Identities." These would play a central role in proving that the many complicated expressions, which lead to infinity in QED, are not independent of one another. Indeed, many of them are identical, and it is these identities that enable the miracle of renormalization to occur. Ward's Identities are the basic foundations on which the entire edifice of renormalization rests.[5]

There remained the problem of the "overlapping infinities"—the missing link that Dyson had left as a conjecture. Dyson had said, "The reader will verify for himself" that these infinities indeed cancel, but even Ward wryly wondered how many tried, as he himself only succeeded after "an immense effort." As we have already seen, Salam did so only after pursuing Dyson for two days and made his name by publishing his solution.

Ward's Identities were a sensation. He was invited to speak about renormalization at Cambridge, which he recalled for two highlights. One was that he slept at Trinity College in a four-poster bed, which had also been used by Elizabeth I, and the mattress "seemed to date from the same period." The other was that it was on that occasion, in 1950, that he made the fateful acquaintance of Abdus Salam.

TRAVELS AND TRAVAILS

As we have seen, in 1950 Salam too was working on renormalization and was about to go to Princeton with P. T. Matthews. Ward had also befriended Matthews. When Ward's Oxford contract came to its end, Matthews and Salam lobbied the authorities in Princeton to good effect: Ward was offered a year's contract at the Institute for Advanced Study during 1951–1952.

At the institute, Ward was given an office and left to get on with his thoughts. Somehow inspiration would have to arrive, and scientific papers written, or the future would be bleak. He found life in Princeton lonely; at least Oxford had had pubs, whereas Princeton seemed stuck in a post-Prohibition puritanical time warp. As for his future, he found the idea of life in a university unacceptable: Theoretical physics was in state of social upheaval.

Up until the 1940s the style of research had been that an individual talent would receive advice from a single professor, working alongside that senior scientist as an assistant. By 1950, aware of the role that physics had played in the recent war, governments poured money into research in the universities. Professors used these funds to hire research students. These same students provided cheap labor for teaching undergraduates, which released the professors from this particular treadmill, enabling them to perform more research. This in turn led to more published papers and further government contracts. The result was inflation, especially in theoretical physics.[6]

The number of professors grew, accompanied by a proliferation of journals that disseminated the results of their research. Some journals charged a fee, known as "page charges," for publishing the papers, and then sold the journals to the libraries of the very universities that had supported the research in the first place. Ward decided that this was not for him and chose to take a job at Bell Laboratories in Summit, New Jersey.

If Princeton had been lonely for a bachelor, relative to Oxford, then Summit was even worse. Once again, Ward decided that he had to get away. Maurice Pryce offered him a five-year position at Oxford, but Ward, looking for security, found an offer of a senior lectureship at the University of Adelaide to be more attractive. So began his second odyssey to Australia.

This was hardly any more successful than the first one. He soon fell out with the head of the department and gave up. Drifting seemingly from one disaster to another, Ward returned to Princeton for a year. Offers of positions at Edinburgh, from Nick Kemmer, and Oxford, from Maurice Pryce, were not followed up, as Ward felt that he was devoid of ideas and could not honorably take a position offered by a friend if he had nothing to offer.

Yet for a man with "no ideas," he had followed up his monumental work on the Ward Identities with further groundbreaking work in other areas of physics. He was internationally recognized as a first-rate theorist. As we have seen, Andrei Sakharov referred to him later as one of the "titans" of QED, comparable to Dirac, Feynman, and Dyson in importance. His singular personality, which underpinned his remarkable ability to produce outstanding pieces of physics, had a downside: Ward was already showing the inability to settle and an antagonism toward personal relationships that would place him outside the mainstream of the physics "family."

FATHER OF THE BRITISH H-BOMB

In his British homeland, powerful players also rated him highly and were noting the singular quality of his work. In Oxford, where Lord Cherwell (Frederick Lindemann) was head of the Physics Department, this was especially so. In addition to his role at the university, Lord Cherwell was science adviser to the prime minister, Winston Churchill. By this sequence of contacts did Ward return to Britain—to work at Aldermaston, developing the British H-bomb.

Churchill had instructed the UK Atomic Weapons Research Establishment (AWRE) at Aldermaston to build an H-bomb in order to have a "seat at the table." Until 1954 British weapons scientists had been able to capitalize on their participation in the Manhattan Project. However, the U.S. Atomic Energy (McMahon) Act of 1946 stopped the U.S. government from sharing its nuclear weapon information with any other country, including Britain. The British had to solve the problem alone, and with the United States and USSR beginning talks on a test-ban treaty, time was pressing. The AWRE was desperate for top-quality theoretical physicists and recruited Ward.

Ward has described taking up his post there as head of the weapons project early in 1955 and "to my amazement" finding that "I was assigned the improbable job of uncovering the secret of the Ulam-Teller invention [the technical concepts of the hydrogen bomb] . . . an idea of genius far beyond the talents of the personnel at Aldermaston."

The thermonuclear programs in the United States and USSR involved scientists of the highest caliber: Bethe, Fermi, Teller, Sakharov, and Ginzburg were all Nobel laureates; von Neumann and Ulam were among the greatest of mathematicians. The UK Atomic Energy Authority did employ some outstanding theoretical physicists in the midfifties. There is no doubt of Ward's quality, which ranked with the very best. Ward's version of events, according to his memoir, is that within six months he had worked out the principles of radiation implosion—the essential feature of the U.S. device—by himself.

In order to put it into practice, a major construction project would be needed. Had it been wartime this might have been done, but there had been bad feeling between Ward and the director of AWRE throughout. Ward subsequently claimed that his idea received the cold shoulder. Dismayed at the lack of support that he was getting, he quit and went back to the United States to the University of Maryland.[7]

Ward's peripatetic lifestyle continued. During his year at the University of Maryland he developed mathematical techniques in statistics, applicable to quantum physics, which have become standard fare today. Ward had "done it again," inspirationally producing new pathways even while he had continued to think that he would "never do anything again."

In 1956 the discovery that the weak interaction provided an absolute distinction between reality and its mirror image—"parity violation"—was sensational. This inspired Ward to return to particle physics, a subject that had flowered since his work at the long-past frontier of QED, just eight years earlier. For once, his morale was high. The enigma of parity violation focused attention on the weak force and inspired Ward to write a note to Abdus Salam about it. The die was being cast for their fateful collaboration.

THE WEAK NUCLEAR FORCE

The sun we see today shines as a result of nuclear reactions that took place in its core more than a thousand centuries ago. The solar fuel consists of protons—the nuclei of hydrogen atoms. The strong force squeezes four protons into tight clumps, and the weak force converts them into the nuclei of helium. This transmutation of hydrogen to helium also liberates energy in the form of photons. These photons then bounce back and forth in the hot inner layers of the sun, gradually working their way to the surface. It takes them about a hundred thousand years to get there, and then they are free. A mere eight minutes later, they reach Earth as daylight.

The feeble nature of the weak force makes the critical conversion of protons into helium hard to achieve. Even after 5 billion years, only half of the solar fuel has been burned. If this time span were likened to a twenty-four-hour day, intelligent life would emerge only a minute before midnight. So had the weak force been just slightly more powerful, our sun would have burned its hydrogen fuel faster, exhausting it long before intelligent life had had time to form. Had the force been feebler, it might have been unable to burn the fuel at all. If that had been the case, stars like the sun would not shine, leaving the universe a dark, soulless place.

The weak force also causes radioactivity. The most common form of this is beta decay, where, as we have seen, the nucleus of one element spontaneously transmutes into that of another by emitting energy, which materializes in the form of an electron—the "beta" particle. Initially, it was thought that the electron preexisted within the atomic nucleus, but today we know that is not so. Instead, it is created from the nuclear energy released in the radioactive decay.

By 1932, with the discovery of the neutron, followed by Wolfgang Pauli's proposal that there also exists a second neutral particle—the neutrino, or "little neutron"—all the pieces were in place for Enrico Fermi to make the first theory of beta decay. Fermi proposed that these four particles—proton, neutron, electron, and neutrino—momentarily coexist at a point in space and time. At that singular location the weak force destroys the neutron, converting it into a proton, thereby changing electric charge from neutral (the neutron) to positive (the proton). In order to conserve electric charge overall, Fermi proposed that a (negatively charged) electron is created at the same time, while the overall angular momentum of the particles is preserved by the simultaneous creation of a neutrino.

Fermi titled his paper "A Tentative Theory of Beta Rays" and submitted it for publication in the journal *Nature*, whose editor duly rejected it on the grounds that it "contained speculations too remote from reality to be of interest to the reader." Years later a subsequent editor would admit this to have been their greatest blunder.

With no electric charge, and interacting primarily by the feeble "weak" force, neutrinos are will-o'-the-wisps that can pass through Earth as easily as a bullet through a bank of fog. More than twenty years would pass before even a single one was successfully captured and identified. This discovery, in the spring of 1956, at last put all the pieces in place for Fermi's idea to be taken seriously.

Fermi had made his model of the weak force by analogy with electromagnetism. The electromagnetic force moves particles around in space, but as they move, they carry their electric charge with them. The weak force, by contrast, shifts electric charges *from* one particle *to* another (Figure 6.1). In Fermi's imagination, the weak force was akin to the electromagnetic but with this transfer of electric charge as well.

However, by 1956, it appeared that the weak force is subtler than Fermi had proposed. In his model, everything happened at a point and had no preference for one direction in space over another—in the jargon it would be a "scalar." Electromagnetic fields, by contrast, have both size and a sense of direction—they are vector fields. Furthermore, beta decay, the archetypal example of the weak force in action, was discovered in 1956 to violate parity—show an absolute distinction between left and right—which is different from electromagnetic forces, which have no such ability. The implication of all this was that the weak interaction is neither scalar, as Fermi had originally proposed, nor vector, like the electromagnetic.

In order to appreciate what happened next, and see John Ward's reemergence, here is a brief detour on how scalar, vector, or other possibilities are decided.

FIGURE 6.1
THE WEAK FORCE TRANSFERS ELECTRIC CHARGE

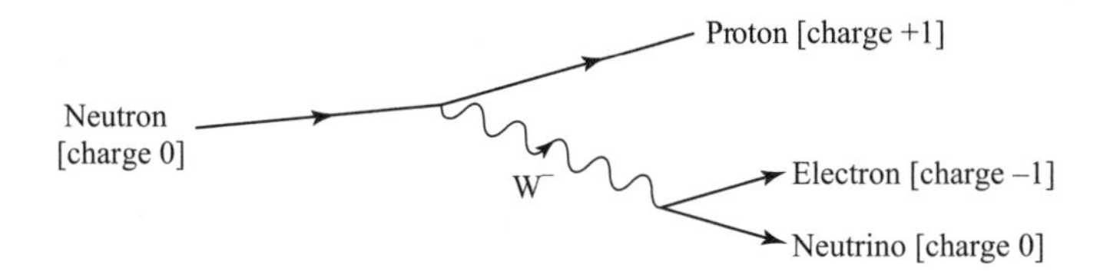

A neutron converts into a proton, along with an electron and neutrino, due to the transient presence of the electrically charged *W*.

If you fire a rifle, the recoil of the butt is in the opposite direction to where the bullet flies. The same is true if a single particle dies, turning into a pair: The progeny fly away back to back. In beta decay, not two but three particles emerge: proton, electron, and neutrino. There is more freedom in their direction of flight than in the case where only two particles appear. In the case of three, a pair of them might follow the same direction; alternatively, the three might be more evenly spaced around the circle. Measuring millions of examples of such "three-particle decays" reveals the "angular distribution" of the progeny. Once this and other properties of the particles are known—such as how they are spinning—it is possible to classify the process into one of five different possibilities. These are labeled SPTVA, which is shorthand for "Scalar, Pseudo-scalar, Tensor, Vector, and Axial." The classification gives profound information about the force that caused the decay.[8]

Processes controlled solely by the electromagnetic interaction turn out always to exhibit the classification *V*, for vector. This is intimately linked to the fact that electric and magnetic fields have not only a magnitude but also, critically, a sense of direction—hence "vector." At the level of quantum theory, the electromagnetic force is transmitted by photons. A photon spins as it flies, with an angular momentum of 1, in units of Planck's quantum. This rate of spin is characteristic of a "vector" particle. So the experimental measurements of electromagnetic phenomena showed them all to be classified in the *V* camp, and deep down this is because the force carrier, the photon, is itself a spin-1 vector particle.

When John Ward first heard that the weak interaction violates parity symmetry, he realized that this would occur if it were classified as a mixture of *V* and *A*, in a combination that has become known as $V - A$ ("*V* minus *A*"). He was not the only one who realized this fact: American Robert Marshak also had the same insight, as Ward soon discovered.[9]

In 1956 Ward was based at the University of Maryland. One day Marshak visited, and as he was a recognized authority on the weak interaction, Ward asked him if $V - A$ could be possible. Marshak immediately asked Ward why he had raised such an idea, to which Ward replied that it would be nice to have some aspect of the weak interaction that could be shared with the electromagnetic. The presence of *V* would ensure that this was the case, and the parity violation then forced the presence of *A* to accompany the *V*.

Marshak explained to Ward that experimental data disagreed with this. "How certain are we of the experimental data?" Ward asked, to which Marshak replied that at least four separate pieces of data would have to be wrong if $V - A$ were to be the answer.

The classification of the weak force appeared a mess, the results of some experiments hinting at *S* or *T*, while others preferred *V* or *A*. Clarity came with a rush. The experiments claiming *S* or *T* turned out to be flawed, and new experiments favored a combination of *V* and *A*. Marshak discussed the matter with several colleagues, including Murray Gell-Mann, a young theorist who would almost single-handedly redefine the frontiers of particle physics. However, that would come later, as we shall see.

Meanwhile, unaware of these developments, Richard Feynman had been traveling in Brazil that summer and also had realized that $V - A$ offered a mathematically tantalizing, though apparently empirically useless, possibility. On his return he asked some of his Cal Tech experimental colleagues to brief him on developments in the weak interaction, and they told him that Gell-Mann thought that the evidence for the *S* interpretation looked doubtful and that the data pointed more toward the correct classification being *V*. For Feynman, suddenly everything clicked into place. Released from the straitjacket of *S* and *T*, he realized that *V* and *A* were possible. Indeed, $V - A$ would be elegantly consistent with everything.

Feynman had come to this epiphany in his own way, having been convinced finally by the news of Gell-Mann's skepticism about the alternative

S interpretation. He was unaware of the discussions that Gell-Mann had had with Marshak, or even of Marshak's interest in the question.

Shortly afterward, Feynman gave a talk at a meeting of the American Physical Society, at which he announced his insight that $V - A$ agreed with all the evidence. He had discussed this with Gell-Mann, and a joint paper by Feynman and Gell-Mann was already entering the literature. As soon as Feynman finished his presentation, Marshak grabbed the microphone "in tears," pleading, to anyone prepared to listen, that "I was first, I was first."[10] Feynman responded, honestly, that all he knew was that he (Feynman) had been last.

Sometimes it is as if the seeds of a great idea are in the wind and flower in several places almost simultaneously. We have seen how Marshak was scooped. Another to have realized the profound opportunities that $V - A$ offered was John Ward; indeed, Ward believed that his question to Marshak in 1956 had set the train in motion.

Ward never published the $V - A$ idea himself, as he never turned his question into an answer with solid-enough foundations—that would be one of Feynman's and Gell-Mann's contributions to history. Instead, Ward saw $V - A$ point toward further profound possibilities for understanding the nature of the weak and electromagnetic forces.

Ward had realized that with $V - A$ established as the classification of the weak interaction, it might be possible to create a theory built along the lines of the successful description of electromagnetism: QED. Unaware of the work of Yang and Mills, Ward intuitively felt that it might be possible to generalize the principle of gauge invariance from QED to the richer forum of the weak force. He had foreseen that if this were so, it would predict the existence of an electrically charged analog of the photon. He put these ideas in a letter to Salam in 1957.[11]

The discovery of the $V - A$ phenomenon had given Ward the tantalizing hint that the weak force, responsible for radioactivity, and the electromagnetic force, which holds atoms together, were like the two faces of Janus, being the manifestations of a single entity—the "electroweak" force. If Ward's recollection of the history is accurate, it was he who, in this letter, first suggested to Salam that gauge invariance might lead to the theory linking weak and electromagnetic forces and innocently set Salam on the course toward a Nobel Prize. Salam's response was positive: Yes, such a generalization of QED was possible, as his student Ronald Shaw had proved in his doctoral thesis. As we have seen, any mention of "Yang-Mills" in the presence of Salam would receive the reminder: "and Shaw."

The ideas of Yang and Mills (with or without Shaw) had largely been ignored since 1954. Suddenly, everything changed. The arrival of $V - A$ as the classification of the weak interaction by 1957 inspired not just Ward but also others to realize that latent within the Yang-Mills equations was a possibility for physics to make a great advance.

However, it would turn out that there are many possible paths through the Yang-Mills jungle. Murray Gell-Mann, who also had recognized the $V - A$ option, spent a considerable amount of 1958 and '59 trying to unite weak and electromagnetic forces but failed to find a model that would fit all the facts.[12] Salam and Ward, however, were relatively slow in getting together and developing the idea. As we shall see, their first try did not agree with nature. By the time they had found the scheme that nature actually uses, they had been beaten—and by a student. Sheldon Glashow, a student of Julian Schwinger, had found a scheme to unite the weak and electromagnetic interactions, following the Yang-Mills idea, and years later his doctoral thesis would qualify him for a share of the Nobel Prize.

Chapter 7

THE MARRIAGE OF WEAK AND ELECTROMAGNETIC FORCES—TO 1964

The first attempts to build a theory of the weak force, which marry it to the electromagnetic force. Introducing Shelly Glashow, who does this first, and then forgets about it. Salam and Ward rediscover the idea three years later.

There may seem little similarity between the colors of a rainbow and the radioactivity of beta decay, but they are in fact different manifestations of a single fundamental mechanism. First, they are both the result of radiation. Electromagnetic radiation had been recognized since Maxwell produced his equations in 1860; as for radioactivity, the very name testifies to its radiant nature. The electromagnetic field is a vector field, whose radiant quanta—photons—are vector bosons. With experiments finally confirming that the mathematical underpinning of the weak interaction too involved vectors, the clues were all there: The radiant quanta of the weak force could also be vector bosons. Here at last was a quantitative similarity between two, hitherto unrelated, sets of phenomena.

Julian Schwinger, who had already played a central role in building Quantum Electrodynamics, now took the first big step toward the eventual unified theory of the electromagnetic and weak forces.

From the very beginning of his work on QED, Schwinger had been impressed with gauge invariance. Its most remarkable implication is that an electromagnetic force necessarily occurs between electrically charged bodies and that this force is carried by an agent—the "vector (or gauge) boson"—namely, the massless photon.

In November 1956 at Harvard University, Schwinger gave a series of lectures that were inspired by such ideas. The equations of electromagnetic forces in QED are the result of gauge invariance built upon numbers—the number of coulombs of electric charge carried by an electron, for example. Yang and Mills had generalized the concept of gauge invariance from numbers to matrices and applied this to the strong interaction. Schwinger realized that matrices provide a natural means for keeping the accounts when electric charge transfers from one particle to another, as in beta decay. He conjectured that it might be possible to derive the very existence of the *weak* force, as well as its properties, by using arguments similar to those of Yang and Mills. And he went further, suggesting that weak and electromagnetic forces might be two manifestations of a unique underlying theory, where for the weak force a hitherto unknown particle—the "*W*" boson—plays an analogous role to the photon of QED.[1]

As the weak force not only moves particles around but also transfers electric charge, Schwinger's *W* would have to carry this charge also. His theory required two possibilities, known as W^+ and W^-, the superscripts denoting the signs of their electric charges, whose magnitudes are the same as that of a proton and an electron, respectively. (Unless it is necessary to distinguish the positive and negative charges, I shall just refer to them collectively as the *W*). In Schwinger's theory, the role of the *W* is thus like that of a photon if it was electrically charged—see Figure 7.1.

FROM SCHWINGER TO SHELLY GLASHOW

Ever since his student days, before the Second World War, Schwinger had been intrigued by the possibility that the electromagnetic and weak forces

FIGURE 7.1
THE *W* IS LIKE AN ELECTRICALLY CHARGED PHOTON

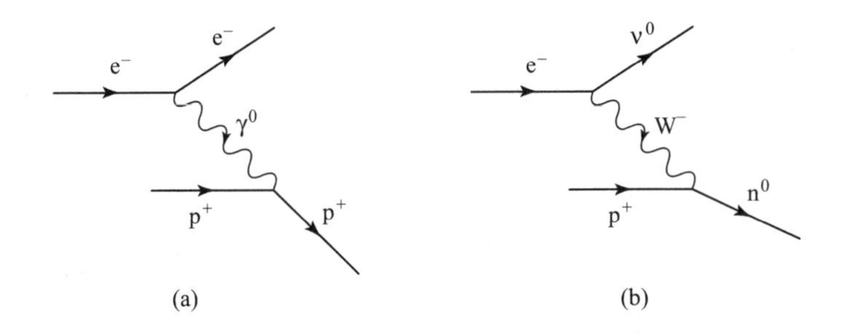

(a) An electron, negatively charged denoted e^-, bounces off a positively charged proton, denoted p^+, exchanging a photon, denoted γ^0. The superscript zero is to remind us that the photon carries no electrical charge. (b) A positively charged proton and negatively charged electron can annul their charges, converting respectively into a neutral neutron (denoted n^0) and a neutrino (denoted ν^0). Consequently, the agent of the weak force would have to transport that charge from the electron over to the proton in order for the neutral siblings to emerge. This negatively charged particle is denoted W^-.

might be unified. His inspiration initially had been little more than numerology, but from this seed has flowered the modern understanding of how nature really works.

As we have seen, calculations in QED are controlled by a number, *alpha*, whose value is 1 divided by 137. Although the student Schwinger had no explanation for its magnitude, he realized that it might be linked to an analogous parameter that governed the strength of the weak force. However, unlike 1⁄137, which is a pure number, the analogous quantity in Fermi's model of beta decay—which is caused by the weak force—has its magnitude expressed in terms of a mass. The value was trifling, being a mere one-billionth of the square of the electron's mass. It was the insignificance of this value, relative to 1⁄137, that had caused the force to become regarded as "weak."

Schwinger's insight was initially rather trivial: Why should it be the *electron*'s mass that is so important in setting the scale of the strength? In beta decay, a lightweight electron is produced, but a much heavier neutron has been converted into a proton. If instead of the electron mass Schwinger chose to set the scale by using the much larger mass of the neutron, the value of the number would rise to about 1 divided by 100,000.[2] This still was small, but no longer utterly negligible. He then realized that if a particle—the *W* boson—mediated beta decay, analogous to the role that the photon plays in QED, the mass of this *W* could set the scale. In such a case, the number could become the magic 1⁄137 if the mass of the *W* was very large. In such a case, the force is not fundamentally weak; it just appears so because of the large mass of the *W* and the consequent difficulty in creating such a beast.

While massless photons can travel across interstellar space, such that we perceive them as starlight, it is unlikely that a massive *W* will manage to cross even an atomic nucleus. Consequently, viewed from afar, the effects of a *W* would indeed appear "weak" in comparison to those of the ubiquitous photon.

When Schwinger first raised this idea in 1941, he was still a student, and J. Robert Oppenheimer, his research adviser, was skeptical about what seemed to him to be nothing other than playing with numbers. As we saw earlier, the mysterious "137" had obsessed Eddington, who had come up with bizarre, and utterly wrong, ideas as to its value. Eddington's fixation with numerology had become something of a joke, and Oppenheimer accused Schwinger of falling into a similar trap. However, by 1956 much had changed. By then Schwinger had become world famous and more confident in his judgment. Also, experiments were beginning to suggest that if a particle carries the weak force, it would have the same spin 1 as a photon. The similarity was too good to ignore: Schwinger published the idea of the *W* boson as the agent in beta decay, analogous to the photon in QED.

From the observed properties of the weak force, Schwinger deduced that the *W* would have to be very massive—at least forty times the mass of a neutron. Later refinements in the calculations elevated this to nearly ninety times, far beyond the reach of any experiment in those days, which explained why it had not been seen. Not until the 1980s would a machine exist capable of doing so, and as Chapter 16 will reveal, the existence of the *W* was duly confirmed in 1984, with a mass of about ninety times that of a neutron.

Exact symmetry between the electromagnetic and weak interactions would have implied that the *W* is massless, like a photon. However, as we have seen, this is not how nature works—there are no massless electrically charged particles of any sort, let alone a *W*. Schwinger was fully aware that the masses spoiled the symmetry of a perfect union, so he carefully avoided the word *unity*, instead describing the weak force as a "partner" of the electromagnetic.

In the 1950s it was technically difficult to study the weak force experimentally, because its effects tended to be swamped by those of the more powerful strong and electromagnetic forces. As a result, there was a lot of contradictory data, much confusion, and a feeling among some physicists that there was little point in building theories of the weak force until its properties were better established. On top of this, or perhaps to some degree because of it, Schwinger's colleagues remained cool about his idea: There was nothing in the data that demanded it; the sole motivation appeared to be that Schwinger had been seduced by the idea of unification.

In this climate, and with little chance of experiment deciding the issue one way or the other, Schwinger turned his attentions elsewhere. However, he did not give up on his pet project entirely and suggested it to a research student, Shelly Glashow. Years later, after Glashow had won a Nobel Prize for what came out of this, he recalled, "Schwinger told me to think about [unifying the weak and electromagnetic forces]. So I did. For two years—I thought about it."[3]

GLASHOW'S STORY

When the Gluchovsky family arrived in the United States as émigrés from White Russia, the immigration official transliterated the name to Glashow. Sheldon Glashow was born in 1932 in Manhattan, the youngest of three boys. His interest in science was awakened during the Second World War by his brother Sam,[4] who was a glider trooper and showed him a life-and-death application of the laws of classical mechanics. Sam had explained that when a low-flying plane drops a load of bombs, they continue to move forward in the same direction and at the same speed as the aircraft. Viewed from the plane, the bombs are falling straight down and will explode directly beneath it, unless the pilot takes evasive action.

Glashow attended the Bronx High School of Science, where one of his colleagues and friends was Steven Weinberg. Glashow and Weinberg went through school together, and next they both went to Cornell for their undergraduate studies, eventually sharing a Nobel Prize in Physics in 1979, though their paths had diverged in the interim. Whereas Weinberg's accolade was for work performed when he had become an established theorist, years later in 1967, Glashow's prizewinning opus was implicit in his doctoral thesis. After graduating from Cornell he had become a Ph.D. student at Harvard, where Julian Schwinger was based.

Glashow spent two years "thinking about" Schwinger's suggestion that the weak and electromagnetic force might be united. When he eventually produced his thesis, Glashow made a remark that a renormalizable theory of the weak force would require the weak force to be linked with the electromagnetic. This comment was based more on youthful chutzpah than profound insight, but Glashow's reasoning was this. The *W* bosons have electric charge, which will enable them to emit and absorb photons themselves; hence, the electromagnetic interactions cannot simply be ignored. While this is undoubtedly true, it is nonetheless a big jump from that to assuming that a renormalizable theory necessarily requires the two forces to be married—and, as events will prove, is not even correct.

Upon completing his thesis, he won a scholarship from the National Science Foundation, which supported his research for two years, from 1958 to 1960. He spent this period working at the Niels Bohr Institute in Copenhagen, building on the work of his doctoral thesis.

Whereas Schwinger had made his model and fitted it to the data—thus inventing the electrically charged *W* bosons to fit the facts—Glashow thought that the ideas of Yang and Mills could be the foundations for a successful theory, which would necessarily imply the existence of the W^+ and the W^-. The showstopper is that Yang-Mills theory predicts the mass of these *W* bosons to be zero—the inconvenient truth that had annoyed Pauli when he heard Yang's talk in 1954. Glashow decided to ignore this. Instead, he modified the equations to take account of the fact that the *W* has a mass. By doing so, of course, he had done away with the very Yang-Mills theory that had initially inspired him. This was a radical first step, but if his hybrid theory were to have any chance of being correct, it would have to be renormalizable. Which is where the problems began.

A quantum field theory of the weak force involving *W* bosons gave absurd answers of "infinity" for much the same reasons as had arisen in Quantum Electrodynamics. However, the trick of removing them by renormalization, which had proved so successful in QED, simply did not work for the weak force. One of the requirements for renormalization in QED had been that the photon has no mass. Hopes of building a theory of the weak force, analogous to QED, but with a massive *W*, appeared to be doomed.

If the Infinity Puzzle was to be solved, some way of accounting for the *W* mass without ruining renormalization would have to be found. It is worth pointing out why this apparently trivial difference—mass—plays such a singular role, because this will be important later when we explore the fine print of the argument in order to find the loophole.

The proof that QED can be renormalized relied on Ward's Identities. These had showed that there are only two independent sources of infinity in the theory, which can be removed by referring to the experimentally measured values for electric charge and mass of the electron. Had there been no such identities, any attempt to remove one infinity would introduce another one somewhere else in the equations; remove that and yet another would appear, and so on ad infinitum. So Ward's Identities lie at the very foundations of renormalization. The critical feature is that they themselves depend on gauge invariance.

Thus, gauge invariance not only generates the electromagnetic force and underwrites its relativistic quantum theory, QED, but also implies that the gauge boson—in this case the photon—has no mass. This latter property is key to the renormalization of QED: Gauge invariance, renormalization, and the massless photon are linked.

There is something akin to the enigma of the chicken and the egg here: Is it the massless photon or gauge invariance that is key to renormalization of QED? As the two are so intimately fused, the question was not central when QED was being used in practice, but it became more relevant when trying to apply similar ideas to the weak interaction.

Gauge invariance is both the engine of Yang-Mills theories and at the root of their renormalization. However, gauge invariance seemingly implies that the gauge boson—in the case of the weak interaction, the *W*—has no mass. Introducing a mass for the *W* into the equations by fiat, as Glashow had done, ruins gauge invariance and, in turn, renormalization. To mix metaphors: Ignoring the massless gauge boson was like throwing the baby out with the bathwater.

Ward was aware of this, as was Salam, who himself had worked on renormalization. Nonetheless, by November 1958 Glashow had convinced himself—erroneously, as it would turn out—that his theory was in fact renormalizable. He wrote a paper that was published in the journal *Nuclear Physics* on New Year's Day, 1959.[5]

During 1958 Salam too had been attempting to unite the weak and electromagnetic interaction. He had responded to Ward's suggestion that they build a theory of the weak interaction by inviting Ward to join him at Imperial College. They had read Schwinger's paper that suggested the existence of a massive *W* boson and were trying to unite the weak and electromagnetic forces themselves. They completed their paper at almost the same time as Glashow, and it was published in February 1959.[6]

Glashow's paper examined a model containing a triplet of "photons," or gauge bosons, the conventional photon being partnered by two electrically charged massive siblings. This was similar to what Salam and Ward were also doing at that juncture.

The peripatetic Ward had left Imperial College on his travels when Glashow came to give a talk in the spring of 1959. Glashow was claiming that his theory was renormalizable. The problem was that Salam and Ward, both experts on renormalization, had been unable to get rid of infinities, which kept emerging from their calculations like mushrooms. Consequently, Salam was astonished when "this young boy" claimed that the theory was viable—renormalizable—after all.[7]

During the days immediately after Glashow's visit, Salam was sufficiently worried that he and colleagues went through the arguments carefully. What could Salam and Ward have missed? In fact, they had overlooked nothing. Glashow's claim that the theory was renormalizable was completely wrong, and if Glashow had done the calculations properly, he would have shown that the infinities were unavoidable, the exact opposite of what he had thought.

Glashow was extremely embarrassed. Years later he recalled the episode as follows: "Anyone competent in quantum field theory could have spotted my error. Nonetheless, Abdus Salam invited me to speak about my work at Imperial College. My talk was well received and afterward Salam had me to his home for a marvelous Pakistani dinner. But when I returned to Copenhagen, two Imperial College preprints awaited me showing that I was wrong. Couldn't Salam simply have told me of my mistake?"[8]

Whatever the truth of this in detail, Salam's reaction was mixed. On the one hand, he must have been relieved that he and Ward had not overlooked something so basic. They were, after all, two of the world's leading experts on renormalization, and it would have crushed their morale to have missed making such a discovery themselves, leaving a graduate student to uncover the great truth. This was their first contact with Glashow, and it left Salam somewhat scornful. Salam was a bit inclined to put people in categories and not move them out again. This episode was an extreme example, Salam claiming that as a result he never again read any of Glashow's papers, although he himself later admitted that this was a mistake.[9]

Glashow wisely dropped his wild claim that the theory was renormalizable, but continued with his pursuit of a unified theory. Fortune came his way the next year, 1960, following a seminar that he gave in Paris, attended by Murray Gell-Mann. Gell-Mann was only three years older than Glashow but already had a huge reputation. By the age of thirty he had introduced the concept of "strange" particles, to explain the bizarre phenomena being discovered in cosmic rays; he was already en route to the grand classification scheme that would group the strong interacting particles—hadrons—into families, eventually explaining them as composed of more basic particles called quarks, and he had developed a method of calculation known as current algebra to account for the behavior of these particles. Also, as we have seen, he had recognized the importance of $V - A$ for encoding the weak interaction. This oeuvre was already more than that of most physicists in a lifetime; for Gell-Mann it was just a prelude. Gell-Mann's ideas seemed boundless, and he too had been musing about how to unify the weak and electromagnetic interactions. At the time he was visiting College de France in Paris, on sabbatical leave from Cal Tech, and invited Glashow to join him for lunch, to carry on discussions from where the seminar had left off.

Gell-Mann ordered fish, which was not normally Glashow's favorite, but he felt unable to refuse.[10] A half century later the memory was fresh, as he told me, "The fish was good!" Fortunately, he enjoyed the meal, and Gell-Mann must have had similar positive impressions of his guest because he invited Glashow to come and work at Cal Tech in the fall of that year. The paper that would eventually lead to Glashow's Nobel Prize matured in the weeks following that lunch and was completed by the time he arrived in California.[11] It was built along the lines that Schwinger had first articulated but with a crucial novel ingredient that took account of mirror symmetry or "parity."

Parity relates the quantum wave of a particle to its mirror image. Empirically, processes controlled by the electromagnetic interaction show a different relation to their mirror image than does beta decay. The latter is said to "violate parity symmetry"—it can distinguish left from right—whereas the electromagnetic interaction respects that symmetry and makes no distinction between left and right. Glashow imposed parity violation on his equations for beta decay—involving W^+ and W^-—in order to agree with the data, but in so doing met an immediate problem: Phenomena involving their neutral partner—the photon—will also violate parity. This is quite unlike the electromagnetic interaction, where parity is conserved.

It is at this point that Glashow might have concluded that here was the proof that his hopes for unification were doomed. He had already boldly ignored one problem, namely, the differing masses of the conventional photon and the charged W partners, by fiat, describing it as a "stumbling block we must overlook," and now his equations implied more nonsense for the mirror symmetry of the electromagnetic interaction. However, instead of seeing this further problem of parity as a threat, he turned it into an opportunity by making his major insight: "We must go beyond the hypothesis of only a triplet [of charged *W* and neutral photons] and introduce an additional neutral [member]."[12]

The missing link, which Glashow had thus provided, was that the electrically neutral and parity-violating partner of the W^+ and the W^- is not the photon. Instead, he postulated that there exists a massive neutral boson with this property, which he named Z^0.

By this stage Glashow is like some drunkard careering from one obstacle into the next. Far from uniting the weak and electromagnetic forces, by introducing the Z^0 he had actually separated them—the neutral sibling to the charged *W*'s was no longer the photon but some hitherto unknown beast. At least he had restored some symmetry. The properties of the weak force as manifested in beta decay implied that the electrically charged *W* had to be very massive, and by inventing a massive Z^0, Glashow had created a set of three massive particles, identical in all respects but for their electric charges.

With this potpourri of ingredients, running counter to all the rules, Glashow had by luck—or inspiration—found the way through the labyrinth. By introducing the Z^0—the particle of "heavy light"—Glashow had stumbled onto his ticket to immortality.

What he had ended up with was far from the equations of Yang and Mills, which had initially inspired him. In Yang-Mills gauge theory, the three gauge bosons have the same mass—zero. Glashow's model, however, contained three massive examples, the W^+, W^-, and Z^0, along with the massless photon of QED, which no longer bears any relation to this trio.

The whole setup looks so arbitrary that it is remarkable that he persevered. Indeed, the original motivation to unite the electromagnetic and weak interactions was long gone: the electromagnetic (photon) and weak (*W* and *Z*) having been cast asunder rather than wedded, as he had intended.

However, all was not lost. His interaction with Gell-Mann helped deepen his understanding of the mathematical principles underpinning Yang-Mills gauge theory. In so doing he discovered how to merge the two forces. Profound properties of quantum mechanics actually link the photon and Z^0.

To see how, we need to take a brief detour.

"ESS-YOU-TWO-CROSS-YOU-ONE"

At the beginning of the 1960s Murray Gell-Mann was becoming interested in the branch of mathematics known as group theory. This classifies sets of common things into families—the groups. The multitude of strongly interacting particles that were spewing out of experiments at particle accelerators suggested to Gell-Mann that many are siblings, their identities and properties being related by the mathematics of group theory. In parallel to this interest in the hadrons, he also investigated the mathematical groups underpinning the equations of Yang-Mills theory.

French mathematician Elie Cartan, at the end of the nineteenth century, had classified mathematical groups, among which were a set known as the special unitary (hence "SU") groups with size N, where N is any integer. The case N = 1 is just the collection of simple numbers. To a mathematician this is so "unspecial" that it is classified simply as U1: QED, with its unique photon, is mathematically a U1 gauge theory. The equations of Yang-Mills and Shaw's original theory, with three gauge bosons, correspond mathematically to an SU2 group.[13] Glashow's model of the weak and electromagnetic forces, with W^+, W^-, Z^0 as well as the photon, is thus known as an SU2 × U1 theory (pronounced "ess-you-two-cross-you-one").

What is the genealogy of these bosons? The electrically charged W^+ and W^- are unambiguously children of the SU2 family. However, identifying the heritage of the two electrically neutral bosons, the Z^0 and photon, is less simple. Quantum mechanics allows them to have mixed parentage, such that both families—U1 and SU2—can lay partial claim to each. Thus, in reality the photon is dominantly from the U1 clan but has some genetic makeup from the SU2 line; conversely, the Z^0 is mainly SU2 while carrying the U1 genes that were lost to the photon. We say that they are each "mixtures" of SU2 and U1, the amount of hybridization traditionally being expressed by the magnitude of a quantity known as the "Weinberg angle," even though Glashow was the first to introduce it.[14]

In September 1960 Glashow completed his paper, which was published in 1961.[15] In it Glashow noted, "Although we cannot say *why* the weak interactions violate parity [symmetry] while electromagnetism does not, we have shown *how* this property can be embedded in a unified model of both interactions." The key, as he presciently described it, was his introduction of the Z^0 as "the price we have to pay" for bringing the electromagnetic and weak forces together. In so doing, he had invested wisely.

Although Glashow had not unified the forces, neither had he utterly cast them adrift. Instead, he had linked them to one another, the "angle" being the measure of this intermingling, and in doing so had set the basis for modern research into a possible grand unified theory of all the forces.[16]

A NEUTRAL WEAK FORCE

The prediction of the Z^0 implied that there exists a form of the weak force that had never been seen before. The key to exposing it is the humble neutrino—the ghostly particle that has no electrical charge and so does not feel the electromagnetic force, but would, if Glashow were correct, respond to the new form of the weak force. The novel prediction that neutrinos could bounce off matter as a result of the Z^0 being exchanged was far beyond any possibility of being measured, at least in 1960, as neutrinos were so hard to detect. There was, however, another consequence of the proposed Z^0, one that unfortunately seemed to eliminate the idea at birth.

Among the plethora of particles being discovered in cosmic rays and at particle accelerators in the 1950s were some that carried a property that Gell-Mann had identified and named "strangeness." Some of these had no electrical charge and, but for having strangeness and different masses, were identical to more familiar particles that are not "strange." For example, there is a particle known as Sigma, which is identical to a neutron but for having about 20 percent more mass and, critically, being strange. According to Glashow's theory, the Sigma could convert into a neutron. However, no such transmutation has ever been seen. Empirically, "strangeness-changing neutral weak interactions" do not exist. Yet in Glashow's theory, they should.

This seems to have been the final straw. Glashow's attempt to unite electromagnetic and weak interactions had foundered originally because of parity violation. He had gotten out of jail by inventing the Z^0 and an angle to allow it to mingle with the photon. But the result was the prediction of a new form of weak interaction, one of whose implications—the existence of strangeness-changing neutral processes—was already ruled out by experiment. In conclusion he added limply, "Unfortunately our considerations seem without decisive experimental consequence."

Glashow dropped this line of work and, having moved to Cal Tech where he became inspired by Gell-Mann's interest in the applications of group theory to hadrons, turned his attention elsewhere. He showed no interest in his SU2 × U1 model for more than a decade.

Abdus Salam's first exposure to Glashow in 1959 had led him, as we have seen, to ignore Glashow's papers thereafter. By 1961 not just Salam but physicists by and large took little or no heed of Glashow's idea. However, the SU2 × U1 theory that Glashow had unwittingly discovered had been the right choice, the scheme that nature actually follows. It would be several years before this was realized, and then due to the work of others.

SALAM AND WARD: ACT 1

Glashow's thesis, in 1958, was not public knowledge outside Cal Tech, and there is no doubt that Salam and Ward's first foray into this area was completely independent of anything that Glashow had done. In December 1958 they submitted their paper titled "Weak and Electromagnetic Interactions" to the Italian journal *Il Nuovo Cimento*, where it appeared in February 1959. They had been inspired by Schwinger, as had Glashow. However, they had set off in a different direction than he did.

Salam and Ward built their model by analogy with Yang and Mills, constructing what is effectively QED but with massless photons that can also carry positive or negative electric charges—in addition to the usual neutral version.[17]

They asserted that "the theory is renormalizable," adding that "it is perhaps the only theory of charged vector [bosons] which can be renormalized." However, there is no indication that they had actually proved this; indeed, given Salam's surprise at Glashow's claims to have proved renormalization of a similar model, it seems certain that they had not. Instead, it seems to have been a conjecture based on the fact that the theory was gauge invariant, with its bosons massless, which was at the core of the proofs that QED is renormalizable.

They then made a bizarre suggestion that the phenomenon of parity violation gives the possibility that the electrically charged "photons" can be massive. They admit that the specter of infinities—lack of renormalization—would therefore appear, and say: "We propose to come back to this in a subsequent paper," although as far as I can discover, they never did. The model had no SU2 × U1 structure and no mixing angle. It has no real overlap with the phenomena that today we know the weak and electromagnetic forces exhibit in practice. In particular, there was no hint of a Z^0 in their theory.

That was the state of Salam and Ward's insight when Glashow first appeared on the scene, giving the talk at Imperial, which led to Salam's ignoring Glashow's future papers. In 2010 Glashow told me that when he visited Imperial College, he was aware already of Salam and Ward's paper and "knew it to be wrong." Although the weak interaction in it appeared to violate parity, "in fact it did not." Glashow described it to me as a "bold step in the wrong direction."[18]

Salam and Ward too seem to have lost faith with their original model, and when Ward visited Imperial College in 1964, they produced their final, more mature, theory together. Salam appears not to have read Glashow's subsequent papers, for in their 1964 work, Salam and Ward cited Glashow's 1959 paper about renormalization, which has only marginal relevance to them, and ignored his 1961 paper, which was very close to their concerns. What's more, it is Glashow's 1961 paper that bought his ticket to the Nobel ceremonies in Stockholm.

In the process, Salam and Ward discovered, like Glashow before them, that a hitherto unrecognized neutral boson was required, which leads to an SU2 × U1 mathematical structure. They called the heavy neutral boson x (today this is conventionally known as the *Z*—for zero[19]—so I shall refer to it as such hereafter). On September 24, they completed their paper with its model of the weak and electromagnetic interactions, incorporating massive *W* and *Z*, but without any understanding of how to avoid the Infinity Puzzle.

So at last Salam and Ward had independently stumbled on nature's choice, albeit three years after Glashow had already developed this model and published it. The breakthrough appears to have been that they found, as Glashow had done, that the neutral interaction in the theory violates parity. This eliminates the photon as being the neutral boson because it is known that electromagnetic interactions exhibit this symmetry. So they introduced a new neutral particle, and an angle—*theta*—that effectively is the same as Glashow had done. As concerns the new neutral, they make a comment about the appearance of the Z that uncannily parallels what Glashow had said three years earlier: It is "the minimum price one must pay to achieve the synthesis of weak and electromagnetic interactions."

Salam and Ward offered a "solution" to the problem of the unwanted strangeness-changing neutral phenomenon. They remarked that the received wisdom about the behavior of strange particles—based on a model by the Italian Nicola Cabibbo—was "incompatible with the present theory," rather than accept that their theory was incompatible with Cabibbo.[20] As Glashow later noted, the Cabibbo model is correct, the conventional wisdom being incomplete, not wrong.[21]

So by 1964, Glashow, Salam, and Ward all have an SU2 × U1 synthesis of the parity violating weak interaction, with the parity conserving electromagnetic interaction, and "the price" being a new parity violating weak neutral force, mediated by "heavy light" in the form of the Z^0. However, Salam and Ward's paper contained nothing significantly more than Glashow had already published.

A final parallel with Glashow's story is that Salam and Ward also quit pursuit of electromagnetic and weak unification. Salam turned his interests to the strong interaction and to gravity, which dominated his published work for several years, and Ward moved away from particle physics. For Salam and Ward especially, there was supreme irony: Unknown to them, at Imperial College that summer, three colleagues—Tom Kibble, Gerry Guralnik, and Dick Hagen—had found the missing link in attempts to unify the weak and electromagnetic interactions.

Within three weeks of Salam and Ward's manuscript being completed, the team of Guralnik, Hagen, and Kibble submitted their seminal paper on "Hidden Symmetry," early in October 1964, explaining how gauge bosons could become massive while maintaining gauge invariance. This paved the way for the eventual solution to the Infinity Puzzle for the weak interaction. But it seems that in the summer of 1964, when Salam and Ward had stumbled on SU2 × U1 while down the corridor their colleagues had found how to escape the straitjacket of massless gauge bosons, no one at Imperial College put two and two together.

A half century later, the concept of "Hidden Symmetry" has become the focus of theoretical physics. Investigating its manifestation in particle physics, where in the public's mind it is associated with the name of Peter Higgs, is an aim of the Large Hadron Collider at CERN. However, it turns out that nature has made use of Hidden Symmetry in a wide variety of phenomena, so much so that symmetry, and its spontaneous breaking, or "hiding," may be included among the great fundamental principles of our universe.

Intermission: 1960

We've reached 1960.

The attempt by Salam and Matthews to make a viable field theory of the strong force has failed, because nature turned out to be richer than anticipated—the solution will not be found until the 1970s, as we see in Chapters 12 and 13.

QED is flawed at very short distances or extreme energy, but there is hope that theories of the other forces may one day somehow provide a resolution of this enigma.

Yang, Mills, and Shaw have discovered a mathematical generalization of QED, with the tantalizing possibility of theories describing the weak and strong forces; the empirical problem, however, is the theory's apparent requirement that the force carrier particles have no mass—like the photon of QED.

The experimental discovery that the weak force exhibits some common features with QED—and thereby possibly involves a "vector" force carrier—has inspired Glashow and Ward to investigate the possibility of constructing a theory uniting the weak and electromagnetic interactions, notwithstanding the mass problem.

Ward has contacted his friend Abdus Salam and suggested that they pursue this line of research.

Chapter 8

BROKEN SYMMETRIES

How nature hides symmetry: in snowflakes, superconductivity, and particle physics. Jeffrey Goldstone discovers a theorem suggesting that hopes for a theory of the weak force involving Hidden Symmetry are impossible. Salam and Steven Weinberg prove this. However, Philip Anderson notices that superconductivity manages to evade the theorem, preparing the way for Peter Higgs to make his name.

We all are aware of symmetry when we see it, even if we are not mathematicians. The symmetry of the Taj Mahal is beauty beyond compare. The west front of Peterborough Cathedral, in my hometown, is an example of broken symmetry (Figure 8.1). Look behind its three Gothic archways, described by Ken Follett as like "doorways for giants," and you find a tower on the north side, but none on the south.[1]

There are two ways of breaking symmetry. Peterborough Cathedral is an example of the most familiar one: where the symmetry was never really there at all. The second example is less familiar but, nonetheless, very common. This is known as "hidden symmetry," where the fundamental laws exhibit symmetry that nature's manifestations do not.

A spiral galaxy is an example. The force of gravity spreads uniformly in all directions: It is "spherically symmetric." Individual stars, such as the

FIGURE 8.1

SYMMETRY AND ASYMMETRY

(a) Peterborough Cathedral. West Front—an example of asymmetry. (Credit: © Arcaid/Alamy)

(b) The force of gravity is spherically symmetric, but large-scale structures can hide this symmetry, as in the case of the spiral galaxy NGC1300. (Credit: © NASA) sun, viewed from afar are themselves spherically symmetric, thus exhibiting this symmetry. However, a spiral galaxy, such as our own Milky Way, is far from spherical, more like a flat plate, with structures in two dimensions but little in the third. In this case, the fundamental spherical symmetry of the force of gravity has been hidden.

This phenomenon of hidden symmetry, though unfamiliar, is widespread. The story of the Higgs Boson is an example, which is currently the focus of attention. Although that saga reached its zenith in 1964, its provenance began much earlier. So, first, let's see some examples in order to appreciate its revolutionary implications in particle physics.

HIDDEN SYMMETRY

What makes a masterpiece? When Pierre-Auguste Renoir painted *The Wave*, he began with a single brushstroke, which on its own would have been indistinguishable from anything that you or I might have done. Even after a few more splashes of color had been applied, it would be hard to tell. However, when large numbers of them are on the canvas, the image viewed close up seems to be an abstract mixture, whereas from afar a beautiful picture emerges—at least, if the artist is Renoir.

The picture emerges as a result of the organization that the impressionist has imposed. The vast numbers of brushstrokes create something beautiful that individual daubs of paint cannot. Analogously, individual electrons and protons are uninteresting—each one is boringly identical to every other. Their mutual electrical attraction can build an atom, which in turn can attract other atoms. Assemble enough of them, and they can form an organized whole, which exhibits properties that individual atoms do not have—such as an ability to think. However, we don't need to enter the mysteries of consciousness to make the point. The property of organization is already apparent with something as familiar as water.

A large number of hydrogen and oxygen atoms, collected into molecules—H_2O—can organize into different phases such as steam, water, or ice. The particular phase depends on conditions such as temperature and pressure. Ice skaters and Inuit fishermen trust the rigidity of hard-frozen ice beneath them, yet a small increase in temperature could make it melt away. Their safety relies on the organization of the individual molecules. When these are locked tightly relative to one another, the composite is solid. Warmth supplies energy, which causes individual molecules to wobble slightly, moving them away from their designated locations in the solid lattice. Initially, these deviations are so small that the assembly remains rigid, but heat them further and their jiggling becomes too violent for the group to survive intact. The constituent molecules break ranks and flow; the solid has become liquid.[2]

Temperature is a measure of the energy in a substance, in particular the kinetic energy of its molecules. The higher the temperature, the greater is the random motion. Conversely, the cooler the liquid, the slower its molecules move, until at zero degrees Celsius—"freezing point"—molecules of water tend to lock together.

In the liquid state, the individual molecules were free to move in all directions without prejudice. Their motion exemplified the rotational symmetry of the underlying laws governing the molecular dynamics. However, at zero degrees Celsius there is reorganization. The atomic jigsaw forms crystalline patterns, as seen in the frost on a winter windowpane. Thus, when water freezes, the underlying laws become hidden by a remarkable change in the way that the molecules are organized.

Any individual snowflake, such as that in Figure 8.2, looks the same if you rotate it though a multiple of 60 degrees, exhibiting "sixfold symmetry" under rotation.

At room temperature, however, the melted snowflake is a drop of water, which appears the same from any orientation. In this situation, water exhibits complete rotational symmetry, which is also the property of the basic laws controlling the behavior of its molecules. In the snowflake, however, where only six discrete rotations survive, the evidence of this fundamental symmetry has disappeared. Somehow the original symmetry has been lost. In the jargon of physics, we say that it is hidden or "spontaneously broken."

"Hidden" describes the fact that if our experiences were limited to subzero temperatures, we would see the discrete sixfold symmetry of the

FIGURE 8.2

SNOWFLAKES AND HIDDEN SYMMETRY

(a) Structure at low temperature, or low energy, emerges from rotational symmetry at high energy. The surface of still water looks the same from all angles—it is rotationally invariant. A snowflake looks the same only under rotations that are multiples of 60°. The example shown is one out of an infinite number of possible orientations. (Credit: © Kenneth Libbrecht/Science Photo Library)

(b) An individual snowflake hides the fundamental rotational symmetry, whereas a multitude of crystals on a windowpane preserve the rotational symmetry overall. (Credit: © Kenneth Libbrecht/www.SnowCrystals.com)

snowflake while the fundamental complete rotational symmetry of its basic molecules would be hidden from us. "Spontaneously broken" refers to the enigmatic change from this complete symmetry, manifested in the liquid phase, to the lesser symmetry of the solid.

This tension between symmetry and asymmetry has been known for centuries. In the fourteenth century, philosopher Jean Buridan, concerned by the dilemma of free will, is traditionally claimed to have contemplated the case of a donkey positioned midway between two identical bunches of carrots.[3] The symmetry of the situation implied that for a donkey that was itself symmetric—and philosophers can imagine such an ideal—there would be no reason for it to select the left versus the right, or vice versa. The result would be that the donkey would starve. It is hard to believe that in practice this would happen, even with an idealized symmetric donkey, but identifying the reasons will lead to deeper truths.

In reality some small perturbation, perhaps a puff of breeze, will break the impasse. Instead of a symmetric situation consisting of two bunches of carrots bisected by a dead donkey, there will be a bunch on one side and a fed living donkey on the other.

A more realistic example, which builds on the donkey's dilemma, is that of a formal dinner where the guests are uniformly positioned around a circular table.[4] Midway between each guest is a table napkin. You are therefore analogous to Buridan's donkey, in that there is a napkin to your left and to your right, and the meal cannot begin until you decide which one to choose. One of the guests, more aggressive than the rest, chooses their napkin. This breaks the symmetry and forces everyone to choose the corresponding napkin around the table. The meal can at last commence.

The napkin example contains one further phenomenon of relevance to this discussion of symmetry breaking. If the guests are very near-sighted, only the pair immediately adjacent to the aggressive guest will see which napkin has been chosen. They make their selection, which in turn forces their neighbors to do likewise. The end result is that a wave of napkin pickups moves around the entire table. This wave is the analog of what is known as a Goldstone Boson, which appears when symmetry becomes hidden in real physical systems[5] and is named after Cambridge theorist Jeffrey Goldstone, who first analyzed this possibility.

To understand the principles, physicists like to reduce a problem to its essentials. For a model we might imagine the shape at the bottom of a wine bottle. This consists of a hump at the center, which is surrounded by a circular valley forming the base, with the vertical walls of the bottle forming the far side of the valley.

Balance a small spherical ball atop the hump. Viewed from above there is complete rotational symmetry: The view from all points of the compass is the same. However, the ball is metastable, and, after the slightest disturbance, it will minimize its gravitational potential energy by rolling downhill. Once on the valley floor, its potential energy is at the minimum; we say that it has reached the "ground state." However, there are infinite numbers of possible ground states corresponding to the different points around the circle at which the ball could come to rest.

Viewed from above we now have asymmetry—a ball at the southernmost point maybe. The original symmetry remains, however, even if it is now hidden. Perform the experiment thousands of times, and keep a record of where the ball falls. After a few trials you will have a wagon wheel with spokes pointing in some directions more than others. After more repeats, the entire circle will gradually fill. Thus, a large number of experiments will leave balls positioned uniformly all around the circular valley, testifying to the rotational symmetry of the initial situation. Any individual will break the symmetry, causing the original symmetry to be hidden.

This example—the hump at the base of a wine bottle—is at the core of ideas that will later be associated with the names of Jeffrey Goldstone and Peter Higgs.

There are three dimensions, each of which plays a role in this model (Figure 8.3). The vertical one determines the amount of the potential energy, which is large at the top of the hump and a minimum in the valley floor. It is the two in the horizontal plane that interest us the most. We can summarize the location of any point by two quantities: One is the radial

FIGURE 8.3
SPONTANEOUS SYMMETRY BREAKING:
THE WINE BOTTLE OR MEXICAN HAT POTENTIAL

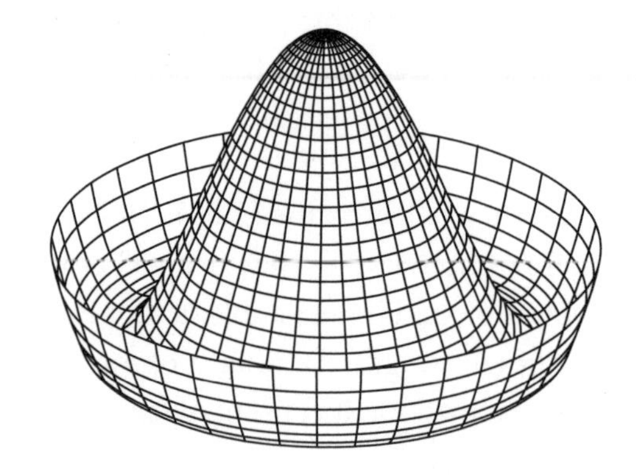

A ball resting at the top of the hill would be metastable. It rolls into the valley to a stable position, which can occur at any angle around the base. Oscillations around this base give the "Goldstone Boson." Oscillations can also occur radially up and down the walls at the side of the valley—"Higgs Boson." See also Figure 9.1.

distance from the center, and the other is its orientation relative to some arbitrarily chosen direction.

When the ball drops off the central hill, it can end up oscillating in the radial direction, up and down the walls of the valley. This finite frequency corresponds to a minimum energy. In contrast, the ball falls at random and ends up at any point around the compass with equal likelihood. There are no forces acting around the valley floor, so there are no angular oscillations.

Wherever the ball ends up around the circle, it has the same energy, namely, the potential energy corresponding to the height of the valley's floor. Each of these positions is a possible "ground state" for the ball in this model of the circular hill and valley. As every one of these infinite possible ground states has the same energy, it costs no energy to go from one to another—after all, this is nothing more than a rotation of perspective.

In quantum field theory, energy is transmitted by particles. In the case here, the energy in rotating from one possible ground state to another is zero, and so the associated particle has zero mass. This massless Goldstone Boson is the consequence of the original symmetry being broken. The memory of that symmetry is the collection of individual broken versions that froze out from the original more symmetric state. The Goldstone Boson is the link connecting them.

For oscillations in the radial direction, there is a minimum energy, which in quantum field theory corresponds to a boson without spin but with mass. In particle physics, this is what has become associated with Peter Higgs.

Although these ideas exploded into particle physics during the 1960s, they were not really new. The earliest analog, though it was not initially recognized as such, is Werner Heisenberg's model of ferromagnetism in 1928.[6] The event that eventually led to particle physicists "discovering" the concept was the explanation of the phenomenon of superconductivity.

SUPERCONDUCTIVITY

In 1911 Dutch physicist Heike Kamerlingh Onnes discovered that when cooled to 269°C, solid mercury suddenly lost all resistance to the flow of electric current. This phenomenon—"superconductivity"—was later found in other materials, such as tin and metal alloys. In a loop of wire made of superconducting material, electric currents can flow for years without needing any voltage to be applied. This astonishing phenomenon defied explanation for decades.

In 1933 German Walther Meissner wondered if the electric currents in a superconductor were carried by known particles—at that time just electrons and ions. In the process of trying to answer this, he discovered a second remarkable property: A superconducting metal expels all magnetic fields from its interior.[7]

It was not until 1956 that superconductivity was explained. There was no need for novel particles; instead, the phenomenon is a result of electrons moving through a crystal lattice of ionized atoms. Superconductivity is a phenomenon that occurs when many electrons act collectively; it is an emergent phenomenon, analogous to the beauty of Renoir's painting emerging from the organized multitude of brushstrokes.

First I will explain how superconductivity arises, and then turn to the profound, and as it transpired inspirational, importance of hidden symmetry in this case.

An electron moving through a lattice of positively charged ions experiences an electrical attraction, which causes a slight distortion of the lattice. As a bell continues to ring after having been struck, so the lattice's distortion may persist for a short while after the electron has passed. A second electron coming through finds a distorted lattice, and interacts with it. If the timing, speed, and spinning motions are right, the two interactions with the lattice cause the electrons to attract one another magnetically. They act cooperatively, like a single particle where the two spins, or individual magnetism, of the constituent electrons have canceled out.

American Leon Cooper was the first to realize this possibility, in 1956, and since then these twins have been called "Cooper Pairs."[8] Their constituent electrons are fermions—particles with half-integer spins, which in quantum mechanics act like cuckoos, where two in the same nest are forbidden. In a Cooper pair, the duo collectively has integer spin and acts like a boson.[9] Bosons, by contrast, are like penguins, where large numbers cooperate as a colony. Bosons can collect together into the lowest-possible energy state—an effect known as Bose-Einstein condensation, after the scientists whose work led to this phenomenon being understood. It is manifested in weird phenomena such as the "superfluid" ability of liquid helium to flow through narrow openings without friction and of superconductivity. In superconductivity, the Cooper Pairs act like bosons, which in concert make a "Bose condensate."

An analogy of the difference between conventional conductors, which have resistance to the flow of current and become hot, and superconductors, which offer no measurable resistance at all, is dancing in a wild nightclub in contrast to what happens when a professional troupe performs a routine onstage.[10] On a packed floor at a nightclub, hundreds of individual dancers are vigorously jiving, waving their arms, rocking from side to side, and bumping into one another. This state is like that of electrons in a metal. To model the electric field, imagine the dance floor is tilted to one side, as on a cruise ship, which is listing slightly. The force of gravity will push the dancers gently toward one side of the room. As they drift across, they continue dancing, the whole ensemble behaving quite chaotically. The more collisions you have, the more energy you waste.

This is how electrons behave in a warm metal when an electric field pushes them in one direction. The electrons move in the direction dictated by the field, meanwhile bumping into one another, losing energy as heat. An overall movement of dancers—in this case electric charge—ensues; electric current flows, but there is a lot of resistance along the way.

The Cooper Pairs in a superconductor are like professional ballroom dancers who are performing as a troupe, rather than as individual pairs. However, in this particular routine, your partner is not dancing with you cheek to cheek, but instead is far across the room, their motion mirroring your own precisely. A large number of dancers may be between you and your partner, each of them in turn being paired with another, somewhere in the crowd. The entire company performs as a coherent whole, a sense of order existing throughout the ballroom.

Any disturbance that would hinder a single dancer in the first example would have to affect the full ensemble of performers in the second case. The collective power of the organized troupe enables it to continue unimpeded; their motion—the electric current for the real case of electrons—loses no energy.

All that is required for this to happen is the existence of the organized pairs. The dynamics of the choreography will determine precisely how they go about it, but the concept itself depends only on the ability of the electrons to pair off. Today we know that this powerful pair bonding is an example of spontaneous symmetry breaking; the symmetry that has become hidden in the real case of electrons in a superconductor is gauge invariance of electromagnetism. This work would lead to two Nobel Prizes.

Having identified the concept of pairing, Cooper and his fellow Americans John Bardeen and John Schreiffer produced their mature theory of superconductivity, which has been known ever since as the "BCS Theory." They won the Nobel Prize for this work in 1972, by which time BCS Theory had inspired physicists for some fifteen years with its beautiful demonstration that nature contains previously unexpected possibilities.

They had discovered a successful choreography—itself a major breakthrough—but not the essential rhythm of hidden symmetry and its profound application to gauge invariance. These features, which give life to the BCS Theory, were revealed by the Japanese American theoretical physicist Yoichiro Nambu in 1959. It would be Nambu's insight, for which he won the Nobel Prize in 2008, that would set the pathway for particle physics for a half century.

SUPERCONDUCTIVITY AND HIDDEN SYMMETRY

Yoichiro Nambu introduced me to the stars in 1971, when we were staying at the newly built National Accelerator Laboratory, today known as Fermilab, near Chicago. He had a powerful collapsible telescope, contained in a small box. Opened up, it stood on a small tripod, which he rested on the hood of his car. This literal looking into the far distance is metaphorically what talking with Nambu can be like. Bruno Zumino, a theorist from Berkeley, once remarked that he had the idea of listening to Nambu so that he could be ten years ahead of the game. However, by the time he finally understood what Nambu had been saying, ten years had elapsed.[11]

Nambu was born in Tokyo, Japan, in 1921 and studied physics at the Tokyo Imperial University. He graduated with a bachelor of science degree in 1942. In 1950 he became a professor at Osaka University, but his talent was already being noticed internationally, and in 1952 he was invited to join the Institute for Advanced Study in Princeton, two years later joining the University of Chicago, where he became a professor.

It was at Chicago that, one day in 1956, he heard Robert Schreiffer give a seminar about the ideas that were about to burst on the world as the BCS Theory. Nambu was impressed by their boldness, but also disturbed by the fact that their theory appeared to disagree with gauge invariance. The BCS paper appeared in print in 1957, but many distrusted it for the same reasons that had disturbed him. After two years of investigation, Nambu appears to have been the first to recognize that gauge invariance does hold true in the BCS Theory but has become *hidden*. He had identified a profound truth: When the temperature gets cold enough, the fundamental patterns of electromagnetism—gauge invariance—may be hidden, as a result of which strange things happen, such as the appearance of the bosonlike Cooper Pairs.

Thus it was Nambu who, by studying the BCS Theory of Superconductivity in 1959, became inspired by the concept of hidden symmetry and then found profound implications in other areas of physics. In so doing, he gave birth to one of the most exciting series of developments of the past fifty years.

In a superconductor, the ground state contains Cooper Pairs. It costs energy to break up any pair, liberating individual electrons. Once liberated, the electrons have higher energy, the difference from their original bonding in pairs being called the "energy gap." The freed electrons receive this energy, which via $E=mc^2$ makes them appear to have gained mass. This gave Nambu an idea: If the universe itself was like a superconductor, could the masses of particles arise by some analogous mechanism?

The way that Nambu investigated this possibility was to suppose that the mass of a proton or neutron is fundamentally zero and that they acquire their masses through the spontaneous violation of some symmetry. To implement this he focused on "chiral symmetry." *Chiral* comes from the Greek for "hand," and *chirality* is a word that refers to the distinction between left- and right-handedness. This is how chiral symmetry relates to mass.

A proton can spin clockwise or anticlockwise, which we may think of as right- or left-handed, like the two possibilities for a corkscrew. Now imagine how that spin appears as you catch up and then overtake it. If it was clockwise as you approached, it will appear to be spinning backward or anticlockwise when you look back after having passed: Its chirality will change. This is fine for a massive proton, but for a massless particle, there is a profound difference. Any massless particle always travels at the speed of light, which is nature's speed limit. Nothing can move faster than this, so if you see a massless particle spinning left- or right-handed, you cannot overtake it and look back: Its chirality stays fixed.

The fundamental rule is that chirality can be conserved for massless particles but not massive ones. Nambu assumed that the fundamental laws governing the strong interactions of nucleons (protons and neutrons) are chirally symmetric, and then he investigated what happens if this symmetry is spontaneously broken.

The result was similar to what he had found in the case of superconductivity: Massive nucleons emerge from the equations. If a nucleon is left-handed, say, its antimatter counterpart is right-handed. In spontaneously broken chiral symmetry, nucleon and antinucleon "condense"—act cooperatively—forming analogs of superconductivity's "Cooper Pairs," with no chirality.

From Nambu's math emerged a massless particle.[12] In this particular case of chiral symmetry, the massless particle is known as a "pseudoscalar" boson, as its quantum wave changes sign if viewed in a mirror[13]—and this change in sign reflects the chiral symmetry that has disappeared.

His paper, published in 1960, was a remarkable success, as a boson with these very properties was already known, namely, the pion, the carrier of the strong force gripping protons and neutrons in atomic nuclei. It is not exactly massless, but is far lighter than any other strongly interacting particle.[14] Thus, Nambu had shown how the breaking of chiral symmetry could give rise to all the basic players needed to understand the atomic nucleus and the strong force that holds it together. Everything looked good.[15]

HIDDEN SYMMETRY AND PARTICLE PHYSICS

Nambu's idea of hidden symmetry had excited particle physicists, among whom were Abdus Salam and Steven Weinberg. Initially independently, they found it inspiring, potentially providing the solution to the conundrum of the tantalizing "nearly" symmetries in particle physics, among which—at least for Salam at that time—was the hope of electromagnetic and weak unification. Any euphoria, however, was short-lived.

The ensuing hiatus came as the result of the paper written by Jeffrey Goldstone, early in 1961.[16] Goldstone is highly skeptical, reluctant to publish until absolutely convinced that he is right and has something to say. That stance ought to apply to all research scientists, but Goldstone has taken it to extremes, and across a long career he has published only a handful of papers, each of which has been profound.[17] In 1961 he argued that the appearance of a massless boson is an unavoidable consequence of spontaneous symmetry breaking. What Nambu had demonstrated for chiral symmetry, with the emergence of the pion, Goldstone was arguing is a general phenomenon.[18]

In his investigation of spontaneous symmetry breaking, Goldstone identified two bosons that played a part: One was massive, the other massless.[19] Both differed from the photon or *W* boson in that they lacked the intrinsic quantum property of spin. While this was fine in the case of the pion, which has no spin, it proved disastrous for the electromagnetic and weak interaction. The carriers of these forces do have spin—they are "vector" bosons—but the real showstopper is that empirical evidence indicated that no *massless*, spinless, electrically charged Goldstone Boson associated with the weak interaction exists. Such a particle would have been easy to see, yet there is no sign of it.

Goldstone had apparently shown that hiding symmetry would imply the existence of a massless particle that empirically does not exist. The implication therefore had to be that Nambu's beautiful idea of hidden symmetry, while successful in the case of chiral symmetry and the pion, was useless for the wider hopes of particle physicists.

That summer, 1961, Salam and Weinberg were at a physics workshop in Madison, Wisconsin, during which, as Weinberg later told me, he had "several conversations with Goldstone."[20] Weinberg was based in the United States, but, following that initial discussion, he spent time that autumn working with Salam at Imperial College in London. Goldstone had given examples but, in the opinion of Salam and Weinberg, no rigorous proof that massless bosons were mandatory.

Salam had left for Zurich when Weinberg found the answer.[21] Weinberg sent an airmail letter to Salam saying, "Just after you left I found a way of proving the Goldstone 'theorem' which is very general and apparently quite rigorous. I hope it is wrong." However, it was right and formed a core part of their paper. Weinberg told me, "Salam and I worked out two rigorous proofs, one of which was heavily based on what I had learned from Goldstone himself."[22] So after drafting their paper, on March 29,

1962, Weinberg wrote to Goldstone, informing him that a manuscript had been sent to *Physical Review* and that although they had given references and acknowledgments to him throughout the paper, "after we finished it, it struck us that your name should be listed as one of the authors."[23]

The paper was by then in proofs, and so Weinberg alerted Goldstone that if he wished to be included, he should reply by cable no later than April 6, as Weinberg was departing that day and Salam by then was already in Pakistan. I found the correspondence in the archives at Trieste, where there is a copy of the Western Union cablegram, containing Goldstone's reply to Weinberg. This arrived at Imperial College just in time, at six o'clock in the evening on April 5, saying, "Make me author consequently change acknowledgements. Thank you. Goldstone." Having had Glashow badger him into producing his first paper on the subject,[24] Goldstone now became a coauthor on the proof of his ideas, courtesy of a last-minute cablegram.

This 1962 paper seemed to kill all hope of using spontaneous symmetry breaking to understand the "nearly symmetries" of nature, not least of building a theory uniting the electromagnetic and weak interactions in which the *W* and the *Z* gain mass as a result of hidden symmetry. A massive *W* might well emerge, but the theory required the existence of a further massless, spinless, electrically charged "Goldstone Boson," which nature does not possess, and this seemed to ruin everything.

Yet within eight years, Salam and Weinberg would be invoking hidden symmetry as the panacea for solving the Infinity Puzzle in the case of the weak force. Within a dozen years, by the early 1970s, the community of physicists had adopted this idea. Much was to happen in the interim.

The way forward was identified by American theorist Philip Anderson in 1962, in a paper published in 1963.[25] His breakthrough originated with a pedagogic example: what happens when electromagnetic radiation encounters some medium, such as plasma, a superconductor, or, we now realize, travels through the quantum foam that fills the vacuum. Anderson considered the first two of these, and showed that gauge invariance could be exact yet allow the associated vector particle—the photon—to act as if it has a mass. Particle physics at the LHC today is investigating the third—the possibility that the *W* and *Z* receive their masses as a result of their passage through the void.[26]

ANDERSON'S PLASMA

More than a hundred kilometers above our heads, solar radiation hits the uppermost parts of the atmosphere, splitting atoms in the electrically neutral air into negatively charged electrons and positive ions. This state of matter is known as "plasma," and this region of the atmosphere is called the ionosphere. Its most famous property is the way it affects the propagation of radio waves.

There is a huge spectrum of electromagnetic waves beyond the rainbow of colors that we can perceive, traveling at the speed of light but vibrating with different frequencies. Infrared and ultraviolet light were discovered first; radio waves soon followed; X-rays and gamma rays came later. These are all electromagnetic waves differing only in the frequency of their oscillations. In the jargon, *ultraviolet* refers to high frequencies, and *infrared*—literally "below red"—refers to relatively low frequencies.

As all electromagnetic waves travel equally fast, the shorter the wavelength is, the more frequently peaks will pass you. Thus, low-frequency oscillations correspond to relatively long wavelength radiation. The vast spectrum of long wavelengths, beyond even the infrared, constitute what are loosely called radio waves. Within these there are divisions illustrating their relative characters: long wave, medium wave, and short wave; low, medium, and high frequency.

When short- or medium-wave radio signals reach the lower edge of the ionosphere, they are refracted, like visible light upon meeting water, or even reflected, like light from a mirror. Having been turned back toward the ground, they hit the surface of the earth and bounce upward, only to be reflected by the ionosphere again. By means of this to-and-fro motion, the wave may skip across many thousands of miles, to the delight of amateur radio enthusiasts.

In the modern era of communication using the Internet and satellites, this is less of a nerd's enjoyment than in the past. However, although these phenomena are most familiar to radio hams, or users of two-way radio communications, most of us at one time or another will have noticed the quality of radio reception vary, from day to night, or during periods of intense solar activity. What is happening is that at night, the atoms in the ionosphere that have been ionized during daylight "heal";

conversely, cosmic rays, meteor showers, or solar storms can all ionize atoms in the atmosphere, changing the structure of the ionosphere, which in turn modifies its effects on radio waves.

The ability of plasma to transmit electromagnetic radiation depends both on the wavelength of the radiation and on the density of the plasma. The dependence on the wavelength is easy to illustrate: Even in the most extreme conditions, when radio communications have been disrupted, the ionosphere continues to be transparent to visible light—we can see the stars even if AM radio waves can't pass through.

As electromagnetic waves travel through space, they interact with whatever lies in their path. It is worth describing what happens when they meet plasma, as the resulting phenomena gave the inspiration that eventually solved the puzzle of how the *W* and *Z* bosons acquire mass and has led to the quest for the Higgs Boson.

An electromagnetic wave consists of intertwined electric and magnetic fields, which jostle electrically charged particles, such as electrons and ions. We know from experience that a force does not accelerate all things the same—it is easier to push a lightweight bicycle than a heavy car, for example. Likewise, electromagnetic waves kick lightweight electrons a lot, whereas heavy ionized atoms are hardly disturbed.

If all of the negatively charged electrons in the plasma are displaced by a small amount relative to the massive positive ions, the attraction of opposite charges will pull them back toward their original positions. However, because of their inertia, the electrons will carry on, overshooting their starting point before being pulled back again. The result is that the electrons in the plasma oscillate back and forth around their positions of equilibrium, causing a wave in the plasma itself. This wave oscillates at a rate known as the "plasma frequency" whose value depends on the density of the plasma. It is the magnitude of this frequency relative to that of the incoming electromagnetic wave that determines whether the incoming wave is reflected back in its tracks, as if at a mirror, or instead continues to travel through the plasma, though modified in intensity and in other ways.

The frequency of visible light is much larger than the frequency of radio waves. So it is possible for the frequency of a radio wave to be less than the plasma frequency, while that of visible light exceeds it. In the latter case the electromagnetic fields forming the light wave change direction so rapidly that the electrons in the plasma don't have time to respond: There is no interruption to the passage of the wave, and hence starlight passes through the ionosphere. Conversely, plasma oscillations overwhelm very low-frequency radio waves, and their energy is reflected.

The magnitude of the plasma frequency is proportional to the density of the plasma. Thus, in low-density plasmas, such as the gaseous ionosphere, this frequency is relatively low: The plasma cuts off radio waves while remaining transparent to starlight. Contrast this with the case of metal, where the density of electrons is high enough that the plasma frequency is in the ultraviolet region.[27] Not just radio waves but now even visible light is reflected, which gives metals their shiny appearance.

The property that only waves of high frequency can pass through plasma, while those below the plasma frequency are cut off, gave Philip Anderson his insight about the subtle ways that photons of light can sometimes act like massive particles.

Imagine living inside plasma. We would only ever be aware of electromagnetic radiation oscillating faster than the plasma frequency. Einstein showed us that the energy of each photon in an electromagnetic wave is proportional to the frequency of the wave. Thus, a minimum frequency corresponds to a minimum energy. A lower limit to the amount of energy that a particle can have is a property of a particle with a mass.[28] So in summary: The presence of plasma impedes the photon and, in effect, gives it inertia—mass.

This is only half the story. The other feature, which Anderson noticed and has further profound implications for the photon gaining mass, concerns the way that waves vibrate.

Maxwell's equations, which describe the properties of electric and magnetic fields, predict that, for light traveling in empty space, these fields vary only in the two dimensions that are perpendicular to the direction

FIGURE 8.4
TRANSVERSE AND LONGITUDINAL WAVES

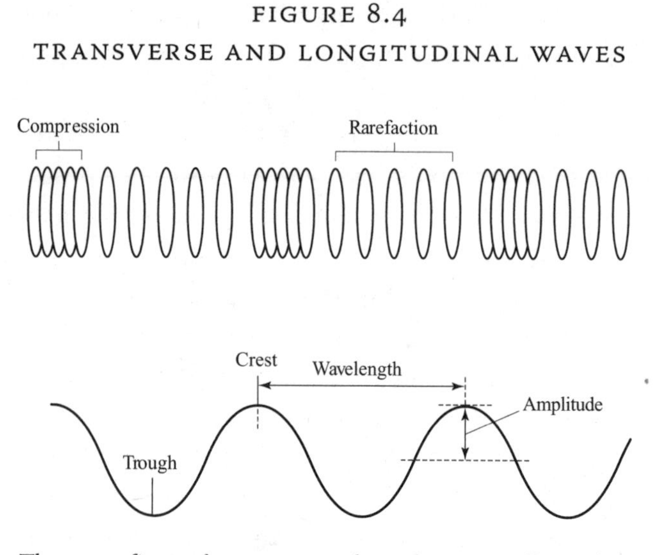

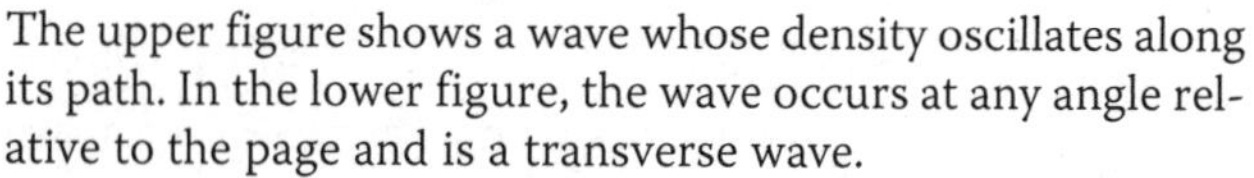

The upper figure shows a wave whose density oscillates along its path. In the lower figure, the wave occurs at any angle relative to the page and is a transverse wave.

of travel, not in all three. As a result, electromagnetic waves in free space are known as "transverse" waves. This failure to use all of the three available dimensions is profound; it is intimately connected to the gauge invariance of Maxwell's equations and the fact that photons have no mass. Had the photon been massive, the waves would have vibrated in all directions, both transverse and parallel to the direction of travel (see Figure 8.4).[29]

A wave that oscillates along its path is called a "longitudinal" wave. This is the normal state of affairs for sound waves, which are the result of alternating regions of high and low pressure in materials such as air; seismic waves, which propagate through rocks after an earthquake; or waves in the sea. However, for electromagnetic waves in a vacuum, the longitudinal wave is absent.

Anderson now realized that within plasma, an electromagnetic wave recovers this "missing" longitudinal component. Suddenly, all three dimensions are being used, and the photons have all the characteristics associated with a massive vector boson. Even more remarkable is that this has happened without spoiling the fundamental gauge invariance of the theory. So if we had lived inside plasma, our experiences of electromagnetic waves would have led us to a gauge-invariant theory where the photons have mass.[30]

With this example at the end of 1962,[31] Anderson had answered a question that Schwinger had raised about gauge invariance and mass. Schwinger had built his theory of QED with gauge invariance at its core. Everyone, Schwinger included, had believed that gauge invariance, and its intimate connection to the conservation of electric charge, automatically implied that the photon had no mass. As this also appeared to be empirically true, this was for several years assumed as a profound reason for the observed, and unique, massless property of the photon. But early in 1962 Schwinger had second thoughts.

His proof had relied on the empirical fact that the electromagnetic interactions of electrons and photons are feeble.[32] Schwinger then tried to find a general proof without making this assumption, but failed to do so. It dawned on him that there is no so such property in general, an implication being that gauge invariance and a massive carrier of a force can coexist. With the example of electromagnetic waves in plasma, Anderson had found one example, but one is sufficient to prove the point: There is no general constraint on the mass of the gauge boson, the force carrier with spin 1 in Yang-Mills theories, if other particles—such as plasma—are present.

The possibility that the *W* and *Z* bosons (at that time still hypothetical) of the weak force could be examples of such a circumstance is, with hindsight, an obvious question. Could an underlying symmetry between electromagnetic and weak interactions lead to *W*, *Z*, and the photon, all massless, and the nature of the vacuum hide this symmetry, analogous to the case of the plasma, but in this example giving mass to *W* and *Z* while leaving the photon massless? The idea that took hold was that although the gauge invariance in the equations of the Yang-Mills theory led one to expect a massless vector boson, nature in practice could override this, leading to a massive boson. As we shall see, it took a while before anyone actually applied this to the weak interaction, as attention initially focused on the strong. But first, there remained the stumbling block of Goldstone's massless spinless boson, which seemed fated to appear, an unwanted guest at the feast.

THE DISAPPEARANCE OF GOLDSTONE'S MASSLESS BOSON

Anderson prefaced the concluding paragraphs of the paper in which he had illustrated how plasma effectively gives a mass to the photon with the general observation that a "Yang-Mills type of vector boson need not have zero mass." He then asked if there was anything that forbids the vacuum in the real world from acting like the examples of plasma and superconductivity. He realized that these examples of spontaneous symmetry breaking seemed to imply the existence of a massless—and unwanted—Goldstone Boson, but then asked the pivotal question: "Is that necessarily always the case?"

Anderson answered, "Obviously no." His certainty was based not on any mathematical theorem but on empirical evidence. The phenomenon of superconductivity contained Cooper Pairs, but these had energy and were not the feared massless Goldstone Boson. Furthermore, the Meissner effect—the elimination of magnetic fields in a superconductor—implies that the photon acts as if it had gained a mass. So superconductivity appears to involve "massive" photons, courtesy of spontaneous symmetry breaking, but without any massless Goldstone Boson. This enigmatic remark about the Meissner effect merits some explanation, as it will be central to future developments.

Different types of materials affect electric and magnetic fields in different ways. An insulator, such as glass or brick, for example, is transparent to them, at least for some range of frequencies: Glass allows the passage of light, whereas a windowless room admits no light but will receive a radio signal. However, this might no longer be true if there is a lot of metal present. Radio reception is poor when your car passes through a tunnel made of concrete, which has been reinforced with metal. This is because metal is a conductor and expels *electric* fields. A normal conductor nonetheless allows *magnetic* fields to penetrate—your radio signal may be poor, but a compass still points toward northern Canada. Superconductors, however, expel both electric and magnetic fields. This "no entry" rule is not instantaneous; the electromagnetic fields can penetrate a small distance, surviving within a thin layer at the surface.

In quantum field theory, the range of a force is inversely proportional to the mass of the particle that carries it. Thus, the fact that a photon has no mass corresponds to the range of the electromagnetic force being potentially infinite: The magnetic field of the earth covers thousands of kilometers, that of the sun hundreds of millions. However, within the superconductor, the range is limited to the thin skin. This short range is understood if the carrier, the photon, has effectively gained a mass.

The BCS Theory explains all of these properties of a superconductor. It therefore is a theory with spontaneous symmetry breaking, with Cooper Pairs, and with photons acting as if they are massive: There is not a massless boson anywhere to be seen. Anderson's conclusion was that superconductivity, which had started all the interest in the first place, had also to be proof that Goldstone's theorem could not be a general truth.

Confident that this was a signpost to the promised land, Anderson in his 1963 paper explored how the theorem was being bypassed by nature. He noticed that in the BCS Theory the photon acted as if it had gained a mass, even though normally it is massless. Anderson then made his proposal. The fundamental theory of QED, which underpins the BCS Theory of Superconductivity, starts with a massless photon, and the mathematics of spontaneous symmetry breaking ought to produce also a massless Goldstone Boson along the way. However, there is no sight of any massless Goldstone Boson in the BCS Theory; indeed, there is no massless particle at all, as the photon itself acts as if it has mass. Anderson conjectured that the two massless entities—the massless photon of QED and the massless Goldstone Boson of spontaneous symmetry breaking—"seem capable of 'cancelling each other out' and leaving finite mass bosons only."[33] In some way, the photon and the massless Goldstone Boson get mixed up, with the Goldstone Boson disappearing and the photon ending up with mass.

We have seen Anderson's example of an electromagnetic wave passing through plasma and acting as if the photon became massive. That pedagogic model also showed how the two transverse oscillations of the massless photon gain the third—longitudinal—mode, which is required in the massive case. Anderson had now built on this with his proposal that the Goldstone Boson, by being absorbed within the photon, provides the "missing" longitudinal oscillation for a massive vector particle.

Although Anderson had identified the way forward, he had not actually identified any flaws in Goldstone's argument.[34] The complete solution had to be found.

There is some irony to the fact that the key to the answer was already in one of Nambu's seminal papers. Foreshadowing even Anderson's insight, in 1961 Nambu, and his collaborator Giovanni Jona-Lasinio, had remarked that in superconductivity there *would* have been Nambu-Goldstone Bosons "in the absence of Coulomb [electrostatic] interaction."[35] In effect, this recognizes that the Goldstone theorem applies only if there are no long-range forces, such as electromagnetic forces, present. Conversely, in the presence of the electromagnetic force, Goldstone's massless boson vanishes. However, it would be three more years before this was finally understood, and a half century before the remarkable implications would be pursued at the Large Hadron Collider.

Chapter 9

"THE BOSON THAT HAS BEEN NAMED AFTER ME," A.K.A. THE HIGGS BOSON

How Peter Higgs—and many others—discover the "Higgs Mechanism" for creating mass. The Higgs Boson—why it is now so important for particle physicists, why it is named after him, and how to become famous in three weeks.

"Gosh! Big's no way to describe it, though it's important in theory [5,5]."

The Higgs Boson is so famous that its anagram appears in Nick Kemmer's favorite crossword: the *Guardian*.[1] Ask why CERN in Geneva is building the Large Hadron Collider (LHC), costing some $10 billion,[2] and the stock answer will be "to discover the Higgs Boson." Yet who is Higgs? With such an expensive boson named after him, were it to be found, the popular wisdom is that a Nobel Prize would surely follow. At least, that was the version put out by the media following "Big Bang Day," when the

LHC was turned on with great hype in September 2008,[3] before an accident then delayed its operation for more than a year. While Higgs is lauded in print and in the broadcast media as the discoverer of what some have disparagingly called the "so-called Higgs Mechanism," predicting Nobel futures is as uncertain as the stock market.

Winners are often those who were carrying the baton over the finishing line after a long relay race. In the present case, the analogy is more that of a steeplechase, in which the hurdles had been erected by Goldstone, Salam, and Weinberg. Schwinger and Anderson had identified the means of clearing the obstacles, before several others reached the tape in a photo finish.

Higgs was certainly among them, as were Belgian François Englert and his American colleague Robert Brout. Abdus Salam confusingly insisted on referring to the "Higgs-Kibble mechanism," which ignores Brout and Englert as well as Tom Kibble's two collaborators—Americans Gerald Guralnik and Carl (Dick) Hagen. Kibble himself has long been embarrassed by this, as a letter from the trio revealed in the *CERN Courier* in 2008. The letter itself had been stimulated by an article in the September edition of the magazine, which had focused on Brout, Englert, and Higgs. The team of Guralnik, Hagen, and Kibble regarded the sole reference to them—"in lectures at Imperial College London, students are told about the Kibble-Higgs Mechanism, in a reference to a later paper published by Guralnik, Hagen and Kibble"—as "rather insulting."[4]

The *CERN Courier*, published on behalf of CERN by the Institute of Physics, is regarded by many as the house journal of the world community of particle physicists. Some saw timing of the article as evidence of a campaign to promote Europeans over Americans for the Nobel Prize in 2009.[5] Such conspiracy theories overlook the fact that Brout, one of those supposedly being "promoted," originates from America, and Kibble, of the injured team, is himself British.

Confusion and misinformation abound. The BBC Web site promoted the version that the "Higgs Boson" was "proposed in 1964 by physicists Peter Higgs, François Englert and Robert Brout."[6] It is indeed true that Brout and Englert beat Higgs into print by just two weeks and were recognized in 2004 when they and Higgs shared the Wolf Prize, widely regarded as the most prestigious award after the Nobel itself; however, as we shall see, although they did introduce the concept of the Higgs Field, there is no explicit mention of any massive "Higgs Boson" in their paper.[7]

In 2010 the J. J. Sakurai Prize, awarded annually by the American Physical Society for work in theoretical physics, was presented to the "Gang of Six" of Brout, Englert, Guralnik, Hagen, Higgs, and Kibble.[8] Never before had more than three people shared this award. In the case of the Nobel Prize, three is the maximum, so speculation has continued to build, exacerbated in 2009 by the start-up of the LHC, the machine that was designed to discover "The Boson"—by whoever's name it be known. Apportioning the credits is unlikely to be without controversy, whoever eventually may be invited to Stockholm. The citation for the Sakurai award mentions the discovery of "the Mechanism for generation of vector boson masses" in relativistic field theory. There is no mention of "The Boson" here.[9]

It may be helpful for what follows to be aware of a widespread misunderstanding about "The Boson," and "The Mechanism." There are two concepts at work, which many articles and media reports have conflated, giving an impression that they are one and the same. Admittedly, they are intimately linked, as are the lines of latitude and longitude spanning the surface of Earth: related but not identical. By analogy, in the model of Jeffrey Goldstone, which first inspired these concerns, if "The Mechanism" for creating mass is what you discover on a journey around the equator, then "The Boson" is the result of a journey along the Greenwich Meridian. In the model example of Chapter 8, and Figure 9.1, the former involves rotation around the valley floor; the latter emerges from the radial oscillations up and down the valley walls.

"The Mechanism" concerns the way that some gauge bosons—vector particles such as the *W* and *Z*, which ought to be massless like the photon—manage to acquire a mass by "eating" Goldstone's massless scalar boson.[10] This is what is widely referred to as the Higgs, or Higgs-Kibble, Mechanism for breaking what otherwise would have been the symmetry between the electro- and weak interactions.[11] However, as the Sakurai Award to six people recognizes, and as Higgs himself has been the first to stress, it has a wider provenance.

FIGURE 9.1
GOLDSTONE AND HIGGS:
ROTARY AND RADIAL OSCILLATIONS

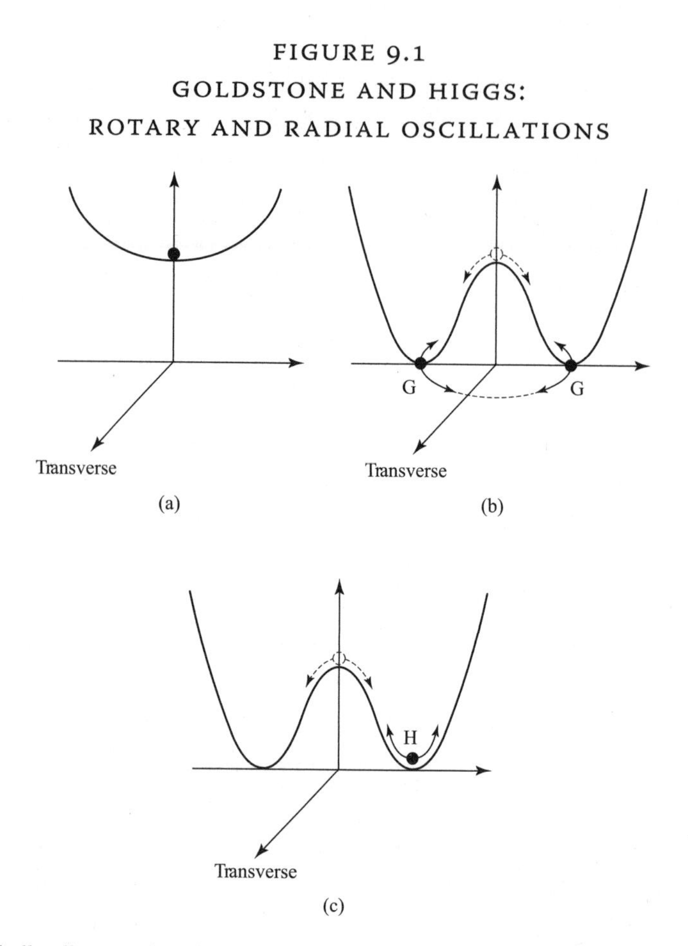

(a) A ball rolling in this shell will end up at the bottom where it has the lowest energy. (b) Suppose some demon pushes the region away from the center downward, so the shell has the shape in (b) and (c). The original point of stability now has higher energy than the valley floor. The ball will roll into the valley at some random point around the base. Oscillations around the valley floor—mode G in (b)—cost no energy. But those up and down the valley walls do—mode H in (c). The mode G is analogous to the massless Goldstone Boson; the mode H is commonly referred to as the massive Higgs Boson.

Although several people discovered The Mechanism whereby *vector-*gauge bosons can acquire mass, Brout and Englert being the first to publish a relativistic demonstration, only Higgs drew attention to the consequential existence of a massive *scalar* particle, which now bears his name. Its provenance is "Goldstone's other boson," which was present in Goldstone's original model.

To be fair to Peter Higgs, it was not he who yoked his name to the particle. He modestly refers to it as "the Boson that has been named after me."[12] Why, how, and when it came to be so named are some of the questions that I shall discuss. The British media, eager for a Nobel laureate, have headlined his name, and "Higgs Boson" has also been a convenient sound bite for those promoting the LHC.

A counterpoint to this adulation has come from Philip Anderson, for whom Higgs was "a rather minor player." Furthermore, he has written that the so-called Higgs phenomenon "was, in fact, discovered in [BCS] theory by me and applied to particle physics in 1963, a year before Higgs' great inspiration."[13]

As far as "The Mechanism" for generating a mass for gauge bosons such as *W* goes, this is indeed true. Anderson is the "A" of what Higgs himself has referred to as the "ABEGHHK'tH" mechanism,[14] the full acronym referring to Anderson, Brout, Englert, Guralnik, Hagen, Higgs, Kibble, and 't Hooft. When we discussed this together, between events at the Edinburgh Festival in 2010, Higgs added, "However, I do accept responsibility for the Higgs Boson; I believe I was the first to draw attention to its existence in spontaneously broken gauge theories."[15] The properties of "The Boson" in particle physics are what the LHC is investigating. While debates about priority for "The Mechanism" may continue, "The Boson" is another issue. So, first, let's meet the saga of "The Mechanism."

GLOBAL AND LOCAL SYMMETRY

If you have read Philip Pullman's *Dark Materials* trilogy,[16] you will be familiar with the cat that disappears through a "hole in the air" in the north of Oxford, reemerging in the center of the city. In the jargon of physics, this is consistent with the "*global* conservation of cats," meaning that the cat exists somewhere on the globe all of the time: in North Oxford at one moment, in the center the next.[17] However, this is not how cats behave. At least, my own tabby, "Ms. Chief," never did, and she came from the same litter as Philip Pullman's. Ms. Chief could get from North Oxford to the center step by step, or in the cat carrier en route to the vet. In so doing, she had moved from one location to the next continuously. In the jargon of physics, that is the "*local* conservation of cats." The same holds true for the atoms that make cats and the particles that make those atoms. Global conservation is true, but local conservation is true also, and much more restrictive—it forbids disappearing through holes in the air, for example.

Both Nambu's model and the examples of Goldstone, such as the ball rolling on the base of a wine bottle, were examples where symmetry was broken globally. The massless Goldstone Boson corresponds to the ball rolling from one point to another around the valley. It is equivalent to your reorienting your perspective such that the entire ensemble turns.

While such global symmetry has a legitimate role in physics, in many cases in practice the underlying dynamics is controlled by *local* interactions. The constraint of *local* gauge invariance for a theory of electric charge implies that electric charge is conserved and that there is force—the electromagnetic force—between electrically charged particles. For example, the electrical attraction between the electron and the proton in a hydrogen atom is, in quantum field theory, the result of a photon being exchanged between the two. First, there is a local interaction between the electron at its particular location and the photon; then the photon travels across the intervening space and transmits an interaction to the proton at the latter's location. The symmetry that controls electromagnetic interactions is therefore a *local* symmetry—one where the rotation of the phase (page 83) can vary from one place to another in space or time, yet the phenomena remain the same.

The electromagnetic force itself is thus a consequence of local symmetry, which spawns the photon. So what happens if this local symmetry is spontaneously broken? Goldstone's theorem applied when *global* symmetry breaks spontaneously; *local* symmetry, however, contains the clue that evades his theorem.

For a theory with local—rather than simply global—symmetry, it transpires that there is no requirement for a Goldstone Boson to exist with zero mass. The local symmetry is what has spawned a massless vector-gauge boson. If this symmetry spontaneously breaks, the massless vector boson absorbs the "would-be" Goldstone Boson and becomes massive itself. The massless Goldstone Boson, meanwhile, disappears from view.

The proof that this happens set the course of research in particle physics for the next half century.

This cannibalism is in effect what Anderson had speculated might happen. However, the demonstration that it does indeed occur in quantum field theory, when the constraints of relativity are imposed, turned out to be rather subtle. Several independent groups of theorists eventually demonstrated different aspects of the proof, including the sextet already advertised. Today, these ideas are part of the staple diet in a student's education, and with the benefit of hindsight might seem unsurprising. However, a half century ago, when the ideas were being conceived, confusion was rife.

Some of today's controversy is a remnant of that confusion, which is itself part of the nature of scientific research. How many ideas are in the subconscious, only to be understood when someone else articulates them? How many remarks, which were indeed obvious but were omitted so as to simplify what at the time was the mainstream argument, haunt one later when they become the center of debate? Only those at the center of the storm can really ever know, and often even they do not. This episode is a paradigm.

MISSED OPPORTUNITIES

As we mentioned briefly in Chapter 8, in 1962 Julian Schwinger asked the prescient question: "Does the requirement of gauge invariance for a vector field . . . imply the existence of a corresponding [vector boson] with zero mass?"[18] The general belief that it does is what had led Pauli to harangue Yang in 1954. By 1962 Schwinger was questioning the orthodoxy, the very concepts that he himself had introduced into physics a dozen years previously.

Schwinger now cited with irony his seminal founding papers on QED, from a decade before, which led him to admit: "[The answer] is invariably in the affirmative." He then continued, "The author has become convinced that there is no such necessary implication." Schwinger, the great architect, had found a flaw in the foundations of his construction. Here was the conception of the route leading to the "ABEGHHK'tH mechanism" for producing massive *W* bosons out of a gauge-invariant symmetry.

In the early 1960s several people around the world took the first step toward the great breakthrough, but faltered, or simply gave up. Schwinger himself either failed to answer his own question or did not pursue it.

Philip Anderson was one who did. As we have seen, his paper in 1963 gave examples of spontaneous symmetry breaking in physics and was one of the first to recognize its importance, but he stopped short of publishing a relativistic field theory demonstrating the breaking of electro- and weak symmetry. Anderson's paper, which showed what happens in a solid medium but ignored the constraints of relativity, was hardly noticed by particle physicists.

Meanwhile, in the Soviet Union, the leading solid-state theorist, Anatoly Larkin, posed a challenge to two outstanding undergraduate teenage theorists, Sacha Polyakov and Sasha Migdal: "In field theory the vacuum is like a substance; what happens there?"

They took up Larkin's challenge and answered it, demonstrating that the quantum of radiation—the gauge boson, which starts out massless in the fundamental equations—*can* become massive. Senior colleagues immediately derided them as naive. Migdal recalls, "We were stomped to the ground at every seminar [where] we tried to present this work. The most disturbing thing was that nobody would even argue on the subject—the mere mention of 'Spontaneous Symmetry Breaking' caused healthy laughter, which ended the conversation."

Migdal's memories are that they were victims of Soviet political philosophy polluting scientific thought. "For the first time since Galileo, the quest for the structure of matter was stopped on political grounds. There is nothing inside [the proton; there has to be] total nuclear democracy—everything consists of everything else. Do not ask whether there was the rabbit inside the hat: you are only allowed to compute how far it will jump and in what direction."[19]

Even physicists of the stature of Vladimir Gribov and Lev Okun, widely respected as liberals and free thinkers, "would not talk to [us] about Yang-Mills theory." Polyakov recalled, "The whole subject was unpopular in Russia and our paper dragged for a very long time."[20] By the time that they had convinced people that they were in fact correct,[21] a half-dozen others—including Higgs in 1964—had discovered the phenomenon independently.

Gerald Guralnik was one of those who approached the finishing line early but failed to publish fast enough.

In 1962, while Guralnik was still a student at Harvard, his supervisor, Walter Gilbert, had been investigating the properties of a mathematical theory of two varieties of massless particle, one of which had the characteristics of a photon, the other having electrical charge but no spin (a "scalar" boson). This showed that if the pair interacted with one another, the photonlike particle became massive. Guralnik talked with Gilbert about this model.[22] Years later, in a public talk, he recalled that "it is a *trivial* step from here to the . . . Higgs model. With any insight, all that was needed to describe the Higgs phenomenon was available at Harvard in 1962."[23]

The following year Guralnik moved to Imperial College, where he was inspired by Abdus Salam and by Tom Kibble. He had long discussions with Kibble "over frequent lunches with vile hard boiled eggs, and dessert—and almost everything else—covered with a yellow custard sauce." They discussed the apparent failure of Goldstone's theorem in superconductivity—the same observation that had inspired Anderson's proposal, though they were unaware of Anderson's work at that time.[24] Through a series of mistakes and misunderstandings, the discussions with Kibble bore fruit as Guralnik eventually stumbled on the truth: Goldstone's theorem[25] need not apply in a theory whose particles are conserved *locally*—like cats—and that took account of Special Relativity. In such a case, the *W* boson, the carrier of the weak force, could turn out to be massive by devouring Goldstone's boson.

Guralnik had first realized this in 1963, but the argument was not watertight, and it was not until 1964 when his old friend Dick Hagen arrived at Imperial College for a visit that they cleaned it all up. By the late summer of 1964 they had "pretty much everything but still had concerns over some fine details."[26] His former mentor Walter Gilbert was lecturing at a summer school at Lake Como, so Guralnik made a special trip to tell him what they had achieved.

As we shall see, Gilbert had just published the paper that sparked Peter Higgs's entrance. At the very time when Guralnik was talking with Gilbert, Higgs's paper was already in production. This would become apparent only later; at the time, and in the absence of telepathy, there was no means of making the connection.

During this period, and before Guralnik, Hagen, and Kibble's paper had been written, Salam and John Ward were completing their work combining electromagnetic and weak interactions in an office just a few doors away. The trio knew nothing of this. Conversely, Salam and Ward also were unaware of the profound insights being unraveled by their colleagues—how to give mass to gauge bosons—which was the final barrier to their own work.

At that time Guralnik was a junior researcher, whereas Ward was a hero. Guralnik told me his impressions of Ward: "Everyone knew about Ward identities and QED. He was in and out. You'd be in the department, like any other day, and suddenly there would be Ward walking down the corridor, en route from Australia or somewhere via Imperial College and Abdus Salam. He was just a different sort of fellow than I was used to talking to about physics." However, he was approachable. "He always seemed a little stern, a little grumpy, but once he started talking, he talked."[27]

Guralnik told me that Ward and Salam were very secretive about their work. The only hint that something exciting was going on was when a case of champagne arrived out of the blue at Imperial College "in anticipation of the prize that they were going to get for their current work" (the Nobel that Salam, but not Ward, would eventually indeed win). Guralnik's colleague John Charap told me a similar tale, as he recalled this occasion himself. He was working in the department late one evening when John Ward invited him into his office to drink some champagne. Charap remembers that Ward had chilled the bottle in one of the solid wastepaper bins, having "satisfied himself that some obstacle in what they [Ward and Salam] were doing had been overcome."[28] Charap did not ask for details:

"We were youngsters and didn't know the great things that were taking place," adding, "I am sure that it happened: I drank the champagne!"

Shortly after this, Guralnik and Ward were having lunch together in a local pub, and Guralnik started to talk about his work—yet to be completed—on hidden symmetry. "I did not get far before [Ward] stopped me. He proceeded to give me a lecture on how I should not be free with my unpublished ideas, because they would be stolen, and often published before I had a chance to finish working on them." As a result of this admonishment, Guralnik did not ask Ward about the work that he himself was doing with Salam.

Guralnik had discovered that the specter of the Goldstone Boson, far from being a problem, could in fact be the holy grail. Unknown to Ward, Guralnik had shown that when gauge bosons, such as the photon or *W*, are present, there is no unwanted Goldstone Boson, and also that the gauge boson can become massive. Salam and Ward, meanwhile, had found the right model for marrying the weak and electromagnetic interactions, at last, but had no idea how to explain theoretically the empirical need for the *W* to have mass.

Guralnik's memory is that at Imperial College in the summer of 1964, all the ingredients were at hand that could have given birth to a full theory of the weak and electromagnetic forces. Today, he muses that on the occasion of that fateful lunch he and Ward between them possessed "enough information to have had a good chance to solve the unification problem on the spot." Had Guralnik and Ward been more open with one another, they might have shared a Nobel Prize, relegating Salam, Weinberg, and Glashow to also-rans.

Unaware of this even more precious jewel that lay just out of their reach, Guralnik and Hagen drafted a manuscript, and showed it to Kibble, asking for any additional input. Fate was about to deal them a cruel blow.

During 1964 Britain had had a series of postal strikes, which had delayed the mail. Their memory of events, as described in several accounts,[29] is that as they were about to send the final version of their manuscript to *Physical Review Letters*, Kibble came in and announced that in the backlog of mail were "three papers, one by Robert Brout and François Englert—(who like Anderson were specialists in the theory of solids)—and two by Peter Higgs." All three also seemed to have discovered how a gauge boson could become massive.

In a slight panic, Hagen and Guralnik read them. When I spoke to them in 2010, I suggested that this must have been one of the worst moments in their scientific careers. Guralnik's response was unequivocal: "You know something, only in retrospect was it a bad moment. The fact is that I did not feel that they were of great weight. My reaction was that Brout and Englert had made uncontrolled approximations and Higgs did not offer the insight that we had developed. I wrote those papers off. That shows how wrong you can be after forty-five years. These were observations made in haste and with the exuberance of youth."[30]

Guralnik, Hagen, and Kibble's work was tightly argued and contained unique insights toward understanding the depth of these new ideas.[31] However, the sad fact was that they had been scooped. They added references to those papers into the text but changed nothing, nor did they add anything as a result of what had happened. They sent the manuscript to *Physical Review Letters*, where it was received on October 12, 1964.[32]

Guralnik continued, "It was silly on my part—but it was honest. In retrospect, I wish we had added the true statement that after this work was finished, it was brought to our attention that related work by, etc." By that oversight, they had implicitly placed themselves as runners-up in a race where the photo finish was never examined.

Guralnik, Hagen, and Kibble are G, H, and K within the "ABEGHHK'tH mechanism" for giving mass to gauge-vector bosons. There is no massive "Higgs Boson" in their paper.[33]

BROUT AND ENGLERT

Robert Brout and François Englert met in 1959, became friends immediately, and have worked together ever since. Brout, born in New York in 1928, was originally a chemist, but by 1959 he had become a well-known theorist, specializing in the physics of solids, or what is today known as "condensed matter." He was already a professor at Cornell University when the twenty-seven-year-old Englert arrived from his native Belgium to become Brout's research assistant. Englert recalled how Brout had collected him from the airport in his "near-century-old Buick." He "took me for drinks which lasted up to the middle of the night. When we left, we knew that we would become friends."

Their styles complemented each other. Where Brout liked to translate abstract ideas into intuitive mental images, Englert has been more comfortable with the formal Franco-European tradition, where images are secondary to formal mathematics. Where for some people such a difference creates difficulties, for Brout and Englert it worked well, enabling their combined insights to be greater than the sum of the individual parts.

They discovered that they had a common interest in the way large numbers of items organize themselves, as in solids and liquids, or, more significant, in the changes of phase from one configuration to another. They became especially interested in ferromagnetism and spin waves.

Electrons spin and act like little magnets. A simple theoretical model of magnetism in a large body imagines an array of spinning electrons evenly spaced along a line (Figure 9.2). Each electron acts like a tiny magnet whose north or south pole is pointing in some chosen direction.

Imagine a clothesline with pegs hanging from it. Each peg is analogous to one of the electrons. In the case of an everyday clothesline, the gravity of Earth defines "down," but in our magnet analogy there is no up-down or sideways—it is like an astronaut's clothesline in outer space. A single peg on the line can point in any direction—the situation is rotationally symmetric.

The same is true for the real example where instead of a clothespin we have a magnet—a spinning electron. Add a second one, which interacts with the first, and then continue to set up a whole line of them: This is the situation that interested Brout and Englert.

Each electron interacts with its immediate neighbors, like two magnets feeling a force depending on their relative orientation. In this particular model,[34] the force was repulsive if the pair spun in opposite senses, attractive if their spins were aligned. The total magnetic energy of the system then depended on the relative orientations of the entire assembly.

FIGURE 9.2

LINE OF SPINNING ELECTRON MAGNETS

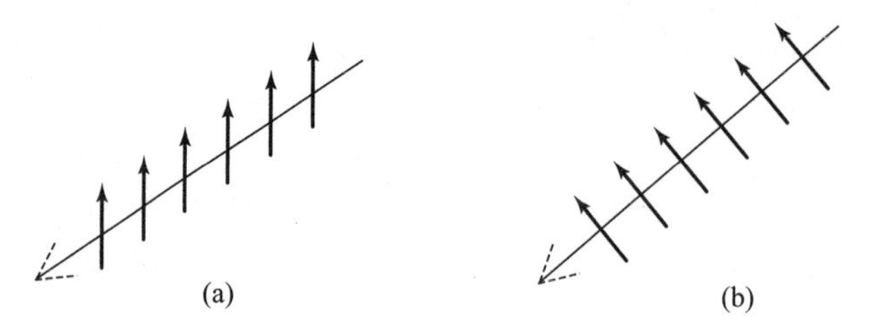

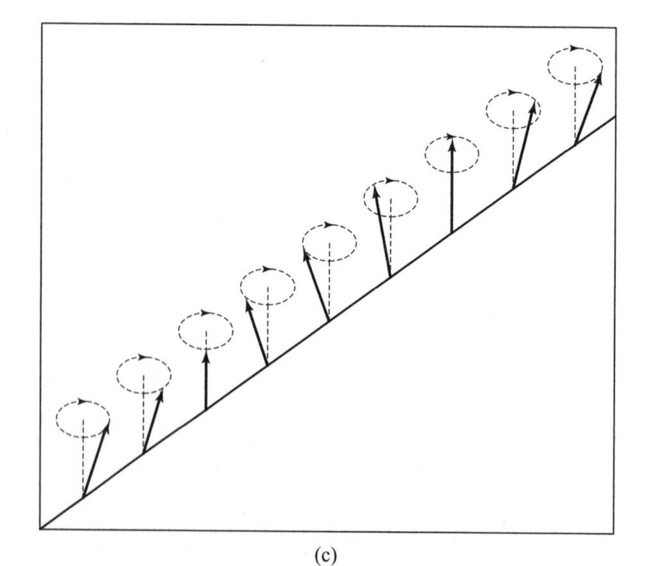

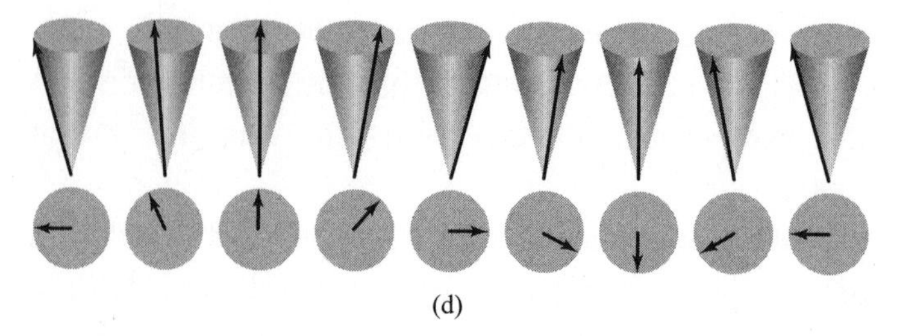

(a) Aligned in a ground state. (b) Aligned at a different orientation but otherwise identical energy. The two examples are simply related by a rotation of perspective—a "global" rotation. They have the same energy. Rotating from one to another is analogous to the mode G in Figure 9.1.

(c) A spin wave with energy—a "magnon"—analogous to the radial motion, mode H, in Figure 9.1.

(d) To illustrate the radial and angular possibilities another way, we can view the magnon from above. This shows the rotation occurring along the direction of the line; this image can be imagined to be rotated out of the page at any angle around that line.

At high temperatures the electrons have so much energy that their spin axes rotate wildly, such that overall the ensemble respects the rotational symmetry. However, as the temperature drops, their mutual interaction tends to align their spins. The lowest energy—the ground state—is the situation in which all of the spins are oriented in the same direction. In this state the rotational symmetry of the fundamental laws is broken.

The direction in which the spins point is arbitrary. Were we to rotate all of the spins—for example, by orienting our line of sight differently—we would perceive another ground state, with the same energy (Figure 9.2b). Here we have again the classic example with all the characteristics that lead to a Goldstone Boson. In this case the boson consists of waves in the spins, where a periodic rotation of their orientations occurs along the line. It costs no energy at all to rotate all of the spins, and so it costs very little to make a gradual periodic change whose wavelength is very large. If the ensemble is infinitely large, then as the wavelength itself becomes infinite, the energy goes to zero. These infinitely long waves of spins are the Goldstone Bosons in this case.

That is what happens if the only interaction is between immediate neighbors. However, if the range of the interaction is larger, the effects can be very different. It turns out that if the range of the force is infinite, the Goldstone Boson is destroyed. Englert and Brout realized that this was a possible analog of what could happen when long-range forces—such as electromagnetism—are present in quantum field theory.

By 1961 their collaboration was very successful, but Englert's scholarship was coming to an end. Cornell offered him a professorship, but his heart was set for Europe. He returned to Brussels, and soon afterward Brout and his wife came over, supported by a Guggenheim Fellowship. The social life and general excellent personal relations won: Brout resigned from Cornell, took up a position in Brussels, and has remained there ever since. Thus, it was in Brussels that they resumed their investigation of broken symmetry.

The question for them was this: If symmetry is spontaneously broken in the presence of a massless vector-gauge boson (such as a photon), which gives rise to a long-range force, does the Goldstone Boson become absorbed into the massless gauge boson, thereby providing the longitudinal oscillation—the analog of Anderson's plasma oscillation—required to convert the massless gauge boson into a massive vector particle? They found the answer: yes! This was the same as Anderson had found with his example of the plasma, but now Brout and Englert had elevated the proof to a full relativistic field theory.[35]

Brout and Englert's analysis successfully demonstrated how it is possible for a massless photonlike gauge boson to acquire a mass while maintaining gauge invariance in relativistic field theory. They were the first to publish this seminal result. They introduced a field, like that of Goldstone, that can seed a massive scalar boson, but did not explicitly mention a particle of the kind that came to be called the Higgs Boson.

ENTER PETER HIGGS

Born in Newcastle in 1929, Peter Higgs is the son of an electrical engineer who worked for the BBC. The family moved to Birmingham, but in 1941, with aerial bombardment a nightly occurrence, his father's employers decided that Bristol would be safer. They moved to the "safety" of Bristol just after its ancient city center was bombed into destruction by German planes.

Peter Higgs was a pupil at the same Bristol school that Paul Dirac had attended thirty years earlier. "I noticed his name many times on the school honours board, and wondered what he had done," he told me.[36] On finding out, this helped develop his interest in physics before he left school. Thus it was, by chance, that Higgs found himself a successor to Dirac, the founder of Quantum Electrodynamics.

Higgs read his father's textbooks, in which he met mathematics and Newton's calculus. "This meant I was a step ahead of the teachers and everyone expected that I would go into engineering" (an uncanny parallel with Paul Dirac, who had himself started out in engineering before moving on to theoretical physics). Higgs, however, shifted his interests toward physics.

In 1945 he had heard Bristol University's two professors of physics—Neville Mott and Cecil Powell, both future Nobel laureates—talk about the scientific background to the Hiroshima and Nagasaki bombs. Attendance was good, and Powell followed this with a series of public lectures about his discoveries of "strange" particles in cosmic rays. His appetite for physics whetted, Higgs entered King's College, London, where he became "the first person to do a newly invented theoretical physics option of the B Sc degree," and upon graduating in 1949 he wanted to begin a Ph.D. in particle theory. Coulson, the professor at King's College, London, unaware that Feynman, Schwinger, Tomonaga, and Dyson had just sorted out the problem of infinity in QED, misinformed Higgs: "The theory is in a very bad state. You may achieve nothing or win a Nobel Prize." As a result, Higgs decided to "play safe" and embarked on a Ph.D. at King's College in molecular physics.

He was working from an office in the same corridor as those of Maurice Wilkins and Rosalind Franklin, the "fourth person" in the teams that shared the Crick-Watson-Wilkins Nobel Prize in 1953 for decoding the structure of DNA. During this period, 1950–1954, Higgs produced his first papers on the helical structures of molecules—though not DNA itself. His first interest, however, remained particle physics, and when he won a postdoctoral scholarship, he was able to move to Edinburgh, where Nick Kemmer was in charge "and a great inspiration."

Higgs moved to Edinburgh in the new year of 1955. It was here that he first met Tom Kibble, who was a final-year undergraduate and attending the research seminars. The following year, however, Higgs won a scholarship to University College, London, and then, inspired by a meeting with Abdus Salam, transferred to London's Imperial College. Finally, by a series of misadventures, he ended up back at Edinburgh, this time for good.[37]

While we were enjoying real ale, Scottish brewed, together in the Edinburgh staff club in 2000, Higgs told me that he had applied for a post in London, for which Kemmer had supplied "an ambiguous reference, full of triple negatives, which created much confusion." There was a post at Edinburgh also, and Kemmer had told Higgs, "Of course it would do no harm if you were to apply for the Edinburgh post," which Higgs took to be Kemmer politely discouraging him. In fact, as Higgs learned later, he had intended it as encouragement.

In 1959, Tom Kibble arrived at Imperial College from Edinburgh, via California, with a fellowship, and Higgs, having been at Imperial College, won the position in Edinburgh. Higgs jokingly told me that this episode was the "real Higgs-Kibble mechanism"—one for creating jobs, as opposed to the widely misnamed Higgs-Kibble mechanism for creating masses. Higgs's involvement in the latter began in those early months at Edinburgh.

One of the perks of being at Edinburgh was taking part in the Scottish Universities Summer Schools, widely recognized as among the most enjoyable advanced schools, not least because of the generous amounts of wine available at dinner.

The 1960 school, which lasted for three weeks, was the first in a series that today spans a half century. At the last minute one lecturer begged off and was replaced by another, who had to come from California. The organizers had budgeted six hundred dollars for his travel, but at such short notice there had been no time to arrange the details. The lecturer arrived one day before the school began and revealed that he had a travel grant from the U.S. National Science Foundation, and hence no need for any reimbursement from the school itself. The treasurer suddenly found himself in the fortunate position of having six hundred dollars to spare, which was a considerable sum in those days, and an urgent need to spend it for the benefit of the school. The solution was brilliant: Use the windfall to provide free wine for the participants at dinner. Thus began the tradition, which has continued ever since. Peter Higgs was given the role of wine steward, which made him popular with all students and lecturers.

These schools are held annually, and I was a student in 1970 when the venue was Middleton Hall, south of Edinburgh. Peter Higgs was away at a conference in Kiev, and it was not until 1973 that I first met him, when I visited Edinburgh to give a seminar. Nonetheless, my first time at a Scottish Summer School holds a special place in my heart.[38] Peter Higgs's first experience of these Summer Schools turned out to be especially memorable for him too, because three future heroes of physics were to meet for the first time.

The venue that year was Newbattle Abbey, a mansion dating to the sixteenth century, built on the site of a former church. With crypt, grandfather clock, and battlements, it would qualify as an ideal location for filming an Agatha Christie murder mystery. There are, so far as I know, no secret passages, but it does contain hiding places for the imaginative. In 1960 among the most creative students were Shelly Glashow and Tini Veltman, who was later to play a seminal role in establishing, with his student Gerard 't Hooft, the renormalizability of the electroweak theory. Of course, in 1960 that was all in the future: Although Glashow had recently completed his paper, Salam, Ward, and Weinberg were not in the frame, and Higgs's contribution was still four years away. However, ideas began to form, courtesy of the wine and the geography of the abbey.

A stairway led down from the entrance hall to the basement, which contained the crypt and the dining hall. The entrances to these rooms faced one another, and a large grandfather clock stood proudly against the wall between them. The crypt had been converted into a lounge, where students and lecturers would meet and socialize.

The staff serving the meals consisted of local people, hired for the purpose. They would allocate three bottles of wine to each table, regardless of whether it was full of adults or participants' families, which included children. Glashow and Dave Jackson, who would later write the all-time greatest textbook on Classical (as distinct from Quantum) Electrodynamics, had realized that by sitting at a table with a family, they could stealthily lower one or two bottles of red wine beneath the table during the meal and then hide them in the bottom of the grandfather clock upon leaving. The front door of Newbattle Abbey was locked at ten o'clock, and the clanging of keys became the signal to retrieve the wine from its hiding place and to meet in the crypt.

Glashow and Veltman were part of a group who met there, enjoying the wine, supplemented by malt whiskey. These discussions—in which Glashow described his new theory marrying the weak and electromagnetic forces—were so well oiled that, as Higgs told me, "they went long into the night, and they were late for lectures." Higgs had had to stay alert as his duties at the school were more than just being wine steward and required him to be up first thing in the morning. "That's how I missed Glashow's model," he explained.

Six years later, on March 16, 1966, Higgs spoke at Harvard about The Mechanism for generating mass. Glashow was present. In their conversation afterward, in which they jovially recalled their happy times in Scotland, Glashow said, "That's a nice model, Peter," but missed the opportunity to realize the potential that Higgs's remarks held for completing his own theory. Both Higgs's and Glashow's papers had by then been published, but neither scientist was aware of the other's insights. As Higgs said to me later—speaking in the third person as if in recognition of the dramatic irony of the situation—"If Peter Higgs and Shelly Glashow had communicated properly, Weinberg and Salam would never have got into it."

So how did Higgs and his mechanism enter into the lexicon of science?

HIGGS'S BOSON

"The portion of my life for which I am known is rather small—three weeks in the summer of 1964. It would have been only two if *Physics Letters* [the European journal to which Higgs had sent the first version of his manuscript] had accepted the paper. But initially they rejected it." Higgs's interpretation of the editor's letter was that "they didn't understand it as it was written in a dead language—the dead language of field theory."[39]

In the early 1960s, "field theory is a dead end" was a widespread belief. However, this rejection of Peter Higgs's first draft of the 1964 paper that would eventually make him famous had profound consequences for the course of physics. Higgs decided that to improve the paper's credibility, he should "add some practical consequences of the theory. That took a week, hence the third week of 1964, and included the [Higgs] boson." So it was that the paper's initial rejection led Higgs to add the feature that helped set him apart from the pack.

Along with Brout and Englert, Guralnik, Hagen, and Kibble, in 1961 Higgs too had become fascinated with Nambu's insight about hidden symmetry and the quandary of Goldstone's massless boson. For him, however, the real story indeed came with a rush in just a few weeks in the summer of 1964.

Recall that Goldstone's theorem asserted that massless, spinless bosons appear when symmetry is broken in any theory that is "Lorentz invariant." Lorentz invariance underwrites Special Relativity and in essence states that the laws of physics appear the same independent of your state of motion. In the jargon: "There is no absolute state of rest" in the universe. This immediately exposes a loophole in Goldstone's theorem and in the explanation of Anderson's observation that there is no Goldstone Boson in a superconductor: The motion of electrons is relative to the superconductor, so there is an absolute state of rest—that of the lump that is the solid superconductor itself—which violates the crucial requirement of Lorentz invariance.

This was widely regarded as *the* loophole and, indeed, led to a widespread belief—wrong, as it turned out—that the presence or absence of a Goldstone Boson was due entirely to the presence or absence of Lorentz invariance, whether the theory is relativistic or nonrelativistic. Widely believed certainly, but no one had actually proved this.

In March 1964 Abe Klein and Ben Lee made the crucial intervention.[40] They demonstrated mathematically that the Goldstone theorem fails in nonrelativistic theories. Had that been all, their paper would have had little more than pedagogic value, finally establishing what was generally believed. However, they noticed that their arguments appeared to apply also to relativistic field theories, which raised the possibility that in such theories, the Goldstone theorem would also no longer apply. As it is these very theories that particle physics is built on, the possibility beckoned that the Goldstone theorem had a loophole.

Ben Lee was Korean and a naturalized American whose interventions during the next ten years, before his untimely death in a car accident, would help determine the future of particle physics. This paper in 1964 was but the first of these critical contributions. What helped start the revolution, however, was that Walter Gilbert claimed that Klein and Lee's paper was wrong.

Gilbert is an American who moved to Cambridge where he enrolled as a research student with Nick Kemmer in 1953, becoming one of the cohort taken up by Abdus Salam when Kemmer moved to Edinburgh. During this period, Gilbert met a fellow American, Jim Watson, who was working with Francis Crick on the structure of DNA. Gilbert was inspired by this, and, having graduated and moved to Harvard where he met up again with Watson, he himself took up molecular biology, eventually in 1980 winning the Nobel Prize in Chemistry for his own work on the molecular content of DNA. He was one of the inspirations behind the Human Genome Project. All this came after his crucial intervention following Klein and Lee's paper, an intervention that launched Peter Higgs.

It is ironic that Gilbert's and Higgs's careers are so counterbalanced. While Gilbert as a student was studying physics, working alongside Crick and Watson, Higgs was working in London, on biophysics, in the same corridor as Maurice Wilkins and Rosalind Franklin. Higgs, trained in biophysics, chose particle physics; Gilbert, having trained in physics, took up molecular biology.[41] Their careers were about to converge dramatically as a result of Gilbert's response to Klein and Lee.

By 1964 Gilbert had been publishing papers in molecular biology for three years. His interest in the relation between gauge invariance and mass had been sparked during 1962, as we saw in Guralnik's history, and led to a paper that briefly interrupted Gilbert's growing oeuvre in biology.[42] On June 22, 1964, he published in particle physics for the last time, with a paper showing that the way Goldstone's theorem is bypassed in nonrelativistic examples does not survive when the constraints of relativity are imposed.[43] This was contrary to Klein and Lee's suggestion that the theorem might fail in relativistic field theory as well.

The journal containing Gilbert's paper arrived in Edinburgh on July 16.[44] By this stage Peter Higgs had been thinking about Goldstone's theorem for two years, so Gilbert's paper grabbed his attention. He quickly realized that, while Gilbert was correct in the way that he had addressed Klein and Lee's argument, there was still a further loophole unaddressed. Within a matter of days, Higgs managed to close it by showing that Gilbert's argument failed for relativistic field theories *if these contained gauge fields*, and hence gauge bosons.

On July 24 he had completed a short note,[45] extending to a mere thousand words, which showed that, contrary to Gilbert's assertion, it is possible to exorcise Goldstone Bosons in relativistic field theory so long as gauge bosons also are present. In the final sentence of his paper Higgs promises that "in a subsequent note" he will show that this results in "qualitative changes in the nature of the particles": in effect, that the massless gauge boson becomes massive as a result of absorbing the Goldstone Boson.

He sent this paper to CERN, where Jacques Prentki, head of the theory group, was also editor of the journal *Physics Letters*. Prentki had the habit of farming out papers to members of the group for comments before deciding whether to publish. This brief paper, criticizing Gilbert's refutation, and showing that Goldstone's massless boson could be exorcised, was accepted and appeared in print on September 15.

However, apart from the enigmatic hint at the end, there is no sign of any mechanism for giving mass to the gauge boson, or of any "Higgs Boson." These would come in his second paper, completed a week later, on July 31. It was this paper that was rejected, the editor saying, "If you develop this work and write a longer paper, you might consider sending it to *Il Nuovo Cimento*."

This suggestion carried mixed messages, as *Il Nuovo Cimento* had a reputation at that time for not using referees at all.[46] Higgs took the first piece of advice, adding some practical consequences, which took his "extra week." He added some sentences at the end, alluding to the presence of scalar bosons, which together with an equation describing their behavior form the first hints of what has become known as the Higgs Boson. Feeling that *Physics Letters* was unreceptive, and disfavoring *Il Nuovo Cimento*, Higgs then sent this revised paper to *Physical Review Letters*.[47]

At the twilight of his career in physics, while transforming to his life as a biologist, Gilbert had unwittingly become the central player in other people's moments of glory or tragedy. Guralnik's written history recalls that he visited Gilbert at Lake Como in July with, if memories are accurate, "pretty much everything" of the means of evading Goldstone's impasse.[48]

Gilbert, meanwhile, had published his paper on this topic only a few weeks earlier, on June 22. When Higgs saw this he had, within forty-eight hours,[49] discovered the loophole for himself, and within weeks had written the two papers that made his name.

If memories of these events are accurate after so long, it is remarkable that at their meeting at Lake Como one or the other of Gilbert and Guralnik did not immediately take the final step. Guralnik's recall of what transpired gives little sense that a Eureka moment was passing by, while he and his wife spent part of an afternoon with the Gilberts at their conference housing.[50]

Guralnik pointed out to Gilbert how his ideas linked to Gilbert's own work two years before with Boulware[51] and outlined "in a fair amount of detail" everything except for some technical contributions that Hagen made later. He told me that Gilbert "asked questions and seemed to understand," and although he was "somewhat non-committal," he had "not reacted negatively" to anything.[52] Guralnik had intended to publish before talking to Gilbert, and although nothing had transpired that bothered or discouraged him, he wanted to pursue further some questions that Kibble had raised about solid-state physics earlier.[53] This took some weeks and seems to have delayed rather than materially affected the paper.

Having been presented with Guralnik's news of the circumstances whereby Goldstone's theorem may be neutered in relativistic field theory, the implications seem not to have hit home with Gilbert. When, later, Gilbert received preprints of Peter Higgs's two papers, on August 27 he wrote to Higgs, first thanking him and then adding, "Unfortunately I believe that your conclusions are wrong."[54] The letter addressed technical issues but gives no hint that Gilbert accepts the essential conclusion—that Goldstone's theorem is bypassed when gauge fields are present.

If work on the mass mechanism had been as mature as memories recall, it is unfortunate that the meeting with Gilbert—one of the seminal authors in the subject at that time—did not lead to more positive outcomes. Instead, Higgs saw the loophole in Gilbert's criticism of Klein and Lee and published, more than two months before Guralnik, Hagen, and Kibble felt able to take the plunge.

At this stage, all members of the Gang of Six had pursued the same quarry, and in each case the recipe was the same: The ingredients contain a massless *W*, which is put into the oven of hidden symmetry, whereupon a massless Goldstone Boson bubbles forth, only to be immediately gobbled up by the *W*. As a result of devouring the Goldstone Boson, the *W* becomes massive.

As we have seen, the editor of *Physics Letters* had received Higgs's first manuscript on July 27, 1964, and accepted it. His second paper was received a week later and rejected, the revised version reaching the editor of *Physical Review Letters* on August 31. Brout and Englert's manuscript had arrived in the office of the editor of *Physical Review Letters* on June 26, 1964, a month before Higgs had completed his first work.

In another of the coincidences in this tale, the editor of *Physical Review Letters* received Higgs's manuscript on the very same day that Brout and Englert's paper was published. When Higgs's paper finally appeared in print, on October 19, Brout and Englert were surprised to see that it included a reference to their own: "He couldn't have seen our paper, so how did he know of it?"[55] Higgs explained: Nambu had been the referee of both papers and "had drawn attention to the work of Brout and Englert. I added a remark about their work."

Their researches were truly independent; in 1964 manuscripts were typed and then submitted by the regular mail. Had the Internet been in existence in those days, Brout and Englert would probably have submitted to an electronic archive, their work being instantly advertised to the world, and the subsequent history different.

There is no dispute that Brout and Englert were first to complete and first to publish. So why is it that Higgs's name is associated with the massive boson and not those of Englert or Brout?

The answer is that so far everyone has been addressing what happened to Goldstone's massless boson. However, there remains the issue of Goldstone's other boson—the massive one. This is what Higgs had uniquely included in his revised paper, the one that appeared in *Physical Review Letters*. Two years later, he developed the ideas in a longer paper, setting in motion the events that would make his name.[56]

I confirmed with Brout and Englert the physical picture that had emerged in 1964 when the various papers appeared. First: Oscillations around the bottom of the wine bottle potential cost no energy and are the massless Goldstone Bosons. Second: In the presence of a long-range force, these are absorbed into the gauge boson, making it massive. Third, and the missing link: The "Higgs" Boson is the result of fluctuations radially from the center of the base—if the Goldstone is the mode around a circular valley at the base of the bottle, the Higgs is the mode oscillating up and down the sides of the valley. This third and special phenomenon is true whether or not there is a long-range force.[57]

I asked them why they had not included any comment about this in their paper. This is their explanation of how they viewed the situation in 1964: "The 'Higgs Boson' is kind of obvious: once there is a field, then in quantum mechanics its fluctuations manifest themselves as particles." For example, we have seen how in quantum theory the electromagnetic field is manifested as particles-photons. Analogously, the Higgs Field can wobble—the base of the valley becoming fuzzy due to quantum uncertainty—and when it does, particles, Higgs Bosons, appear.

Englert continued: "To us this was obvious and we felt not necessary to make it explicit after we introduced [what is often called] 'Higgs field.'[58] In fact, the 'Higgs boson' was already known in other areas of physics. The 'Higgs boson' has indeed nothing to do with gauge invariance: It is a general property of global symmetry breaking."[59]

It is ironic that Brout and Englert made no mention of what is today central to modern particle physics. Higgs too agrees that the massive bosons are obvious—when I challenged him on this he immediately exclaimed, "I agree!"[60] Indeed, it was so obvious that, but for the referee's intervention, which led him to add some practical implications—notably the massive boson—he might not have mentioned it either.

From conception to birth, Higgs's inclusion of the boson had come in the three weeks interregnum between his first and second papers. I was present at the banquet in 2000, held a few months after Peter Higgs's seventieth birthday, at which he recalled his creation of what has become the

"Higgs Boson." He summarized, "The amount of labor was rather small, and I am staggered by the consequences."

GURALNIK, HAGEN, AND KIBBLE

So it was that Brout, Englert, and Higgs had all published papers, while Guralnik, Hagen, and Kibble were writing their own paper, ignorant of developments due to the British postal strike until the fateful moment when Kibble announced the news.

The widely held attitude that this area of research was a by-water—the "field theory is a dead end" mantra having spread far and wide—might also help to explain the sequence of missed opportunities that would continue to haunt those involved.

Guralnik subsequently talked about their work in several places, only for it to be dismissed as nonsense: "I was told in no uncertain terms that I did not understand electromagnetism or quantum field theory." Most audiences' reactions, on hearing a talk about the phenomenon of hidden symmetry, was like that when you see a magician levitate: You know it is an illusion even though you cannot explain how it is done. To many hidden symmetry seemed to appear out of nowhere, as if by sleight of hand. Its results appeared suspect. Either you were being tricked, or, if the speaker genuinely believed what they were doing, then they were fooling themselves. Such reactions were reminiscent of what Migdal and Polyakov, the Russian "Schoolboys," had themselves been told.

At a conference in Germany, "Heisenberg and other famous people thought these ideas were junk, and made it clear that they felt that way. I was young, so this rebuttal was scary."[61] The experience with Heisenberg was typical of the general negative reactions to these ideas. Julian Schwinger, his former professor, had been at that conference. Schwinger said nothing about Guralnik's talk, but afterward offered him a ride in his brand-new Iso Rivolta sports car: "Julian remembered from my Harvard days that I loved cars and would be very interested in the wonderful machine which cost a noticeable part of the amount of money that he would receive with his Nobel Prize later in the year." Edward Teller's wife joined them and sat in the front seat. Schwinger was driving very skillfully, changing gears adroitly as the car accelerated out of the bends. For Guralnik, this offered some redemption after the experience with Heisenberg; for his part, Schwinger was obviously delighted with his new toy; Mrs. Teller, however, merely said: "In the U.S., if you get a car this expensive, it has an automatic transmission."

When Guralnik's London scholarship came to an end, he won a temporary contract at Rochester in New York State. The leader of the team there was Robert Marshak, who had earlier made several major contributions to the development of weak interactions, not least with recognizing the significance of $V - A$ that had started the revolution. Guralnik told Marshak of his own plans, which were "in retrospect leading to what would become the unified model of weak and electromagnetic interactions." Marshak's reaction to Guralnik's work was skeptical, echoing the negative reactions that had been the staple response to such ideas at that time. Marshak was also acutely aware of the intense competition for tenured posts in universities (then as now) and advised Guralnik that he "had to work on something else if [he] wanted to stay in physics."

Forty years later, Guralnik's reaction is philosophical: "Since he was an expert on weak interactions and the job market, I obeyed. I am still sure he was correct." This is indeed a sanguine perspective because, some years after that conversation, Marshak met Guralnik at a conference (at of all places Shelter Island, where the Infinity Puzzle had been born). With great decency Marshak "publicly apologised to me for stopping my work on symmetry breaking" and, as Guralnik recalled Marshak's words, "probably stopping you [Guralnik] from getting a Nobel Prize."

FORTY-SIX YEARS LATER

Around 1964, the "Gang of Six," and also the "Russian Schoolboys," discovered how mass can appear in what, at the outset, was a relativistic field theory without mass. This result transcends any particular model. The challenge today is for particle physics to determine by experiment how, and to what extent, The Mechanism is actually used by nature.

Tom Kibble recalled: "As a matter of fact [in 1964] the Higgs Boson was not seen by anyone as a particularly important feature. The interest [then] was in the way the gauge vector bosons acquired a mass, eating up the Goldstone [massless] Boson on the way. Now of course it's important because it has still to be observed, and its observation would confirm the validity of the theory."[62]

In 1966 Higgs followed up on his earlier work and wrote a longer paper in which he considered the decay of the massive boson.[63] This showed that if vector bosons have acquired their masses as a result of spontaneous symmetry breaking, the more massive the vector boson is, the greater is its affinity for the Higgs Boson. In 1967, Steven Weinberg recognized that the electron too can acquire mass by this mechanism. If spontaneous symmetry breaking gives mass to all fundamental particles—whether they are the massive *W* and *Z* bosons, or fermions such as the electron, its more massive sibling the muon, and even heavier tau[64]—it will be the decay pattern of the Higgs Boson into these various particles that will prove it.[65] According to the theory, the Higgs Boson will tend to produce the massive flavors of a given family more readily than their lightweight counterparts. It is this pattern that the LHC seeks, as we shall see later.

Kibble's perception of the perceived unimportance historically is in accord with the memories of others, which were aired when the six shared the Sakurai Prize in 2010. For Englert and Brout, "This was our first paper in field theory and we were somewhat isolated. We didn't know about Glashow's work [on the SU2 × U1 model of the weak and electromagnetic forces] at all." They had successfully eliminated Goldstone's *massless* boson. Their expertise was principally in condensed matter physics, whereby they knew of Goldstone's *massive* boson in much the same way as did Anderson, regarding it as "so obvious" that they made no comment about it.[66] Higgs alone both included it in 1964 and later, in 1966, made some limited investigation of its phenomenological significance for particle physics. Which, for many particle physicists today, is the nub.

During the International Conference on High Energy Physics at Berkeley in 1966, Ben Lee prominently referred to the "Higgs Boson" and the "Higgs Mechanism." Hagen was present, and following the conference he "sent a letter to about twenty of the most well known participants in which he urged reconsideration of that name. Regrettably, no copy is currently available of that letter."[67] As the GHK team, including Hagen, had not completed their paper until after the appearance of Higgs's original papers, and as Higgs had built on this further during 1966, with a study of how the massive boson decays, Lee's assessment at that time is perhaps not unreasonable. In any event, Hagen's letter seems to have had little effect. As we shall see in the next chapter, the following year, Steven Weinberg produced his seminal paper, which uses these ideas to build what is now confirmed as a viable theory of the weak and electromagnetic forces.[68] Weinberg's paper, which cites Higgs prominently in pole position, eventually became the most highly quoted paper in theoretical particle physics.

Within the community of particle physicists it is Higgs's name that is freely associated with the "Boson that has been named after [him]." That is how it is likely to remain. A historian of science might argue, as some have, that misnomers pollute this particular part of the record. The massless boson attributed to Goldstone is perhaps more justly credited to Nambu, and indeed is often referred to as the Nambu-Goldstone Boson. The massive boson, which in particle physics is named for Higgs, may be traced to Goldstone's original paper. Tom Kibble recalled a suggestion that the Higgs Boson "should be called the Goldstone boson, while the Goldstone boson should be called the Nambu boson—though that would be very confusing!"

The words on the tomb of President Kennedy will always be attributed to him, though it was Ted Sorensen who wrote them. Their impact and resonance through the years come from the writer and the orator both. So perhaps will be the legacy with this boson. It will be attributed to Higgs, if only because its discovery will be in a particle-physics experiment and that is the name by which that community knows it.

However, there is a danger here of diminishing the profound nature of the phenomenon. Spontaneous symmetry breaking occurs throughout nature, transcending any specific discipline, and is one of the most beautiful examples of the universality of science. The anticipation is to see precisely how nature makes use of it in breaking the symmetry of electroweak into the individual electromagnetic and weak forces, giving structure to matter and enabling perhaps life itself to exist. This is like the Kennedy oratory, whoever may have produced the words.

Intermission: Mid-1960s

We've reached the middle of the 1960s.

A theory uniting the electromagnetic and weak forces has been achieved, and the earlier worries about the apparent need for massless force carriers assuaged.

This has emerged out of ideas on symmetry being hidden, which had been known in other areas of science, and then applied to relativistic quantum field theory–particle physics. Originally, a theorem due to Jeffrey Goldstone had been thought to show that this could not happen. The loophole in his theorem, which led to the possibility that mass can emerge spontaneously in theories where, initially, there was no mass, has been established independently by six people, who published their work within a few weeks of one another in the summer of 1964. One of the sextet is Peter Higgs, whose name today has become associated with this development, and is best known for its—as yet unproved—consequence: the existence and properties of the "Higgs Boson." While this is a central focus of particle physics investigation today, later chapters will show that in 1964 the concepts were widely regarded as an interesting mathematical discovery, awaiting some realistic application.

Chapter 10

1967: FROM KIBBLE TO SALAM AND WEINBERG

Tom Kibble turns the Higgs Mechanism into a useful tool and teaches Salam the idea, who then incorporates this into the Salam-Ward model of the weak and electromagnetic forces. Weinberg also uses The Mechanism, and publishes a paper, which leads to his Nobel Prize. Salam sees Weinberg's paper and realizes he's been scooped. Meanwhile, almost everyone else ignores these ideas.

SPRING

By the summer of 1964, all the ingredients for the theory uniting the electromagnetic and weak forces were to hand. At Imperial College, London, all that remained was for Salam and Ward to complete their theory by incorporating the discovery that Kibble, Guralnik, and Hagen had made. However, the penny didn't drop. I asked Tom Kibble, "Were people talking to one another?" He laughed, admitting, "Not as much as they ought to have been."[1]

Kibble's impression is that, in 1964, Salam hadn't really noticed the papers on symmetry breaking, or at least not appreciated them: "Ward himself was a particularly secretive person, suspicious, with a tendency to think you were likely to steal his ideas. So he didn't talk to other people very much. Salam, however, talked to other people a lot, though I'm not sure that he always listened."

Whereas in 1964 Salam appears to have been deaf to what Kibble and his colleagues had achieved, in 1967 everything dramatically changed. In the intervening three years, Kibble had persevered with spontaneous symmetry breaking, plumbing its depths and trying to link the ideas with the realities of particles and forces. After all, while The Mechanism had shown the basic principles of how mass could emerge for the gauge boson, the only gauge boson known in the 1960s was the photon, which is massless. The massive *W* and *Z* bosons were predicted, but far beyond the reach of experiment at that time. What rules determine which bosons become massive and which, like the photon, remain untouched?

Having been one of those who first realized how to give gauge bosons a mass, Kibble managed to generalize the theory toward the needs of the real world. As Yang, Mills, and Shaw, a dozen years before, had generalized the ideas of gauge invariance by using the richer pastures of mathematical "SU" groups, so now Kibble did the same for the "mass mechanism." Some of what he achieved was implicit in Brout and Englert's original paper, but the full theory—containing all the ingredients for applying spontaneous symmetry breaking to the real world of particles and forces—was worked out by Kibble in 1967.

His success was to show how nature can be selective when endowing masses to gauge bosons. His equations expressed the mathematics of how this occurs. To help the reader he even described a simple model, which is almost, but not quite, that actually used by nature when it filters a massless photon from the soup of the massive *W* and *Z* bosons. Kibble seemed more inspired by the mathematical beauty than any actual phenomenology.

As we sat in his office at Imperial College in 2010, Kibble told me that he had had "one or two quite lengthy conversations with Salam about the problem in 1967.[2] I think that for whatever reason that got through to him in a way that the earlier work had not."

With these private tutorials, Salam began to appreciate the beauty and power of the idea. These conversations with Kibble made such an impact that, years later, Salam always referred idiosyncratically to the "Higgs-Kibble mechanism"—a feature that Kibble freely admits has always embarrassed him.

What had excited Salam so much in 1967 relative to his uninterest in 1964? To understand this, we need to appreciate where Salam's real interests lay.

His oeuvre shows a lifelong interest in group theory. In particular, he had always been intrigued by Yang-Mills theory and haunted by the memory of his student Ron Shaw, who had shown him the powerful implications of adding group theory to QED. Salam had also been fascinated by searches for symmetry groups that might encode the properties of the strong interaction. Under his aegis, in 1961 Israeli theorist Yuval Ne'eman had identified the group SU3 as a way of describing the properties of the hadrons—a feature independently discovered by Murray Gell-Mann. At one point in 1965, Salam thought that he had found the ultimate description of strong interactions, marrying SU3 with the theory of relativity in a grandiose mathematical scheme known as "U-twiddle-12." This was paraded in the British media as the ultimate theory of everything, but experimental evidence was already undermining its foundations, and it died. Group theory was indeed close to Salam's heart.

It is group theory that I believe caused Salam's awakening in 1967. Whereas spontaneous symmetry breaking had made no impact on Salam in 1964, Kibble's incorporation of group theory into the ideas in 1967 resonated for him. The model of the weak and electromagnetic interaction that he and Ward had built was itself deeply rooted in group theory. Salam, after tutoring by Kibble, realized how the mass mechanism could be grafted into those ideas.

Kibble, meanwhile, went away on sabbatical, first to Brookhaven during the summer of 1967 and then to Rochester. So he was absent when a series of unrelated events took place, in Imperial College and in Belgium, that summer and autumn. The events at Imperial that year, which would provide Salam his Nobel ticket a dozen years later, have long been shrouded in mystery and controversy; we will decode them later. The more immediate and public breakthrough centered on a conference in Belgium in October and a paper by Steven Weinberg in November, which would eventually lead to his own share of that Nobel Prize.

SUMMER

Weinberg spent the summer of 1967 at Cape Cod, working on a new theory of the strong interaction of pions, which was based on ideas from spontaneous symmetry breaking.

Nambu's original investigations had dealt with each type of pion in isolation. He had not examined how the three versions, with positive, negative, or zero charge, shared their common "pion-ness" when interacting with one another. The commonality, or symmetry, is mathematically expressed by the group SU2, which we met on page 80.

Weinberg had been using this symmetry as his point of departure in an attempt to build a theory of the strong force, where nucleons interact with pions. The proton and neutron—nucleons—themselves form two "faces" of the underlying SU2. In addition, each of them can behave like a left- or right-handed screw as they move. The empirical behavior of the interactions of pions and nucleons showed that the underlying symmetry of SU2 tended to act on the left- or right-handed possibilities independently. The mathematical structure of the resulting equations, which Weinberg was using, is called SU2 × SU2.

However, in nature this symmetry is not perfect. The imperfections are because nucleons have mass. Had they been massless, they would have traveled at the speed of light, the left- and right-handed possibilities acting truly independent of one another: In such a case, the symmetry could have been exact. However, the nucleons do have a mass, so that left and right get mixed up, and the chiral symmetry is broken. This is in effect what Nambu had investigated in 1959 but without the added richness of the SU2 × SU2 mathematical structure.

Weinberg had been investigating this more sophisticated theory around 1965, initially with considerable success. He derived theorems, which explained the behavior of pions and nucleons interacting at low energies, such as in nuclear physics. He was able to predict how pions behaved when they scattered from one another. By supposing spontaneous symmetry breaking to be the reason the mathematics of SU2 × SU2 was only an *approximate* symmetry of the strong interactions, Weinberg led the way in developing a whole theory of hadronic physics at low energies.

This proved so successful that it tended for a while to reduce the interest in what Brout, Englert, Higgs, and the Imperial team had done. In 1964 they had shown how to get rid of the unwanted massless Goldstone Bosons, but Weinberg's work was convincing him, and many others, that the pion itself was the Goldstone Boson of a spontaneously broken SU2 × SU2. When Higgs's first paper appeared, Weinberg thought, "Well that's nice; he's found a way of getting rid of Goldstone bosons but now I've become convinced that the pion is one."[3] Far from being unwanted, the Goldstone Boson had turned up at center stage.

Nonetheless, Higgs, Kibble, and the others had all demonstrated that the vector bosons of a Yang-Mills theory could become massive without spoiling the fundamental gauge symmetry. This opened the possibility that a complete Yang-Mills theory of the strong interaction might after all be possible. In addition to the well-known pion, examples of massive vector particles that feel the strong force had been found, notably the "rho" meson. Like the pion, this too occurred in the ubiquitous three charged varieties: positive, negative, and zero. Superficially, these rho-mesons had all the hallmarks of being the gauge bosons of strong interactions. And they are massive: Was this the theater for applying spontaneous symmetry breaking after all?

The idea, which at first sight seemed so promising, failed to fit the data. For some phenomena, the symmetry empirically is broken, but for others it works perfectly.[4] When these patterns were incorporated into the maths, the rho-meson stubbornly remained massless, contrary to reality. The problem with which Pauli had berated Yang years before—"Where are these massless vector mesons?"—refused to go away.

Higgs, also, had been trying to construct a theory of the strong interaction, built upon spontaneous symmetry breaking, but had decided it was impossible. En route to a conference at Rochester, he visited

Brookhaven. His diary, and a letter written to his wife a few days after his arrival,[5] records that it was August 25.

The journey was memorable for Higgs. He had flown from Scotland with Icelandic Airways via Keflavik, where he had been delayed, not arriving at JFK until 11:15 p.m. A taxi, train, and a second taxi finally brought him to Brookhaven about 2:30 in the morning on Friday, August 25. For his internal clock, this was already breakfast time, and after some sleep he went into the laboratory.

Kibble had already left Brookhaven. Higgs recalled, "I joined a discussion including Weinberg and others on how to understand the [strong interactions], or rather, how not to. I told them how I had failed to do it with spontaneously broken gauge symmetry. Weinberg had been having no success with something similar. Maybe this encounter helped him to realise a few weeks later that he 'had been applying the right ideas to the wrong problem' as he remarked in his Nobel Lecture."[6]

Weinberg has no recollection of this meeting, or of being at Brookhaven in 1967.[7] In any event, it seems that Higgs and Weinberg had reached a similar impasse. But in the middle of September, while driving his red Camaro to work at MIT, Weinberg had the epiphany of "applying the right ideas to the wrong problem."[8] Instead of the strong interactions, for which the ideas refused to work, the *massless* photon and the hypothetical massive W boson of the electromagnetic and weak interactions fitted perfectly.

Weinberg needed a concrete model to illustrate his general idea. The strongly interacting particles—"hadrons"—were, for him, a quagmire, and so he restricted his attention to the electron and neutrino. The electron can spin like a left- or right-handed screw, whereas the massless neutrino is only left-handed. The left-right symmetry, which had been a feature of the strong interaction, was gone. Instead of two SU2, the mathematics now only needed one, the second SU2 being replaced by simple numbers—in the jargon, "U1." So Weinberg set up the equations of SU2 × U1—the same "ess-you-two-cross-you-one" as Glashow, Salam, and Ward, had done, unknown to him. His theory, like theirs, required two massive electrically charged bosons, the W^+ and W^- carriers of the known weak force, and two neutral bosons, the massless photon and a massive Z^0.

What set Weinberg's breakthrough apart was that, with Kibble's ideas as a guide, and by incorporating the mass-generating mechanism in the simplest way, he was able to predict the masses of the *W* and the *Z*. The feature, however, which would eventually set the theory apart from all others, establishing it in the pantheons of science, was the fact that spontaneous symmetry breaking had seeded a theory that is renormalizable.

Weinberg had written a viable description of electromagnetic and weak interactions.[9] The one missing link in Weinberg's construction was that he did not know this.

WEINBERG'S MODEL

"The history of attempts to unify weak and electromagnetic interactions is very long, and will not be reviewed here." So began the first footnote in Steven Weinberg's seminal paper,[10] (Figure 10.1) which was published in November 1967 and led to his Nobel Prize in 1979. Weinberg's footnote recorded Fermi's primitive idea of 1934 and also a model "similar to ours by S Glashow [in 1961]." There is no mention of work by Salam and Ward.

This omission is a significant commentary on the attitude toward the weak interaction at that time and to Salam's contributions. The sea change would come later. During the 1970s, as momentum toward a Nobel award grew, the names of Weinberg and Salam became synonymous with the newly united "electroweak" theory; Glashow's name reappeared on the scene only at the last moment, while Ward's all but vanished.

That, however, is for the future; in 1967 Weinberg had started his paper by articulating the challenge as both an opportunity and a threat. He focused on the leptons—those fermions, such as the electron and neutrino, that do not feel the strong interactions. "Leptons interact only with photons, and with the [weak] bosons that presumably mediate weak interactions. What could be more natural than to unite these spin-one bosons [the photon and the weak bosons] into a multiplet," he pondered. That is the opportunity. The threat is that "standing in the way of this synthesis are the obvious differences in the masses of the photon and [weak] boson." He then suggested a solution: Perhaps "the symmetries relating the weak and electromagnetic interactions are exact [at a fundamental level] but are [hidden in practice]." The problem here, as Goldstone had shown, is that such an idea "raises the specter of unwanted Goldstone bosons." Weinberg then draws attention to the ideas of Higgs, Brout, and Englert, and Guralnik, Hagen, and Kibble, citing them in that order, and uses these to build his model.[11] This kills the Goldstone specter and gives masses to the *W* and *Z*. A further important insight is that Weinberg also shows how this mechanism leaves the photon massless.

His opening paragraph ended with the prescient observation that "the model may be renormalizable." The argument upon which this remark is based appears at the very end of the paper, though with somewhat less confidence than the promise hinted at the beginning. He begins the final paragraph with the question: "Is this model renormalizable?" The extent of his intuition is revealed in his argument: Although the presence of a massive vector boson hitherto had been a scourge, the theory with which he had begun had no such mass and, as such, was "probably renormalizable." So, he pondered, "The question is whether this renormalizability is lost [by the spontaneous breaking of the symmetry]." And the conclusion: "If this model is renormalizable, what happens when we extend it . . . to the hadrons?"

By speculating that the model may be renormalizable, he was hugely prescient, as 't Hooft would prove four years after. As a lot will hang on the question of renormalizability, it is worth backtracking two weeks before the paper was submitted. A chance encounter at the Solvay Congress in Belgium, during October, may have helped convince Weinberg that he was on the right track.

SOLVAY CONGRESS: OCTOBER 1967

By the end of September, Weinberg had his ideas in place, as he set off to Belgium to attend the Fourteenth Solvay Congress on Fundamental Problems in Elementary Particle Physics, held in Brussels, October 2–7.

Weinberg did not speak about his forthcoming paper but did make some remarks after other talks. The summary report shows that at the conclusion of the talk by Hans-Peter Dürr, a theorist from Munich, on the afternoon of October 2, Weinberg was the first to comment. Dürr's title had been "The Goldstone Theorem and Its Possible Application to

Elementary Particle Physics." Weinberg first remarked about the strong interaction, recalling Nambu's work, that had started all the interest, and then said, "We can also ask whether these ideas can be applied to unify the electromagnetic and the weak interactions."

Weinberg then made his seminal remark, which the proceedings recorded, "This raises a question I can't answer: Are such models renormalizable?" He continued with a similar argument to that which later appeared in his paper, ending with, "I hope someone will be able to find out whether or not [this] is a renormalizable theory of weak and electromagnetic interactions."

There was remarkably little reaction to Weinberg's remarks; as he himself recalls, there was "a general lack of interest."[12] The only recorded statement came from François Englert, who insisted that the theory is renormalizable; then, remarkably, there is no further discussion.

Englert and Robert Brout, then relatively junior scientists, had both attended the conference. The Solvay Congress is a closed meeting, restricted to a select band of world experts, but by tradition people connected with the field at Brussels University could participate, although generally they did not intervene. However, Englert reacted to Weinberg's comment, stood up, and, according to the published record of the meeting, said that he and Brout had shown that the mathematics describing the massive vector mesons is such that "the answer to Weinberg's question is that these massive vector gauge fields constitute a renormalizable theory."[13]

The published proceedings contain no further discussion on this issue, and Englert's assertion apparently remained unchallenged. Weinberg had raised the question of renormalizability, and Englert had made a response to the effect that the issue was in fact settled.

After four decades some forget, or remember with advantages. The occasion was so singular for Englert, and his collaborator, Brout, that their memory of the occasion was vivid, richer in detail certainly than the brief summary in the proceedings.

Englert recalls that he was new in the field, nervous, but at the same time quite confident in what he and Brout had achieved with their work on "The Mechanism" for generating masses. He remembered his reaction as Weinberg, "asked something like—Why does a renormalizable massless theory become non-renormalizable after breaking?—I do not remember the exact words but it seemed to me that he suggested that the acquired mass of the gauge vector boson still left a theory that was *not* renormalizable. I reacted because I believed that we had serious indications that the broken phase should be renormalizable."[14]

This impression differs from Weinberg's perception of the renormalization question. He recalls, "I thought there was a good chance that it was renormalizable." He explained that because spontaneous symmetry breaking manifests itself in phenomena at low energies, his intuition told him that this symmetry breaking "shouldn't matter when you ask questions about high momenta which is what comes up in questions about renormalizability. But [my intuition] didn't go beyond that."[15] As we shall see later, Weinberg had a manuscript of his forthcoming paper with him at the conference, in which his speculations about the renormalizability of the theory were spelled out.

Englert's confidence at that time was based on the mathematical formulas referring to a quantity known as the "propagator" of the vector boson, which had emerged in his and Brout's analysis.[16] In the formulas that had previously appeared, the propagator contained a term that depended on the ratio of the energy of the boson to its mass. As the energy of a virtual particle can range all the way to infinity, this ratio can also, and this in turn spoiled renormalizability of the theory. However, in their formulation of the problem, a result of "The Mechanism" was that this term did *not* grow when the energy itself was large. The value was in effect a number: the ratio of the energy to the energy itself, not to the mass. If this form of the propagator was correct, then the possibility of a viable theory remained alive.

A small debate then ensued between Englert and Weinberg, which the proceedings did not record. In retrospect, their discussion may have been based on mutual misunderstanding.

Englert explained that the equation for the propagator in their work contained the energy of the boson, rather than simply its mass, and recalled that Weinberg said, "No. The propagator has [energy divided by *mass*] and is not renormalizable." Englert said that he repeated that his and Brout's propagator did not contain mass in that way. Then, in the memories of Englert and Brout, Weinberg said, "Someone has made a mistake somewhere."

Englert explained that at that time he and Brout were new in the field, and Weinberg's reaction made him worry that perhaps he was not correct. So much so that "when the secretary came to me and asked me to write [my comment] for the proceedings, I said that I didn't want to. So I wrote nothing and was convinced that the proceedings didn't have anything about my intervention."

However, Brout wrote an account, a fact that Englert was unaware of for several years. "Then years later, maybe ten years later, [after 't Hooft and Veltman had proved the renormalizability] Bernard De Wit said to [Englert] 'How could you say that the theory was renormalizable in 1967?'" So Englert replied, "Power counting [the fact that energy was divided by energy and not mass] as a consequence of gauge invariance. I said that it *should* be renormalizable, *not* that I had proved it. And then I said [to De Wit]—but how do you know what I said?" And De Wit replied, "'It's in the proceedings.' Well, that was the first that I knew of it."

Only then did he check and found that Brout had written it. I asked Brout about this, who confirmed, "I wrote it because I thought it was relevant!"[17] Brout was convinced that the theory was renormalizable and said, "I wrote it up in that sense."

Years later, it seemed to Englert and Brout that they were all talking at cross-purposes.[18] Weinberg was unaware of the details of their work at that stage, and the standard form of the equation for the description of the *W* boson depended upon a ratio of its energy and mass.[19] Once the form of the equation in Englert and Brout's approach was known, the *possibility* that the theory was renormalizable became more likely. However, four years would pass, and much work would have to be done, before anyone managed to prove it.

WEINBERG'S PAPER

At some point during the Solvay conference Weinberg presented a handwritten draft of his paper to Hans-Peter Dürr, the speaker after whose talk Weinberg and Englert had made their remarks. Forty years later I obtained a copy by a roundabout route. Weinberg himself had not seen it in all that time and thought that all record of his Nobel-winning manuscript had been lost.[20] The manuscript is notable for there being no sign of second thoughts, or editing, which suggests that it was a provisional final draft of an idea that had been worked through in the preceding days (see Figure 10.1). The only hint of modification after the first draft had been written is a memo squeezed in at the end of a reference to Higgs, to include references to Brout and Englert, and to Guralnik, Hagen, and Kibble, for the idea of spontaneous symmetry breaking, on which the paper was based. Weinberg's intuition about the renormalizability of the model is already present in this manuscript and is identical to what appears in his article in the *Physical Review Letters.* There is no mention of Glashow's SU2 × U1 model in the draft, but this is included in the version that was published in the *Physical Review Letters* the following month. This is the only substantial difference.

This manuscript was submitted to the editors of the *Physical Review Letters* on his return to the United States and received by them on October 17. It appeared in print on November 20 and was available on library shelves by December. This chronology will turn out to be significant later when we trace the origins of Salam's route to his share of the Nobel.

There is another citation in Weinberg's paper, with significance for understanding Salam's further entry into this history. It is of the paper published by Tom Kibble at that time.[21]

The basic idea of mass generation without the specter of Goldstone Bosons was due to the assorted theorists mentioned above, in the summer of 1964. However, a crucial feature of Weinberg's model was the trick of being able to give masses to the *W* and *Z* while leaving the photon massless. This extension of the mass-generating mechanism was due to Kibble. Weinberg recognizes this and credits it. Weinberg's genius was to assemble the various pieces of a jigsaw and display the picture. Arguably, it was the technical mechanism discovered by Kibble that was central to this, no less than the generic ideas that had preceded it.

As for reaction to these ideas at the time, they were underwhelming. We have already seen that Weinberg's recollection of the Solvay meeting was that he mentioned his ideas "to a general lack of interest." The record of that meeting shows that, following the interjection of Englert, there was no further discussion of this topic. I consulted other delegates, and, apart from Brout and Englert, no one could recall Weinberg's comment or Englert's response. Even Weinberg himself had no memory of it.

In the world at large, the impact was no greater. Weinberg's paper appeared in November 1967 to a deafening silence: "Rarely has so great an accomplishment been so widely ignored."[22] At the risk of veering into cliché, the events of 1967 bring to mind the Buddhist idea of a tree falling and no one hearing it.

Six months after the appearance of Weinberg's paper, Salam gave a talk in Göteborg in May 1968 outlining how spontaneous symmetry breaking could be combined with the model of Salam and Ward (see Figure 10.2 for a draft version of the original Salam and Ward paper). Kibble's tutorials had inspired him, though three years too late. As Weinberg's paper was already published, and Salam's talk contained nothing additional to that, it too made no impact at the time. Furthermore, it appeared in obscure proceedings of the Swedish Academy. Few appear to have read it even today, yet within five years it would become a plank in Salam's name being joined with Weinberg in what became known as the "Weinberg-Salam model," culminating in Salam's sharing the Nobel Prize in 1979.

This single contribution apart, Salam himself appears to have done nothing more with these ideas for several years. Weinberg too seems not to have been inspired by his own creation, as his oeuvre for the remainder of that decade continued to concentrate on the strong interactions on which he had been focused throughout.

Today, Weinberg's paper has been cited more than 7,100 times. Having been cited but twice in four years from 1967 to 1971, suddenly it became so important that researchers have cited it three times every week throughout more than forty years. There is no parallel for this in the recorded history of particle physics. The reason is that in 1971 an event took place that has defined the direction of the field ever since: Gernard 't Hooft made his debut.

FIGURE 10.1
WEINBERG'S ORIGINAL MANUSCRIPT

A Model of Leptons

Steven Weinberg*
Department of Physics, Massachusetts Institute of Technology
Cambridge, Massachusetts

Leptons interact only with photons, and with the intermediate bosons that presumably mediate weak interactions. What could be more natural than to unite[1] these spin-one bosons into a multiplet of ~~gauge~~ vector fields? Standing in the way of this synthesis are the obvious differences in the masses of the photon and intermediate meson, and in their couplings. We might hope to understand these differences by imagining that the ~~group of~~ symmetries relating the weak and electromagnetic interactions are exact symmetries of the Lagrangian but are broken by the vacuum. However this raises the specter of unwanted massless Goldstone bosons.[2] This note will describe a model in which the symmetry between the electromagnetic and weak interactions is spontaneously broken, but in which the Goldstone bosons are avoided by introducing the photon and the intermediate boson fields as gauge fields.[3] The model may be renormalizable.

*On leave from the University of California, Berkeley, California

1. The history of attempts to unify weak and electromagnetic interactions is very long, and will not be reviewed here. Possibly the earliest reference is E. Fermi, Z. Phys. 88, 161 (1934).

2. J. Goldstone, Nuovo Cimento 19, 154 (1961); J. Goldstone, A. Salam, and S. Weinberg, Phys. Rev. 127, 965 (1962).

3. P. W. Higgs, Phys. Letters 12, 132 (1964); Phys. Rev. Letters 13, 508 (1964); Phys. Rev. 145, 1156 (1966)
Hagen et al. Brout & Englert

4. See particularly T. W. B. Kibble, Phys. Rev. 155, 1554 (1967). A similar phenomenon occurs in the strong interactions; the ρ-meson mass in zeroth order perturbation theory is just the bare mass, while the A1 meson picks up an extra contribution from the spontaneous breaking of chiral symmetry. See S. Weinberg, Phys. Rev. Letters 18, 507 (1967), especially footnote 7; J. Schwinger, Phys. Letters 24B, 473 (1967); S. Glashow, H. Schnitzer and S. Weinberg, Phys. Rev. Letters 19, 139 (1967), Eq. (13) et. seq.

5. T. D. Lee and C. N. Yang, Phys. Rev. 98, 101 (1955)

6. This is the same sort of transformation as that which eliminates the non-derivative π couplings in the σ-model; see S. Weinberg, Phys. Rev. Letters 18, 188 (1967). The π reappears with derivative coupling because the strong interaction Lagrangian is not invariant under chiral gauge transformation.

7. For a similar argument applied to the σ-meson, see S. Weinberg, ref. 6.

8. R. P. Feynman and M. Gell-Mann, Phys. Rev. 109, 193 (1957).

Volume 19, Number 21 PHYSICAL REVIEW LETTERS 20 November 1967

taken very seriously, but it is worth keeping in mind that the standard calculation[8] of the electron-neutrino cross section may well be wrong.

Is this model renormalizable? We usually do not expect non-Abelian gauge theories to be renormalizable if the vector-meson mass is not zero, but our Z_μ and W_μ mesons get their mass from the spontaneous breaking of the symmetry, not from a mass term put in at the beginning. Indeed, the model Lagrangian we start from is probably renormalizable, so the question is whether this renormalizability is lost in the reordering of the perturbation theory implied by our redefinition of the fields. And if this model is renormalizable, then what happens when we extend it to include the couplings of $\vec{A}_\mu$ and B_μ to the hadrons?

I am grateful to the Physics Department of MIT for their hospitality, and to K. A. Johnson for a valuable discussion.

*This work is supported in part through funds provided by the U. S. Atomic Energy Commission under Contract No. AT(30-1)2098).

†On leave from the University of California, Berkeley, California.

[1]The history of attempts to unify weak and electromagnetic interactions is very long, and will not be reviewed here. Possibly the earliest reference is E. Fermi, Z. Physik 88, 161 (1934). A model similar to ours was discussed by S. Glashow, Nucl. Phys. 22, 579 (1961); the chief difference is that Glashow introduces symmetry-breaking terms into the Lagrangian, and therefore gets less definite predictions.

[2]J. Goldstone, Nuovo Cimento 19, 154 (1961); J. Goldstone, A. Salam, and S. Weinberg, Phys. Rev. 127, 965 (1962).

[3]P. W. Higgs, Phys. Letters 12, 132 (1964), Phys. Rev. Letters 13, 508 (1964), and Phys. Rev. 145, 1156 (1966); F. Englert and R. Brout, Phys. Rev. Letters 13, 321 (1964); G. S. Guralnik, C. R. Hagen, and T. W. B. Kibble, Phys. Rev. Letters 13, 585 (1964).

[4]See particularly T. W. B. Kibble, Phys. Rev. 155, 1554 (1967). A similar phenomenon occurs in the strong interactions; the ρ-meson mass in zeroth-order perturbation theory is just the bare mass, while the A_1 meson picks up an extra contribution from the spontaneous breaking of chiral symmetry. See S. Weinberg, Phys. Rev. Letters 18, 507 (1967), especially footnote 7; J. Schwinger, Phys. Letters 24B, 473 (1967); S. Glashow, H. Schnitzer, and S. Weinberg, Phys. Rev. Letters 19, 139 (1967), Eq. (13) et seq.

[5]T. D. Lee and C. N. Yang, Phys. Rev. 98, 101 (1955).

[6]This is the same sort of transformation as that which eliminates the nonderivative $\vec{\pi}$ couplings in the σ model; see S. Weinberg, Phys. Rev. Letters 18, 188 (1967). The $\vec{\pi}$ reappears with derivative coupling because the strong-interaction Lagrangian is not invariant under chiral gauge transformation.

[7]For a similar argument applied to the σ meson, see Weinberg, Ref. 6.

[8]R. P. Feynman and M. Gell-Mann, Phys. Rev. 109, 193 (1957).

Volume 19, Number 21 PHYSICAL REVIEW LETTERS 20 November 1967

Dear Dr 't Hooft

This letter suggests (though it does not prove) the renormalizability of the massive Yang-Mills models discussed in your recent preprint, in which the vector meson mass arises from spontaneous symmetry breaking.

S.W.

A MODEL OF LEPTONS*

Steven Weinberg†

Laboratory for Nuclear Science and Physics Department,
Massachusetts Institute of Technology, Cambridge, Massachusetts

(Received 17 October 1967)

Leptons interact only with photons, and with the intermediate bosons that presumably mediate weak interactions. What could be more natural than to unite[1] these spin-one bosons into a multiplet of gauge fields? Standing in the way of this synthesis are the obvious differences in the masses of the photon and intermediate meson, and in their couplings. We might hope to understand these differences by imagining that the symmetries relating the weak and electromagnetic interactions are exact symmetries of the Lagrangian but are broken by the vacuum. However, this raises the specter of unwanted massless Goldstone bosons.[2] This note will describe a model in which the symmetry between the electromagnetic and weak interactions is spontaneously broken, but in which the Goldstone bosons are avoided by introducing the photon and the intermediate-boson fields as gauge fields.[3] The model may be renormalizable.

We will restrict our attention to symmetry groups that connect the observed electron-type leptons only with each other, i.e., not with muon-type leptons or other unobserved leptons or hadrons. The symmetries then act on a left-handed doublet

$$L=[\tfrac{1}{2}(1+\gamma_5)]\binom{\nu_e}{e} \quad (1)$$

and on a right-handed singlet

$$R\equiv[\tfrac{1}{2}(1-\gamma_5)]e. \quad (2)$$

The largest group that leaves invariant the kinematic terms $-\bar{L}\gamma^\mu\partial_\mu L-\bar{R}\gamma^\mu\partial_\mu R$ of the Lagrangian consists of the electronic isospin $\vec{T}$ acting on L, plus the numbers N_L, N_R of left- and right-handed electron-type leptons. As far as we know, two of these symmetries are entirely unbroken: the charge $Q=T_3-N_R-\tfrac{1}{2}N_L$, and the electron number $N=N_R+N_L$. But the gauge field corresponding to an unbroken symmetry will have zero mass,[4] and there is no massless particle coupled to N,[5] so we must form our gauge group out of the electronic isospin $\vec{T}$ and the electronic hyperchange $Y\equiv N_R+\tfrac{1}{2}N_L$.

Therefore, we shall construct our Lagrangian out of L and R, plus gauge fields $\vec{A}_\mu$ and B_μ coupled to $\vec{T}$ and Y, plus a spin-zero doublet

$$\varphi=\binom{\varphi^0}{\varphi^-} \quad (3)$$

whose vacuum expectation value will break $\vec{T}$ and Y and give the electron its mass. The only renormalizable Lagrangian which is invariant under $\vec{T}$ and Y gauge transformations is

$$\mathcal{L}=-\tfrac{1}{4}(\partial_\mu\vec{A}_\nu-\partial_\nu\vec{A}_\mu+g\vec{A}_\mu\times\vec{A}_\nu)^2-\tfrac{1}{4}(\partial_\mu B_\nu-\partial_\nu B_\mu)^2-\bar{R}\gamma^\mu(\partial_\mu-ig'B_\mu)R-\bar{L}\gamma^\mu(\partial_\mu-ig\vec{t}\cdot\vec{A}_\mu-i\tfrac{1}{2}g'B_\mu)L$$

$$-\tfrac{1}{2}|\partial_\mu\varphi-ig\vec{A}_\mu\cdot\vec{t}\varphi+i\tfrac{1}{2}g'B_\mu\varphi|^2-G_e(\bar{L}\varphi R+\bar{R}\varphi^\dagger L)-M_1^2\varphi^\dagger\varphi+h(\varphi^\dagger\varphi)^2. \quad (4)$$

We have chosen the phase of the R field to make G_e real, and can also adjust the phase of the L and Q fields to make the vacuum expectation value $\lambda\equiv\langle\varphi^0\rangle$ real. The "physical" φ fields are then φ^-

1264

KX520 1-3

The cover page and references of Weinberg's manuscript, as given to H.-P. Dürr at the time of the Solvay conference, which may be compared with the paper in published form (Courtesy S. Weinberg and M. Veltman). The origin of the arrow on the manuscript adjacent to reference 3 is unknown, but may be a reminder to complete that reference: the citations of Brout and Englert and "Hagen et al." (Guralnik, Hagen, and Kibble) have been added later, possibly following discussions at that meeting. The published version comes from a copy sent by S. Weinberg to G. 't Hooft, following the 1971 proof of Weinberg's conjecture about renormalizability. This paragraph has been highlighted in the published version.

FIGURE 10.2
SALAM AND WARD MANUSCRIPT

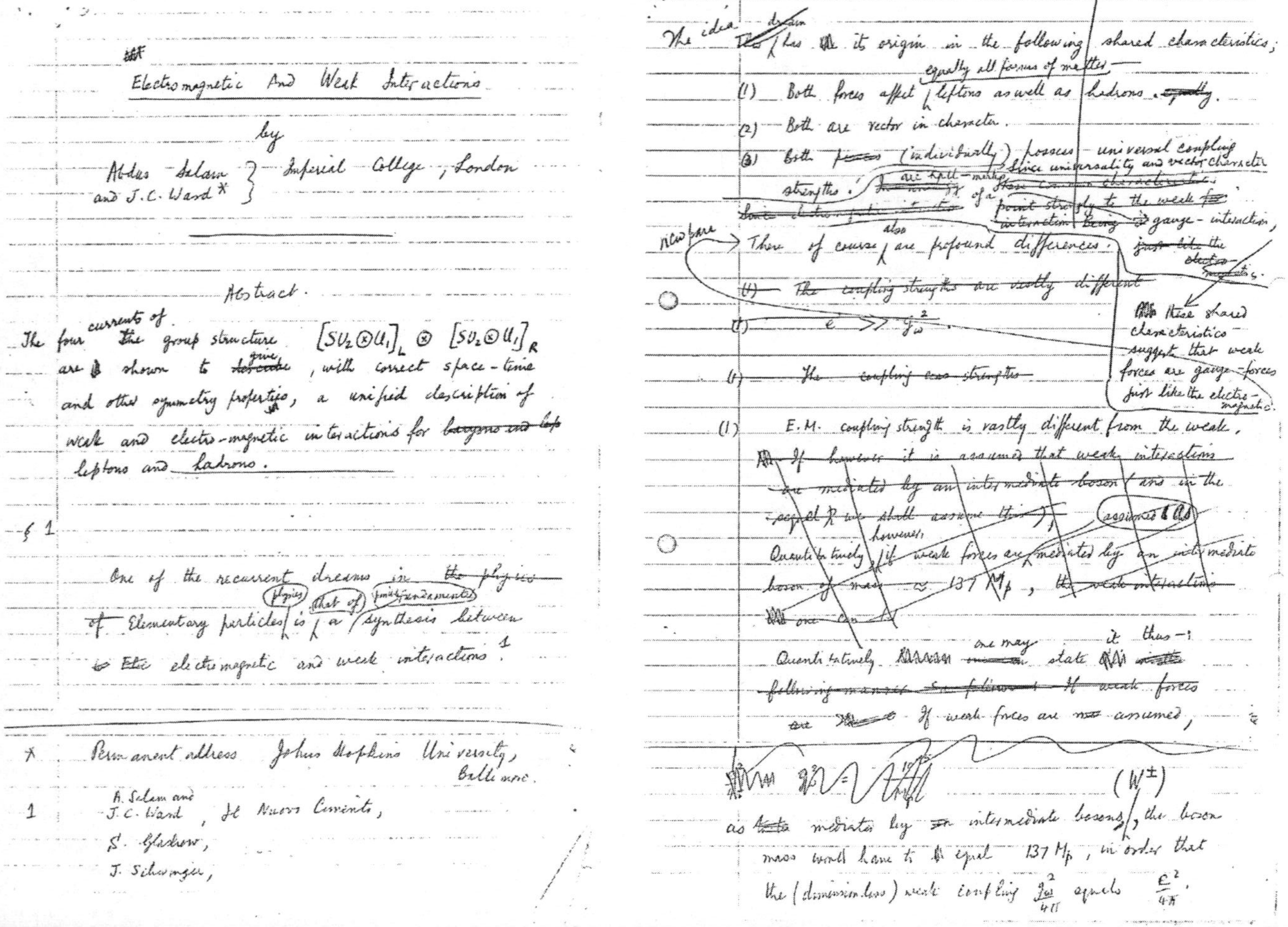

Electromagnetic And Weak Interactions

by

Abdus Salam and J.C. Ward* } Imperial College, London

Abstract.

The four currents of the group structure $[SU_2 \otimes U_1]_L \otimes [SU_2 \otimes U_1]_R$ are shown to give, with correct space-time and other symmetry properties, a unified description of weak and electro-magnetic interactions for leptons and hadrons.

§ 1

One of the recurrent dreams in physics of Elementary particles is that of a synthesis between electromagnetic and weak interactions.[1]

* Permanent address Johns Hopkins University, Baltimore.

1 A. Salam and J.C. Ward, Il Nuovo Cimento, S. Glashow, J. Schwinger,

(2)

The idea draws its origin in the following shared characteristics;

(1) Both forces affect equally all forms of matter leptons as well as hadrons.

(2) Both are vector in character.

(3) Both (individually) possess universal coupling strengths.

There also of course are profound differences.

(1) E.M. coupling strength is vastly different from the weak.

If weak forces are assumed ... as mediated by intermediate bosons $(W^\pm)$, the boson mass would have to equal $137\,M_p$, in order that the (dimensionless) weak coupling $\frac{g_W^2}{4\pi}$ equals $\frac{e^2}{4\pi}$.

Salam and Ward's 1964 paper marrying weak and electromagnetic interactions. The three pages of manuscript suggest that the original idea had been to write a long paper for a journal with an abstract. The script also shows that it is often hard to construct the start of a manuscript, but that the central arguments flow more easily. Note also the redrafting, which is a reminder of what creative writing was like in the days before word processors. (Courtesy of Louise Johnson)

(3)

In the sequel we assume just this. For the outrageous mass value itself ($M_W \approx 137 M_p$) we can offer no explanation. We seek however for a synthesis in terms of a group structure such that the remaining differences viz

(2) Contrasting space-time behaviour (V for EM vs. V and A for weak)

(3) Contrasting ΔS and ΔI behaviour, both appear as ~~facts~~ aspects of the same symmetry.

~~The interaction Lagrangian is~~
Once the group-structure is known, ~~the interaction Lagrangian is [illegible] would be obtained~~ ~~By~~ the gauge principle would ~~give the [illegible]~~ unambiguously give the interaction Lagrangian. For hadrons at least we also require that the group-structure include SU_3.

We ~~succeed~~ succeed partly in the attempt.

ELECTROMAGNETIC AND WEAK INTERACTIONS

A. SALAM and J. C. WARD *
Imperial College, London

Received 24 September 1964

One of the recurrent dreams in elementary particles physics is that of a possible fundamental synthesis between electro-magnetism and weak interactions [1]. The idea has its origin in the following shared characteristics:

1) Both forces affect equally all forms of matter-leptons as well as hadrons.
2) Both are vector in character.
3) Both (individually) possess universal coupling strengths. Since universality and vector character are features of a gauge-theory these shared characteristics suggest that weak forces just like the electromagnetic forces arise from a gauge principle.

There of course also are profound differences:

1) Electromagnetic coupling strength is vastly different from the weak. Quantitatively one may state it thus: if weak forces are assumed to have been mediated by intermediate bosons (W), the boson mass would have to equal 137 M_p, in order that the (dimensionless) weak coupling constant $g_W{}^2/4\pi$ equals $e^2/4\pi$.

In the sequel we assume just this. For the outrageous mass value itself ($M_W \approx 137\, M_p$) we can offer no explanation. We seek however for a synthesis in terms of a group structure such that the remaining differences, viz:

2) Contrasting space-time behaviour (V for electromagnetic versus V and A for weak).
3) And contrasting ΔS and ΔI behaviours both appear as aspects of the same fundamental symmetry. Naturally for hadrons at least the group structure must be compatible with SU_3.

Lepton interactions define both the unit of the electric charge and (from μ-decay) the (bare) value of weak coupling constant. Leptons therefore must be treated first.

There is only one genuine lepton multiplet (in the limit $m_e = m_\mu = 0$) which really treats the neutrino field on the same footing ** as μ and e. This is the Konopinski-Mahmoud multiplet.

$$L = \begin{pmatrix} \nu \\ e^- \\ \mu^+ \end{pmatrix} \tag{1}$$

In terms of SU_3 generators ***, the electric charge clearly equals:

$$Q_l = \begin{pmatrix} 0 & & \\ & -1 & \\ & & +1 \end{pmatrix} = -2U_3 = -\tfrac{2}{3}\sqrt{3}\,(I_0 - V_0) = 2(I_3 - V_3), \tag{2}$$

while the weak interaction (with no neutral currents) has the unique form †

* Permanent address, John Hopkins University, Baltimore.

** There are other schemes where one postulates multiplets consisting of a two-component neutrino field together with a four-component electron or muon. These do not satisfy even the most elementary requirement of a genuine group-structure, i.e. that in some limit at least, the particles concerned should be transformable, one into the other.

***
$$[T^i, T^j] = i f^{ijk} T^k$$
$$\{T^i, T^j\} = \tfrac{1}{3}\delta^{ij}\mathbf{1} + d^{ijk} T^k$$
$$I_3 = T^3,\ I^\pm = \tfrac{1}{2}\sqrt{2}\,(T^1 \mp iT^2),\ I_0 = T^8,\ [U_0, I] = 0$$
$$U_3 = \tfrac{1}{2}\sqrt{3}\,T^8 - \tfrac{1}{2}T^3,\ U^\pm = \tfrac{1}{2}\sqrt{2}\,(T^6 \mp iT^7),$$
$$U_0 = \tfrac{1}{2}\sqrt{3}\,T^3 + \tfrac{1}{2}T^8$$
$$V_3 = \tfrac{1}{2}\sqrt{3}\,T^8 + \tfrac{1}{2}T^3,\ V^\pm = \tfrac{1}{2}\sqrt{2}\,(T^4 \mp iT^5),$$
$$V_0 = \tfrac{1}{2}\sqrt{3}\,T^3 - \tfrac{1}{2}T^8$$

Note

$$Q_h = T^3 + \tfrac{1}{3}\sqrt{3}\,T^8 = \tfrac{2}{3}\sqrt{3}\ \ U_0 = \tfrac{2}{3}\sqrt{3}\,(I_0 + V_0) = \tfrac{2}{3}(I_3 + V_3)$$
$$Q_l = T^3 - \sqrt{3}\,T^8 = -2U_3 = -\tfrac{2}{3}\sqrt{3}\,(I_0 - V_0) = +2\,(I_3 - V_3)$$

Explicitly,

$$Q_h = \begin{pmatrix} \tfrac{2}{3} & & \\ & -\tfrac{1}{3} & \\ & & -\tfrac{1}{3} \end{pmatrix},\quad Q_l = \begin{pmatrix} 0 & & \\ & -1 & \\ & & +1 \end{pmatrix}$$

Q_h is the conventional hadron charge operator, Q_l gives lepton-charge.

† Define

$$\psi_L = \tfrac{1}{2}(1 + \gamma_5)\,\psi \qquad \psi_R = \tfrac{1}{2}(1 - \gamma_5)\,\psi$$
$$(A^+B)_L = \tfrac{1}{2}A^+\gamma_4\gamma_\mu(1 + \gamma_5)B$$

Chapter 11

"AND NOW I INTRODUCE MR. 'T HOOFT"

Introducing 't Hooft and Veltman. Veltman believes in Yang-Mills theories and builds a computer program to calculate Feynman diagrams. His student 't Hooft goes to Cargese, meets Ben Lee, and realizes how to build a viable theory of the weak force using the Higgs Mechanism. 't Hooft and Veltman prove that the theory is renormalizable. 't Hooft makes his debut in Amsterdam.

What was it that set Gerard 't Hooft apart from the rest of us, and from a generation of already established scientists? Many years later I asked him: "If at the start when you set out to prove the renormalizability of the weak force you had realized the enormity of what you were trying to do, would you have begun?" His answer was revealing: "At that time I felt I could do anything. I was so absorbed by the problem that I didn't care [about the difficulty]. This was the problem that I wanted to solve."[1]

In 1967 when Weinberg's paper was published, I was a new graduate student at Oxford. Some of my colleagues noted his idea, but no one took it to be anything remarkably different from many other short-lived attempts to confront the weak interaction. It made little impact. Across the

North Sea, a couple of hundred miles away in Holland, 't Hooft also was starting on his career. I, like most of my colleagues, had chosen pragmatically, settling on a thesis with a good chance of making inroads, enough to get onto the first rung of the research ladder in a climate where tenured positions were becoming increasingly rare. The strategy was to establish oneself enough to help cement a brick here and there in the wall of knowledge; 't Hooft, however, was prepared to risk everything in the ambitious hope of constructing the foundations for a whole new edifice.

This slightly built, trim, clean-cut man, with a tenor voice, hardly seemed someone who would stamp authority on the masses. Our research supervisors each had huge reputations in their fields—for 't Hooft this was Tini Veltman, focusing on field theory, contrary to the received wisdoms of the time, and in my case at Oxford, Dick Dalitz, who first saw the potential reality of quarks as seeds of protons, neutrons, and a host of ephemeral hadrons that had been discovered in high-energy physics experiments in the 1950s and '60s. The recognition of the quark layer of reality would help to revolutionize physics in the subsequent decades, as later chapters will reveal, but it would be the work of 't Hooft on the weak interaction that would turn out to be so central to the impending revolution. In Oxford we focused on the strong interaction and on building models based on quarks, a few brave souls choosing to skirt around the edges of the puzzling weak interaction. In Utrecht, unknown to any of us then, 't Hooft—by circumstances, by chance, but above all by virtue of his intense personality—took up the challenge head-on. As we saw above: That was the problem he wanted to solve.

For my part, I remembered the problem that as a child I had wanted to solve, indeed, that I thought I had solved: Fermat's Last Theorem. Every youngster learns Pythagoras's Theorem and knows that there are many integers that will satisfy the equation $a^2 + b^2 = c^2$ (for example, 3, 4, and 5). French mathematician Pierre de Fermat, in 1637, asserted that he had proved that this was no longer possible with integers for $a^3 + b^3 = c^3$. A few trials show you cannot do it, but do not prove it to be an impossibility in general. Fermat, however, claimed to have found a proof, which he omitted to include in the margins of his notebook. For three centuries mathematicians wrestled with this, and failed. I first saw the problem mentioned in a national newspaper. I was astonished that the statement of the problem was so easy that I could understand it, yet it was supposedly impossible to prove. What excited me was that I immediately saw the answer and wrote to the editor.

However, unknown to me there was a typo: The equation had been written incorrectly as $a^3 \times b^3 = c^3$. I proudly informed the editor that if $c = a \times b$, the problem was solved. At that time, it never occurred to me as suspicious that a child of ten had seen what generations of the ablest minds had overlooked. When the editor told me of the typo, I was astonished; not at my naïveté (I was still too naive) but that rotating a multiplication sign through 45 degrees to turn it into "+" had such profound consequences.

That defeat convinced me there was no point in continuing with Fermat's puzzle. A few years later, Andrew Wiles came across Fermat's theorem in a book at his local library. Like me, he was then ten years old; like me, he decided to solve it; unlike me, he succeeded—though it took him more than a decade of dedicated work some thirty years later. 't Hooft had the perseverance of Wiles, or at least more than most. Fermat's Last Theorem had tantalized mathematicians for three centuries, whereas the problem of building a viable theory of the weak interaction had existed for only three decades. However, in some popularizations at the time, this challenge drew comparison with the one that Fermat had set for generations of mathematicians.[2]

The element of chance that enabled the drama to come to its successful climax was that 't Hooft came under the wing of Tini Veltman.

TINI VELTMAN

As I said at the beginning: Tini Veltman is a contrarian. In the 1960s, while almost everyone was ignoring field theory, Veltman's intuition told him that they were wrong. What he had realized was that a theory built along the lines that Yang and Mills had invented could be the answer to the weak force. No one else seemed to be paying these ideas much heed. It was an inspired decision; today we recognize Yang-Mills Theories to be the answer to the electromagnetic, weak, and also the strong forces.

His story starts in 1962 when Yang and T.-D. Lee, interested in the possibility that electrically charged *W* bosons are the agents of the weak interaction, investigated their electrical properties. The amount of electric charge on a *W* is specified—it exactly matches that of a proton or an electron—but its magnetism was not. The *W* can also have a more complicated electrical property, known as a quadrupole moment. Lee made a calculation of this in an analogous manner to Schwinger's calculation, fifteen years earlier, of the electron's anomalous magnetic moment.

Veltman became interested. However, when he tried to extend Lee's calculations, he quickly encountered difficulties. The algebra had "up to 50000 terms" in some parts of the calculation.[3]

"Necessity is the mother of invention": To solve this, Veltman pioneered algebraic calculations by computer.

In 1963 he had spent time at Stanford in California, where SLAC—the Stanford Linear Accelerator Center of electrons—was being built. Veltman was housed in the workshops where the construction was in full swing; the noise was constant. Across from his office, James "B. J." Bjorken and Sid Drell were working on their magnum opus—the text of the book that for years became the standard work on relativistic quantum mechanics (but omitted Yang-Mills theory). Their discussion on how to present the arguments competed with the din of the engineering. Veltman "escaped to the computer centre," where he wrote his program: "Schoonschip."[4]

Veltman characteristically "called the program *Schoonschip* among others to annoy everybody not Dutch."[5] The name translates as "clean ship" and is a Dutch naval expression for "clearing up a messy situation." Schoonschip was well named, becoming the first practical computer routine for making complicated symbolic manipulations and enabling calculations that previously had been beyond an individual's algebraic ability.

In January 1964, Veltman was visiting T.-D. Lee in New York and told him about Schoonschip. Veltman recalls that Lee "barely reacted," but later Veltman heard that immediately after he left Lee's office, Lee "wanted one of the local physicists to develop an analogous program." Veltman never forgot this.[6]

When, with the aid of Schoonschip, Veltman competed the calculation of the *W*'s quadrupole moment, he noticed an intriguing feature. He had calculated the answers for a range of possible values of the magnetic moment. The expressions diverged, meaning that their numerical values would become infinity, but for one particular value of the magnetic moment, the algebraic expressions were simpler: Nearly all of the divergences disappeared. At the time, he didn't know what to make of it, but "it remained in my memory."[7] This was fortunate, as five years later he would see the same result in a completely different place.

During the 1960s, Veltman visited CERN regularly and discussed physics with Irish theoretician John Bell. In 1968 Bell wrote a paper that gave Veltman an epiphany: Bell's ideas suggested to Veltman that the weak interaction might be described by a Yang-Mills theory. So he started to learn about this theory.

It was while doing so that he had a surprise. He noticed that the Yang-Mills equations lead directly to the "magic" result for the *W*'s properties that he had noticed five years before. He felt intuitively that this was important and might be the seed for a viable description of the weak interaction. He decided to concentrate on the renormalizability of Yang-Mills theories. The calculations turned out to be even more complicated that anything he had dealt with before.

As we have seen, Yang and Mills considered the proton and neutron as a pair of twins (distinguished by their electric charge). In QED the source of the electromagnetic force is electric charge, which exists in positive or negative amounts, but the accounting involves no more than arithmetic: Add up the number of positive charges, subtract the number of negatives, and you have calculated the total amount of charge. The rules are the simple manipulation of numbers. Whereas QED described the proton and neutron independently, the equations of Yang and Mills united them: In addition to the proton or the neutron being examined in isolation, Yang-Mills theory generalized QED to take account of what happened when one turned into the other,[8] and in this case the accounting involves matrices, which are mathematical generalizations of ordinary numbers.

For several years few theorists looked at Yang-Mills theories; they are technically complex, and as we have seen, there was no clear evidence that nature made use of them: nice mathematics but wrong physics.[9] Those who decided to press ahead, regardless, then hit further barriers: The theory was haunted by unphysical "ghosts"—weird particles that have a probability to exist that is less than nothing.

By 1968, when Veltman started along this route, others, who had briefly dabbled with Yang-Mills Theories and been haunted by the ghosts, had quit. However, Veltman took a contrary view: As long as these transitory phantoms occurred only deep within the unobservable quantum engineering that made the physical machine operate, there was nothing wrong in principle: Ghosts, after all, do not exist. If they were exorcised before all the accounts were audited, so that they disappeared in the final reckoning, he felt that there need be no contradiction with common sense.

As for renormalization: In QED the fact that the photon was massless had been crucial; for a theory containing a massive *W*, the techniques that had worked so well in QED failed. The essence of why masslessness is so important is implicit in our example of pixels. For a theory to be renormalizable, the behavior of a particle remains the same at all resolutions—'t Hooft describes this nicely as "a dressed particle follows the same laws as a naked particle when seen under a microscope."[10] This immediately highlights the problem: The range of a force will appear to increase when seen in the microscope—unless the range is infinite. An infinite-range force requires its carrier, the gauge boson, to be massless. This is the case in QED, with its massless photon: QED is renormalizable. By contrast, a massive boson, such as the *W*, cannot satisfy renormalizability on its own.

Veltman felt that at least something in Yang and Mills ideas was right. They seemed to have the possibility of being renormalizable, though they implied the existence of unwanted massless vector bosons. In Veltman's view, it was just a matter of getting a mass into the equations somehow.[11]

Veltman was not completely alone in realizing that Yang-Mills theories might describe the weak interaction. However, he was unique in his single-minded refusal to give up attempting to tame the Infinity Puzzle.

In 1961 Glashow and Murray Gell-Mann at Cal Tech had been investigating mathematical group theory, and Gell-Mann had become interested in the way these techniques lead to Yang-Mills theory. Along the corridor, Feynman was trying to make a finite quantum theory of gravity; Gell-Mann suggested that he first look at Yang-Mills theory as a warm-up exercise.[13]

As the gauge bosons of Yang-Mills theory are massless, and the *W*—assuming that it exists—has mass, Feynman inserted extra terms into the Yang-Mills equations to allow for this mass. This spoils the gauge invariance, but it can be done in such a way that this occurs in just one term in the entire set of equations. Feynman then used his famous diagrams to calculate what happened when the various particles interacted with one another.

Feynman discovered that he could make the calculations simpler by adding "ghost particles"—weird entities that may be created during the process, only to die before the reaction is complete, having no real life of their own. Although this made the calculations less complicated, it left the theoretical foundations more like sand than solid rock.

Feynman's calculations started off all right, if the mass of the *W* was very tiny, but as he continued, going beyond the first approximations with which he had begun, the equations became increasingly complicated. Even for tiny masses, these complications would have to be solved in a complete theory; for a massive boson, which empirically is the case for the *W*, they ruined the whole edifice almost from the start. He gave some lectures about this stillborn work at a conference in Poland in 1962,[14] and that was that. He never formally published anything, nor developed what seemed to be a dead end, with or without ghosts.

Veltman started off along a similar route to Feynman. Like a mountaineer trying to conquer Everest before Hillary and Tensing finally succeeded, Veltman began in the foothills, following the route that Feynman had charted.[12]

I have a childhood memory of the conquest of Everest in 1953. A film of the expedition was shown at school, and as the entourage of sherpas hiked through mountain forests along innocuous trails, I wondered what all the fuss was about. Then the real drama began as they met the more frightening challenges of the high country, set up base camp, and headed toward the peak. Ever since I have loved hiking, an activity that seems remarkably popular among physicists, but have no serious ambitions for Everest. The analogy with the attempts to make a viable theory of the weak interaction is similar.

The way that Feynman had introduced mass looked tantalizingly harmless to Veltman, and the first steps suggested that the theory was indeed renormalizable. However, to prove that would require entering the territory where Feynman had been defeated. There was nothing especially difficult. In our Everest analogy, there was no metaphorical confrontation with vertical walls of rock or ice; rather, it was the lengthy equations that showed the journey would sap one's stamina. At first progress was good, but soon the goal seemed to recede as fast as he moved forward. The number of separate computations grew more and more until thousands of separate calculations were needed.

The difference between this and the assault on Everest is that for Veltman, the challenge was to see the view from the top of the mountain; no one cared much how you actually got there. So, instead of trying to climb it using the standard tools, why not have a machine help you up the slope? That is what Veltman had with Schoonschip.

The machine succeeded where Feynman had been defeated. It metaphorically climbed Mount Massive Yang-Mills and found the answer. That's the good news. Unfortunately, the answer turned out not to be what Veltman had hoped for. After all of this effort, he found that the infinites did not cancel out; the theory was *not* renormalizable. In short, the "massive Yang-Mills theory" is not the answer to the theory of the weak force.

Or, to be more precise, the theory was not viable if masses were introduced into the equations in the way that Feynman, and now Veltman, had done. If Yang-Mills theory, in its pristine symmetric form, with a triplet of massless "gauge bosons," is renormalizable, then some way had to be found of sneaking mass into the equations without destroying this critical feature: exact local-gauge invariance.

The stage was set for the arrival of Gerard 't Hooft.

THE ROUTE TO CARGESE

As part of his undergraduate program, 't Hooft had to write an essay about "anomalies." These are quirks of quantum theory, where theorists[15] had discovered some quantities that, when calculated in two different ways, give two different answers. That something can have two different answers, depending on how you choose to do the arithmetic, is nonsense, or, at least, "anomalous"; if an anomaly shows up, the implication is that the underlying theory is incomplete. The reason anomalies occur, and the techniques for avoiding them, is subtle, and beyond the needs of this narrative. However, one of the criteria for any successful theory is that it must not give any anomalies. 't Hooft learned this from his undergraduate project, unaware that it would soon turn into an important part of his mathematical tool kit.

He knew that he wanted to study particle physics but had no special project in mind when he came under Veltman's wing. By the late 1960s Veltman had become convinced that Yang-Mills theories are important and so assigned his students the task of reading about them. Thus it was that 't Hooft began to study field theory, at a time when the received wisdom was that Veltman and his students were "dusting an old and deserted corner of physics."[16]

't Hooft was reading the classic papers on Yang-Mills theory, whose mathematics he found very beautiful, as had Yang, but the apparent refusal of nature to use massless gauge bosons was the obvious block for applying them to physics. Another student, Jon Ubbink, had been given the paper of Brout and Englert to read, which showed how to give mass to the gauge boson.

When Ubbink presented his talk, 't Hooft was there as a young student, not really understanding much of the arguments, but something stuck in his mind.[17] Clearly, mass can be generated in some formal way: "There was something called Higgs' mechanism," but it "came across as abstract and nothing to do with the real world," and made no permanent impact on him at that juncture.

This attitude toward The Mechanism was quite prevalent at that time. The fact that this narrative is focusing on the weak and electromagnetic interactions is because we have the benefit of history. As we have seen, initially the general hope had been that the Yang-Mills ideas would prove to be the panacea for the strong interaction. It was the empirical absence of massless strongly interacting scalar bosons that had, first, frustrated Goldstone, Salam, and Weinberg, and then inspired the work of Higgs and the Gang of Six. However, by 1967, attitudes to the Goldstone paradox, and the need for "the mass-generating mechanism," had matured. Steven Weinberg's experience typifies the changing attitude.

Throughout the 1960s, Weinberg had been interested in phenomena that are almost, but not quite, symmetric, and he had been wondering under what circumstances the blemish is due to spontaneous symmetry breaking. Nambu's work on the spontaneous breaking of chiral symmetry in nuclear physics had discovered that when a massless nucleon (proton or neutron) acquires a mass, Goldstone's boson is manifested in the form of pions. Weinberg developed these ideas, eventually constructing a whole theory of strong interactions applicable at low energies. The pion was central to this entire construct. So when Weinberg first came across Higgs's paper, as we saw earlier, he thought, "Well, that's nice; he's found a way of getting rid of Goldstone bosons, [which had so concerned Goldstone, Salam, and me in 1961], but I have [now] become convinced that the pion is one."[18]

The pion is so nearly massless (at least on the scale of other particles that feel the strong interaction) that Goldstone's theorem appeared to be a great success. It was not until late in 1967 that Weinberg had his insight that the ideas of spontaneous symmetry breaking could be applied to the weak and electromagnetic interaction with potentially even greater success. Until that time he, and others, had come to love massless Goldstone Bosons. It was within such an environment that 't Hooft also found no obvious need for the "Higgs Mechanism" in particle physics.

However, as 't Hooft read and thought more about the Yang-Mills theory, he rediscovered "The Mechanism" as a route to making massive gauge bosons. The presentation of his fellow student Ubbink came to him subliminally, so he makes no claim to have discovered the trick independently. In any event, with this additional insight, he was by chance well prepared for a piece of fortune: His application to attend a summer school in Les Houches was rejected.

In 1970 the mountain resort of Les Houches, near Chamonix, hosted a summer school in field theory. 't Hooft had applied but too late: All the places had been allocated. His second choice was a school in Cargese on the island of Corsica. It was here that fate gave 't Hooft his lucky break.

French theorist Michel Levy had established the Institut des Sciences at Cargese, adjacent to a beach. That already made a memorable environment. He had also worked with Murray Gell-Mann, and they had invented the "pi sigma model," which built on the original ideas of Nambu and went beyond the original ideas of Matthews and Salam (see Chapter 4). This model was a theme in the lectures at the school, which included a series by Benjamin Lee and also German theorist Kurt Symanzik.

The novel feature in Gell-Mann and Levy's model was the presence of the "sigma." The sigma is a particle with no spin—a "boson"—and, as we said in Chapter 8, when there are several of them, they cluster like penguins. In quantum theory, bosons can collect together into the lowest-possible energy state, forming a "Bose condensate."

In Gell-Mann and Levy's trial theory, the sigma bosons experienced Bose condensation. The effect is that the pions and nucleons feel the presence of this condensate, which is like a physical medium impeding the free motion of other particles. It is the effects of the Bose condensate that hide the fundamental symmetries of the underlying theory; the presence of the condensate is the source of spontaneous symmetry breaking. The intriguing feature of the pi-sigma model was that it is the seed of a field theory of strong interactions that is renormalizable.

The model was not, and is not, a complete theory of strong interactions. Hadrons exhibit far more phenomena than the pi-sigma model describes. But that was not the main issue in 1970. The mathematical features of the model excited theorists as being a prototype for a more mature approach, which was yet to be found. At Cargese, the lectures that interested 't Hooft were those of Lee and Symanzik, which demonstrated how the theory could be renormalized—purged of infinity.

A critical feature of the pi-sigma model that enabled this happy circumstance was that the particles get their masses from spontaneous symmetry breaking—from the presence of the Bose condensate of the sigma field. This resonated for 't Hooft, who had already mused that the "Higgs Mechanism" might be key to renormalizing Yang-Mills theories.

I asked him the order of events: Had he stumbled on the Higgs Mechanism himself independently before Cargese or been turned toward it as a result of Lee's and Symanzik's lectures? 't Hooft made clear that he had "rediscovered the Higgs mechanism, before Cargese."[19] Peter Higgs himself had even said to me that if he himself hadn't discovered The Mechanism, 't Hooft would have. 't Hooft however doesn't make such a claim: "I don't claim any originality there; I don't know how much I had picked up from the student who had read those papers [by Brout, Englert, and Higgs]. Indirectly his presentation came back to me."

"Was this a subliminal awareness of the mechanism then?" I asked.

"Exactly," he replied. "That's how things often work."

I imagined 't Hooft in 1970, aware of The Mechanism before Cargese, having the key to the puzzle in hand. I put it to him that "it's one thing having the idea, but it's still a big journey from being aware of a possibility to actually doing something with it. After all, both Weinberg and Salam were both aware but didn't succeed. So is the Cargese experience the big difference for you; before and after Cargese for you were two different universes in a sense?"

On hearing Lee's and Symanzik's lectures about the mass mechanism and pi-sigma model, 't Hooft confirmed, "It strengthened me in my belief that the mechanism would show these same features [for Yang-Mills theories]. Then I asked the question to both Ben Lee and to Symanzik: 'What you have been doing for the linear sigma model, should one be able to do the same for a Yang-Mills theory?' They both gave me the same answer—they didn't know; they had not looked at Yang-Mills!"[20] He remarked also on the irony of something else that they had said: If they had been a student of Veltman's, they would have asked *him*!

My experience of research is that it is like climbing up and over a hill. The going gets tough, but there comes a point when you realize that you can indeed carry on to reach the peak so long as you are confident that it's within range. I suggested to 't Hooft that the experience at Cargese told him: "Now I know what the answer is, I can proceed to find a way to get from where I am now to where I want to be." He agreed: "There may be a lot of truth in that. Of course, you can always ask a 'what if' question:

What if I hadn't gone to Cargese; would I have succeeded? I don't know. It's very hard to tell. I think 'yes' but if someone said 'no,' I would have no good arguments to contradict them."

Whatever might or might not have been, the fact is that Lee and Symanzik sowed the seeds. 't Hooft confirmed that "after Cargese I was convinced that this was the way to solve the problem. I was encouraged by what Ben Lee and Symanzik had said in Cargese to believe that this scheme would work."

'T HOOFT CLIMBS EVEREST

't Hooft was now sure that he was on the right track and that spontaneous symmetry breaking could introduce masses without destroying the gauge invariance of Yang-Mills theory and, most important, that it would give finite answers—be renormalizable. If only fools rush in where angels fear to tread, the portents for 't Hooft were not good: At least two other leading theorists had given up. 't Hooft described the challenge to me: "Salam had a grand overall view including the idea that the theory would be renormalizable, but he didn't know how to prove it properly and was unable to solve the technical part of the proof. Weinberg also expressed his opinion that these theories had to be renormalizable. He set a student to work on the problem, but the student couldn't succeed in making it work. So Weinberg turned away, thinking it was a nightmare."

't Hooft succeeded where others had failed because he used the mathematical techniques of "path integrals." This is the extension to quantum field theory of the techniques that Feynman had originally invented for quantum mechanics (see Chapter 2). It is only with these tools that the renormalization of Yang-Mills theories has been proved.

In 1942 Feynman had invented a novel approach to the quantum mechanics of particles—his path-integral formulation. He had concentrated on their trajectories. He was not really enthused by quantum field theory, perhaps because of the plague of infinity. Like Dirac, who also disliked the renormalization ideas, Feynman applied the path-integral approach to the dynamics of particles, but not to field theory.[21]

As a student, Steven Weinberg had heard lectures about path integrals from John Wheeler—Feynman's research supervisor. Weinberg's memory is that this experience put him off path integrals. He recalled, "Wheeler was very poetic. He never actually showed that the path-integral approach was a consequence of ordinary quantum mechanics. Instead, it came across as an independent approach to physics, which is the way that Feynman had thought of it originally. I didn't like this dualistic view [of applying the ideas to particle trajectories while ignoring field theories]. I hated it."

This prejudice against path-integral methods was so deep that when Weinberg first heard about 't Hooft's proof of renormalization, which had used these techniques, he "was skeptical at first." He told me, "I didn't trust anything that came out of the path-integral method. So at first I didn't trust what 't Hooft had done. That's not a criticism of him, but of myself. I should have known better."[22]

Veltman himself had become converted to path integrals only in 1968. He was visiting the university at Orsay, near Paris, for the academic year 1968–1969. Having learned of Feynman's work on Yang-Mills theory, Veltman realized, "There was no escape—I had to learn path integrals."[23] In order to teach himself, he decided to give a series of lectures. Ben Lee was present and became interested—happenstance that would have far-reaching implications later.

The lectures were a success, notwithstanding that the students were more interested in making revolution that year. Even so, when all was done, Veltman decided that he did not yet understand the concepts well enough, and so, on returning to his home institution in Utrecht, he set up a course on the subject. He asked his student Gerard 't Hooft to make notes of the lectures.

Veltman finally decided that he understood path integrals, but never felt completely comfortable with them. Resonating with Weinberg to a degree: "I distrust them." By contrast, "'t Hooft had no such emotional ballast, and he became an expert in path integrals."

A RENORMALIZABLE THEORY

In order to demonstrate that Yang-Mills theory is a viable description of the electroweak force, two things are needed. One is that it is renormalizable—freed of infinity—and the other that it is unitary. The latter means that there are no ghosts, that is, entities with probability less than zero, and that the total probability of something happening at all is 100 percent.

To check these you must calculate the quantum amplitudes of the basic theory, which in turn requires you to choose a gauge—the basic definition of the various quantities, such as the measure of potential energy felt by the particles. There is one particular choice of gauge that exhibits the unitarity of the theory very easily; this is known as the "unitary gauge." However, there is no free lunch: The unitary gauge makes the question of finiteness of the theory opaque. It is possible to assess this by using a different gauge, similar to what Englert had used when he made his remarks to Weinberg at the 1967 Solvay conference.

While Englert was undoubtedly correct about the theory being finite, that alone does not establish that it is also unitary. The key part of any proof is to show that the theory is gauge invariant, such that if you check finiteness in one gauge, and unitarity in another, you can then translate between different gauges to establish that results remain true, independent of the choice of gauge. Only by showing that the results are independent of the accounting schemes will you have established that the theory itself is viable.

In QED the translation between gauges had presented no special problems. However, in Yang-Mills theories, the extra mathematical complexities make it impossible—at least, if you use the canonical formalism. This appears to have been the insuperable barrier.

Salam's notebooks show no signs that he made serious efforts, let alone any inroads toward solving the problem.[24] Weinberg also tried and failed; he assigned the problem to a research student, Larry Stuller, who also had no success. To this day no one has managed to complete a proof in the canonical formalism; there are deep reasons it is impossible unless you use path integrals. By using that technique, 't Hooft was able to find a gauge where it was obvious that there is only a finite number of infinite expressions, which can all be removed by relating them to measured quantities, such as charge and mass of fundamental particles. He was then able to translate between this and a gauge where unitarity was also established. The gauge invariance was preserved throughout, and 't Hooft established that the answers are identical: The theory had all the properties needed to be a consistent description of nature.

That is why 't Hooft eventually succeeded where others had failed. However, when he returned from Cargese, although he had a plan of attack,[25] much work would be required before the task was complete.

First, he had to work out how the pure Yang-Mills theory—with massless bosons—is renormalized. Others had already worked out the rules for calculations: American Stanley Mandelstam had done so, as had, in Leningrad, Ludwig Faddeev and Victor Popov. However, there were problems: The rules of the Russians were not identical to those of Mandelstam, and in addition the expressions of all three disagreed with those that Feynman had developed by a factor of two. Feynman had introduced a mass for the boson by hand, whereas the others had remained strictly massless. Feynman himself thought that this disagreement was a minor problem—"Who cares about a factor of two?" he remarked[26]—but it turned out to be critical; Feynman's theory, which had introduced a mass for the boson by hand, was fundamentally different and was a reason that he had been unable to complete the calculations.

't Hooft looked at the case when all particles were massless. Many among the cognoscenti believed the massless Yang-Mills theory to be renormalizable, but 't Hooft felt that no one had actually proved the fact.[27] He completed the first draft of a proof at the end of 1970 or early in 1971.

Veltman was working in Paris that year and visited Utrecht several times. 't Hooft felt it was hard to get Veltman's attention. 't Hooft's memory is that he had "big fights with Veltman," who was not convinced that 't Hooft had gotten any further than he himself had earlier.[28] Veltman was right: 't Hooft had shown that there was no problem with infinity, but had not shown that the total chance of something happening came out as 100 percent; in the jargon, he had not "satisfied unitarity." In effect, he had not proved that the theory contained only physical particles and not ghosts. 't Hooft eventually managed to get rid of the ghosts, but Veltman was still not convinced. He was worried that "anomalies" lurked within the calculations. If there are anomalies, a theory is not gauge invariant and, hence, is wrong. The anomaly problem was more difficult, but 't Hooft eventually satisfied Veltman of this, too.[29] His paper, which was the first explicit demonstration that massless Yang-Mills theories are renormalizable, appeared in 1971.[30] Veltman thought that although "it was perhaps not earth-moving, his path integral manipulations were nice and useful."[31] As we have seen, and Weinberg also remarked, it would be this technique that proved seminal in ultimately making the breakthrough.

This was fine for a warm-up exercise, but said nothing about the real challenge: What happens with massive particles? Veltman had spent so long on trying, and failing, to find a renormalizable theory with massive electrically charged *W* bosons that he was becoming convinced that it couldn't be done. That was until one day in 1971 when they had a memorable conversation.[32]

't Hooft was excited at having solved the problem for the case when everything was massless, but Veltman tried to dampen his enthusiasm by reminding him of the real challenge, which involved massive particles. Veltman then guided 't Hooft by saying, "All you need is just one example [of such a theory with massive particles] that is renormalizable." Once that was done, Veltman told him, some "freak" who likes building models could fix it to look like the real world.

Veltman was saying these things rhetorically, convinced that it was impossible, but 't Hooft replied, "I can do that." Veltman stopped and said, "What?" to which 't Hooft repeated, "I can do that." 't Hooft's memory[33] is that Veltman later recalled the occasion thus: "We were having a walk, and at that moment I nearly bumped into a tree." Veltman today recalls it more soberly as, "Write it down and we shall see."[34]

't Hooft explained that he had identified a missing ingredient in Veltman's previous unsuccessful attempts. 't Hooft's inspiration from Cargese had been the role that the sigma and Bose condensation played in the renormalizability of the pi-sigma model. In his proof of renormalizability of the massless theory, gauge invariance had been crucial.[35] He knew that if he included an analog of the sigma field in the equations, this could introduce masses while preserving the gauge invariance. 't Hooft was convinced that when this was done, the infinities would disappear.

He started to tell Veltman about The Mechanism for generating masses in the equations, but Veltman wasn't interested in all of these details. His single-minded focus was on the goal, and he interrupted, saying, "Just give me the equations that you think work." 't Hooft wrote the equations on a notepad, and Veltman, still skeptical, said, "It looks crazy, adding this weird particle without spin or charge, but I'll take these equations and put them in my computer to see what happens."[36]

All his program codes were on the computer at CERN in Geneva. In those days there was no Internet where you could log in from afar, so he had to go to Switzerland himself to do the calculation.

My own memories of computing in those days are of programs punched out on paper tape or cards. The trick was not to tear the tape, or drop the pack of cards—worse still to jumble them up en route from one's office to the computer center. If you made it without mishap, you delivered your program to the administrator's desk, and then the wait began.

The huge computer filled a room the size of a warehouse. This was the state of the art in 1971, but such machines were often less powerful than a modern laptop. Your program would share time with several others, and it could be many hours before the results appeared. Often the printout informed you that the program had failed. An inadvertent typo, such as putting a comma where there should have been a period, could be enough to abort the calculation, causing you to start all over again. A turnaround time of twenty-four hours sounds awfully frustrating today, but in 1971 was wonderful: It certainly beat the alternative of working out the answer by hand, which could take months, assuming it was possible at all.

Veltman had run many programs calculating Feynman diagrams, so he was expert in modifying his codes to test 't Hooft's equations. Memories of what followed differ in detail, but not substance. 't Hooft's memory is that the results, which arrived within a few hours, showed that most of the infinities had vanished; nonetheless, a few stubbornly remained. Veltman then called 't Hooft in Utrecht to say that his equations had nearly worked, but not quite. They then realized that Veltman's program had failed to copy a factor 4 in front of one term: "I couldn't understand why you put that crazy factor of 4 there," Veltman supposedly said.[37] But then Veltman duly inserted a free parameter at that point in the calculation, before going through the whole procedure once more. This time, he discovered that if you assume that that free parameter equals 4, miraculously, *all* of the infinities disappeared! 't Hooft recalls, "By then he was as excited about it as I had been."

Veltman's recollections differ in the details. He wrote:

> Basically I ran a number of little tests, which showed that the work of 't Hooft was right. In the past I had done much of that stuff, and I quickly saw that with the new approach all pieces fell into place. It is not true that first things went wrong and that I called him, and then things went right. I knew all about that [numerical factor] before I departed. Everything went right from the start, with the rules as given in the manuscript that I had with me. I went to CERN to verify his work, and everything came out dandy.[38]

Veltman added: "We did not discuss any precise computational details in that phone call for the simple reason that he did not know what I was doing."[39] The main point that concerned Veltman was analogous to the case of the "overlapping infinities" that had hindered the proof of renormalizability in QED (see page 68). In modern jargon these arise in Feynman diagrams containing "two-loops." Veltman had studied this extensively, and already written two papers about it[40] in the simplest example—Yang-Mills theory with an SU2 mathematical structure. There had appeared to be a barrier in a particular piece of the two-loop calculation, and Veltman wanted to see how that worked out in 't Hooft's approach. He recalls, "That was easy to see and I understood that quickly. It was for my own satisfaction—after all, if 't Hooft's article was wrong, I for sure would get the blame."[41]

Veltman gave Bruno Zumino—a theorist at CERN—'t Hooft's manuscript for a critical reading. Zumino then told Veltman about Weinberg's article, as one of 't Hooft's models was the same as Weinberg's that combined the electromagnetic and weak forces.[42] Veltman then made a phone call to Utrecht and told 't Hooft about it, and recalls, "At that moment he told me on the phone that he had more or less produced a model for the weak interactions, which looked like the Weinberg model."[43]

't Hooft had originally been inspired by what he had heard at Cargese and had focused on a model of the strong interaction. His Yang-Mills particles would have been the positive, negative, and electrically neutral versions of the rho-meson, but as these have the same spin as a photon, he had included both the rho and the photon in his mathematics. As a result of this chance, his mathematical model had SU2 × U1 as its foundation—coincidentally the same structure that, unknown to them, the electroweak theory empirically requires.

Veltman recalls, "It was a bit of a let-down to see that the right model existed already, as you can imagine, but we swallowed that and continued."[44] This is in contrast to 't Hooft who remembers this as a moment of great excitement, saying that Veltman "was as excited about it as I had been."[45] These two reactions typify the personalities of the pair and their experiences to this point. For 't Hooft, the renormalization was the achievement, the empirical realization secondary. For Veltman, who had pursued renormalization for so long, the triumph of the final proof was dampened by the fact that the model itself had been anticipated. In any event, no one had settled whether or not Weinberg's model was renormalizable: Veltman and 't Hooft knew that they had stumbled on something of immense importance.

"AND NOW I INTRODUCE MR. 'T HOOFT"

We have now reached the point with which we started the Prologue: Veltman realizing that the big conference planned in Amsterdam for August would provide the perfect opportunity for him to launch 't Hooft and this theory. Veltman had been given a wide brief, the only constraint being that he organize a series of theoretical presentations. He couldn't resist devoting it to the problem of renormalization and, without revealing his hand, inviting T.-D. Lee and Abdus Salam to speak.

Following what Veltman has referred to as Salam's "baloney," and then Lee's description of his attempts to solve the puzzle by including particles with weird properties, the moment had at last arrived. Veltman's memory is that he said, "And now I introduce Mr. 't Hooft, who has a renormalizable theory every bit as good as Quantum Electrodynamics."

I have talked to Veltman, 't Hooft, and a Dutch colleague, Chris Korthals-Altes, about that day. Memories differ. If at the time people had realized that they were witnessing history being made, perhaps someone would have recorded it. Korthals-Altes recalled Veltman's having said: "as elegant as anything we have heard before," which is the version that I used in the Prologue. When I asked 't Hooft, he could not remember Veltman's words as: "I was totally focused on the talk I was about to give." Veltman, however, is sure: "It was a previously cooked up carefully formulated statement. Everybody else was hearing it for the first time and not knowing what was coming to them." I chose Chris's version in the Prologue as it gives the sense of the occasion and of its impact on people. Veltman added, "I think nobody truly noticed the importance of what I said except Sidney Coleman, T.-D. Lee, Ben Lee, and perhaps Salam."[46] After forty years the precise words may be lost, but their significance is not.

't Hooft recalls that the room in Amsterdam was not very big. Those were the days before modern, computerized PowerPoint presentations. The style was to write on transparent acetates, which were placed over a light box and the image projected onto a screen. You could use colored pens that left "permanent" traces—removable by alcohol but not by water—or "washable" ones whose marks could be removed merely by rubbing with a sheet of tissue paper. The latter were favored in case last-minute changes were called for. However, their danger was that they had a tendency to be diluted and spoiled by water vapor, or—all too often—sweat from the fingers of a nervous speaker. 't Hooft had given informal talks in Utrecht, but this was his first appearance on an international stage. He recalled that his "transparencies were badly prepared [with the] wrong kind of pens—[and] smudgy."

Veltman had been given an hour for his session. Lee and Salam, each having twenty-five minutes, allowed only ten minutes to squeeze in 't Hooft's talk, which was supposed to announce the technical solution to a problem that has defeated some of the world's greatest theorists. I said to him, "I am trying to get a picture in my mind. Here you have in effect solved the Fermat's Last Theorem of particle physics . . . ," to which 't Hooft modestly interjected, "Sort of." I continued, "and there are people in the hall who have worked on it unsuccessfully themselves, and you are to describe the solution in ten minutes? Could you do more than make a series of unproven assertions?"

He answered, "It was presented in such a way that 'the ultimate proofs are still to be given.' I didn't say that this had been proved and finished, as it wouldn't have been true. There were still some [loose ends to tidy up]. We had gone a long way understanding dimensional regularization [a technical innovation that had played a significant part in his breakthrough], but that certainly hadn't been completed yet."

In ten minutes he explained how these theories were based on gauge invariance, which allowed you to calculate in different gauges. In one gauge, the "unitary gauge," it was straightforward to prove that probabilities are sensible—there are no chances less than zero, and the total chance that something happens at all is 100 percent. "From there you can make a gauge transformation to another gauge where it is easier to see that the theory is renormalizable, [and then] prove that the [Feynman] diagrams give the same [answers]. You can say that in ten minutes."

With understatement, 't Hooft said to me that after his presentation, "There wasn't time for a lengthy discussion." It was "in the break afterwards that people started to talk." He recalled that Ben Lee was very interested. It was Lee whose lectures at Cargese in 1970 had inspired 't Hooft, and it would be Lee who, following 't Hooft's lecture, would translate 't Hooft's novel mathematics into a form that others were more familiar with. To some degree, as Dyson in QED had shown how Feynman's techniques linked with the more familiar ones of Tomonaga and Schwinger, so did Lee bring 't Hooft's work to mass recognition.

With hindsight, this session at Amsterdam marks the moment when field theory was reborn as the golden path for understanding the fundamental forces. For Veltman, it was also a triumph to follow Salam's talk, not least when accounts of the talks appeared months later in the conference proceedings. Salam had added a paragraph to the end of his own paper: "Finally I welcome the renormalizable G 't Hooft's theory of weak interactions. . . . [T]he same theory was proposed by S. Weinberg [in 1967] and earlier by J. Ward and myself [1964]. See also the Proceedings of the Nobel Symposium in Gothenburg [1968]."

These are Salam's first comments following 't Hooft's breakthrough. Salam is claiming priority for the SU2 × U1 model, which is justified, in particular citing the role of Ward. These remarks, and the inclusion of Ward, will weigh heavily in what later transpires. At this juncture, the most singular feature is that 't Hooft has proved renormalizability of the theory, something that none of the aforementioned managed to do.

His talk had lasted just ten minutes, stuck on at the end of a session in a side room to the main conference. Copies of 't Hooft's paper were distributed at the conference, and news spread by word of mouth. The editor of the journal to whom 't Hooft submitted it for publication asked my Dutch colleague Chris Korthals-Altes at Oxford to advise on whether it should be published. Chris didn't know what to do with it, as it was so technically complex, containing pages of painstaking mathematics. So he consulted J. C. Taylor, the leading expert on field theory at Oxford. Taylor verified that it was correct, realized its significance, and advised that it should be published immediately.

What none of them thought was to try to apply the same ideas to the other force, the strong force. The folk wisdom was that the literal strength of the strong force placed it outside the realm of field theories. However, experiments at SLAC in Stanford were about to show that that force is not always as strong as it appears. The explanation would eventually involve the very same ideas that 't Hooft had just presented as explanation of the weak force.

Nor did anyone then realize how this breakthrough would change the direction of two thousand years of science. Ever since the philosophers of ancient Greece, the quest had been to find the basic seeds of matter. What 't Hooft and Veltman had done was to show how to understand the *forces* of nature. It soon became clear that this was the first step toward construction of the long-sought unified theory. However, it would be the work of an American, James Bjorken, that would turn this theoretical promise into confirmed law, leading to Nobel Prizes for Glashow, Salam, and Weinberg in 1979, and for Veltman and 't Hooft in 1999.

In what follows we shall see how these breakthroughs have defined the direction of particle physics ever since. The first thing that happened was that 't Hooft missed out on the possibility of a second Nobel, for there was a terrible irony: 't Hooft didn't realize that in the course of his work, he had also stumbled on the key to the strong interaction.

We will come to this in Chapter 13. But first, it is worth summarizing what has been achieved, because from this moment competitors began to maneuver for the Nobel Prizes that were sure to follow. Some have been awarded; others may yet be; some never will, because the subjects are no longer alive.[47]

THREADS

The media love heroes and simple story lines. The remarkable emergence of 't Hooft, a mere student solving the puzzle that had defeated great masters, gained its own life. Over the years, Veltman's role tended to be overlooked.

What 't Hooft had achieved was indeed remarkable, but no one, least of all himself, would claim that he did it alone. My comparison with Fermat's Last Theorem is an example of the media oversimplification, not least because Andrew Wiles solved Fermat on his own; 't Hooft's triumph, by contrast, was the culmination of a vast effort.

In our analogy of climbing Everest, one could say that Veltman had made a map of the route, prepared the equipment, and almost reached the summit before 't Hooft started. When Veltman, nearing the summit, had been faced with an impassable crevasse, 't Hooft found a route, but even then it was Veltman's tools that established this to be the way and got them successfully through.

As for the route: The phenomenon of hidden symmetry proved crucial. 't Hooft rediscovered this, but even here subliminal awareness probably played some role. As for the SU2 × U1 model, which formed the skeleton of their theory, Weinberg had already both seen its promise and worked through its implications, together with speculating that hidden symmetry was the way to reach the summit. Glashow, Salam, and Ward before him had also stumbled on SU2 × U1 and identified the *Z* boson as a heavy analog of a photon.

Yet no one before 't Hooft and Veltman seems to have truly believed that these were the Commandments of nature. Glashow, Salam, and Ward, and then Weinberg, had each ignored their own creation, choosing to work on other areas, turning their sails into the wind only after 't Hooft's appearance. And not just these few; the world at large had paid little attention to them. Following Weinberg's paper in 1967, no one other than Salam referenced it for two years; however, after 1971, every day, someone somewhere has had need of it, such that today it is the most highly cited paper in theoretical particle physics.

Salam himself published only once he realized that Weinberg had rediscovered the ideas of Salam and Ward and moved them on by invoking hidden symmetry. And following 't Hooft's proof of the viability of such an idea, Salam quickly positioned himself in the Nobel race by referring to his own 1968 paper. His work with Ward, on which this was based, received little publicity and was largely unknown to the hordes of newcomers flocking to the field. Instead, as we shall see in later chapters, they were hearing a lot about the "Weinberg-Salam model."

We shall now see three main threads emerge from 't Hooft's entrance. First, not only had the weak interaction been solved, but 't Hooft had the key to the strong interaction too and failed to realize it. Second, a whole program of experiments was developed, which first tested and then confirmed the predictions of the theory with remarkable success. These have formed the base to today's conceit that, at the Large Hadron Collider, the final proofs of the whole edifice will be achieved. Third, as those experiments began to confirm the theory, maneuvering for the Nobel stakes began.

Since a maximum of three people can share a Nobel Prize, speculation about the successful trio was rife. While 't Hooft's name was highly publicized, Veltman's role was less advertised outside the field; the "Weinberg-Salam" model captured media attention, which ignored Glashow and left Ward forgotten. There was some speculation that the common conjunction of the names Weinberg and Salam, together with the prominent focus on the role of 't Hooft, would make them the choice. The Nobel committees dug deeper. Eventually, 't Hooft and Veltman shared an award in 1999, while Weinberg, Salam, and Glashow shared the award in 1979. Ward, however, was bypassed.

The breakthrough discovery that Yang-Mills equations describe the electromagnetic and weak interactions had been the result of theory. In parallel to this, experiments had made remarkable discoveries about the strong interaction in the latter half of the 1960s that would prove no less singular than those we have just seen in the case of the weak interaction. These would eventually show that Yang-Mills theory is the key to all of these forces. However, what with hindsight may appear obvious took a long while to be realized.

Intermission: Early 1970s

We have reached the start of the 1970s.

The "Higgs Mechanism" for generating mass in theories that initially had no mass has been generalized by Kibble and in turn exploited by Weinberg in 1967 to build a theory uniting weak and electromagnetic interactions. Abdus Salam too has incorporated Kibble's ideas into the similar theory that he and Ward had constructed earlier, in 1964.

Hardly anyone takes any notice, because there is no demonstration that these theories are viable—that is, renormalizable.

Gerard 't Hooft in 1971 proved that they are, in his thesis, using tools forged by his supervisor, Veltman.

By 1972 there is universal agreement that physics at last possesses a viable theory describing the electromagnetic and weak interactions. The key has been the use of Yang-Mills theories (Shaw's name has been forgotten) and The Mechanism of Higgs and Kibble (the rest of the sextet being similarly overlooked by many).

The enigma of the strong interaction, however, remains—'t Hooft, meanwhile, is unaware that he has the solution to this also within his equations.

PART 2

REVELATION

Chapter 12

B. J. AND THE COSMIC QUARKS

Birth of the quark model: George Zweig is told it's complete rubbish; Murray Gell-Mann says quarks are unphysical; James "B. J." Bjorken has a big idea, which leads to their discovery. Feynman invents the "parton model." The strong force acting on quarks appears to be a paradox: The closer you look, the feebler it appears to be.

In 1966 Jim Bjorken was a thirty-two-year-old professor of physics, working at SLAC, the Stanford Linear Accelerator Center, in Stanford, California. An intelligent, shy man, who loved climbing in the Sierra Nevada, or hiking in the hills that separate San Francisco Bay from the Pacific Ocean, he is universally known as "bee-jay." The name came from his days as an undergraduate at MIT, when a neighbor in the dorm, also named Jim, and he got tired running for the hall phone, far away, when someone yelled "Jim." His friend went by the moniker "J. S.," and, "I did not like JB, so I permuted the letters and adopted the nickname BJ."[1] It managed to spread from the dorms to the MIT physicists who were working at Stanford, and from there to the world. In the forty years that I have known him, I have never heard him called James, or Jim; it is "always

BJ-this or BJ-that." He even signs himself: bj.[2] His stature in physics matches the physical: he is about two meters tall.[3] Even today, in his seventies, with a full head of trim hair and spectacles, he looks like a clean-cut young student.

That year, 1966, was when Bjorken had his big idea: how to find quarks, the seeds of protons, neutrons, and many other ephemeral hadrons. Although none of this seems remarkable today, Bjorken was going out on a limb in 1966. For the idea that quarks might be concrete particles—real particles—had been widely disparaged at the time, not least by Murray Gell-Mann, one of the creators of the quark model.

That the proton and the neutron are not the ultimate seeds of the atomic nucleus was already suspected, but the hypothesis that quarks might, in a sense, be the seeds of those seeds, as one Russian doll lies within another, was already highly controversial. Gell-Mann was skeptical, but the independent coinventor of the quark model, George Zweig, was adamant that quarks were real. In 1964 Gell-Mann dismissed Zweig airily: "The concrete quark model—that's for blockheads!"[4] Four decades later, Zweig recalls, "I can still hear Murray's voice." I too remember how much it depressed me to be told by Gell-Mann himself, in 1968, that quarks were "just a mnemonic," without physical reality.

No one suspected how dramatically perceptions were to change. When I met Gell-Mann, he was en route to the international conference in Vienna in 1968, at which the first evidence for Bjorken's idea on how to prove the reality of quarks would be announced. Little did I realize how the consequence of that event would influence my own life, taking me to Stanford and working alongside "B. J." This chapter will tell the story of how Bjorken's ideas in the latter half of the 1960s led to the discovery of quarks and provided the means that would turn the theoretical ideas of Higgs and 't Hooft into a golden age of science.

QUARKS

Bjorken's big idea involved bouncing beams of high-energy electrons off protons and then watching how the electrons recoiled. In so doing, he believed, one would be able to determine what the interior of a proton looked like.

Electrons hadn't counted for much outside Stanford fifty years ago. True, people knew that they had to pay for them in electric current and that QED described their intimate terpsichore and affinity for light, but few thought them to be a useful tool to illuminate the dark recesses of the proton. Stanford, however, had an electron accelerator in the 1950s, small by modern standards but a leader at that time. In 1956 several physicists migrated west to Stanford in order to exploit its unique facilities. That was the year when Bjorken too moved there, but "it was the mountains and the general Stanford reputation, not electrons, that was the main draw for me."[5]

Away from California, attention had focused almost entirely on smashing protons into other protons and then examining the debris. There was a logical basis to this. Electrons were best for studying electromagnetic forces, but the success of QED showed that these were understood; the weak force, as the name suggests, was weak and hard to isolate, so it was thought that the strong interaction was likely to be solved first—and the best tools to achieve this were believed to be beams of protons, which, unlike electrons or photons, feel the strong force.[6]

That the strong interactions between hadrons, such as protons and pions, would be "solved first" is what the Ph.D. prospectus at Cambridge University promised when I surveyed the field in the long-ago days of 1967. Oxford University, by contrast, had become a center for research into the novel idea of quarks.

The leader of this program was Dick Dalitz, a short, stocky Australian who had been head-hunted to Oxford from Chicago. He had arrived together with an American car of the grandiose style favored in the 1960s. Fondly referred to by students as his "acre of real estate," its presence filling the driveway of the old house where the theoretical physics department was then based, it showed if Dalitz was in residence.

Dalitz had taken the idea of quarks seriously and shown how it could be successfully extended far beyond the original aims of Gell-Mann and Zweig. However, to do so he had to ignore some basic rules of quantum mechanics. Today we understand why this worked empirically, as we shall see later, but at the time it divided opinion. Few outside Oxford believed in quarks, and even some in the department were skeptical about Dalitz's version of the quark model.

In the latter half of the 1960s, quarks were more synonymous with Dalitz than with Gell-Mann or Zweig. It was to Oxford that I went in 1967, though at the time hardly anyone else in the world was taking the quark model seriously.[7] With hindsight, it is remarkable how slow we were to recognize the seeds of change. That summer in a major international conference at Stanford, Bjorken was describing what would happen if an electron beam struck and scattered from a proton that consists of three quarks. Today we all recognize Bjorken's insight as a remarkably accurate description of what turned out later to be reality, but, under pressure, he disowned his creation almost as soon as he had made it. Why he did, and the consequences of it, are for later; in summary, in the summer of 1967 protons were the tools that the majority of physicists expected would reveal the new reality. And hardly anyone expected that the reality consisted of quarks.

Other students at Oxford were working on the strong interactions using traditional ideas beloved in Europe, with enigmatic names such as Regge theory, bootstraps, and current algebra. It was from students working on current algebra that I learned of Weinberg's work on the strong interaction: His epiphany to convert to the weak force came a few months later. Some in Oxford were themselves trying to solve the mysteries of the weak interaction. Here I first heard of the *W* boson—the theoretical beast that did for the weak force what a photon does for the electromagnetic. However, at the time this did not stand out as something special: The theories of the weak interaction involving the *W* were plagued with infinity, and some attempts to stop infinities erupting throughout the calculations had led to the unsuccessful prediction of hordes of other particles: *X,Y*, and *A,B,C*, though not, as far as I recall, *Z*, which would later become a key player.

One day I heard a talk by one of Dick Dalitz's assistants, Gabriel Karl. Karl explained how all of the myriad of hadrons could be explained as built from just three smaller particles—the "up, down, and strange" quarks. The idea that all of the data, which would take years to assimilate, could be reduced to just remembering the names of three quarks and a few simple rules of how they combined seemed too good to be true. I chose to focus my thesis on the quark model.

THE QUARK MODEL

The idea of quarks had been born in 1964, independently by two Americans, George Zweig and Murray Gell-Mann.

By the 1960s, experiments with cosmic rays and at accelerators had revealed scores of hadrons. In 1962 Gell-Mann found a way of gathering the mushrooming hadrons into families, most famously containing sets of eight. He poetically named his scheme the Eightfold Way after the Buddhist path to truth. The mathematics behind this involved group theory and the particular group known as SU3.[8]

In March 1963 Gell-Mann gave a talk at Columbia University about his new SU3 theory. A couple of weeks earlier another theorist, Gian Carlo Wick, had given an introductory seminar about SU3; upon hearing it, Robert Serber realized that in addition to families of eight and ten, which had already been discovered, there should be a basic family of three (as in SU "three") and, moreover, the octets and tens could be built up as composed of groups of these more basic entities. As Serber later recalled, "The suggestion was immediate; the [hadrons] were not elementary but were made of [what we now call] quarks."

A fortnight later, Gell-Mann was in town. During lunch at the Faculty Club, Serber explained the idea to him. Gell-Mann asked what the electric charge of the basic trio is. Serber had not looked into this, so Gell-Mann figured it out on a table napkin. The answer turned out to be +⅔ or −⅓ fractions of a proton's charge, which was an "appalling result," as no such charges had ever been seen. Gell-Mann mentioned this in the colloquium and said that such things would be a "strange quirk of nature." Serber remarked later, "Quirk was jokingly transformed into quark."[9]

George Zweig was a graduate student at Cal Tech, which was Gell-Mann's home base, and aware of the Eightfold Way scheme for classifying hadrons. Zweig's research supervisor was Feynman, whom he would visit every Thursday afternoon from "1.45 to 4.30 when we would adjourn for tea."[10] Each week Zweig would try to find something interesting to discuss. In April 1963 he saw the report of an experiment, which showed that a particle known as the "phi meson" seemed unnaturally stable. Zweig found this remarkable and chose it for the Thursday afternoon discussion.

Feynman, however, was not impressed. Recalling his own experiences with experiments during the "$V - A$" saga (Chapter 6), he warned Zweig that experiments could be unreliable. Zweig, however, could not get the phenomenon out of his mind and was convinced that something important was being revealed. This led him to the idea that the phi-meson and other particles that feel the strong force are made of constituents.

With this insight, and independent of Serber or Gell-Mann, Zweig too had realized that the mathematical success of SU3 could be the clue to a deeper layer of reality within hadrons. What Gell-Mann would call quarks, Zweig called "aces." Today they are traditionally referred to as "quarks," and so I shall use that name from here on. The difference is that Zweig regarded these quarks as physical objects that flow along within hadrons and capable of swapping from one hadron to another. With this assumption he discovered that he could explain the ways that various hadrons were created or died; in particular, in his picture the phi-meson is the lightest example built entirely of the strange "flavor" of quark, which leads to its abnormal stability. During 1964 he was visiting CERN, and it was while he was there that he wrote two papers: a short one outlining the basic idea, and a long one showing the implications and tests of the model.

The head of the theory group at CERN in those days was Leon van Hove, a tall, austere Belgian with strong opinions. The theory group secretary had typed the two papers before Zweig spoke with van Hove, who regarded Zweig's idea as "complete rubbish."[11]

Next on the agenda was to get the papers published. Zweig wanted to submit them to *Physical Review.* No European journal at that time was as highly regarded as the American *Physical Review*, and for Zweig, as an American citizen, this was the natural destination for his work. To cover the production costs, the journal charged scientists a fee to publish their papers. Van Hove blocked that, telling Zweig that CERN would not pay.

So Zweig responded, "I have a research grant from the US NAS-NRC [National Academy of Sciences, National Research Council] that pays my salary and has already paid CERN $1500 for publication costs." Van Hove told Zweig that all papers leaving the Theory Group at CERN must be approved before they could be sent out for publication and that Zweig's papers would not be approved unless he submitted them to European journals. He then instructed the Theory Group secretary, who had typed the two preprints, not to type any more of Zweig's papers.[12]

The editorial conventions for CERN preprints differed from those at *Physical Review*, and so Zweig's wife retyped the second—long—preprint in a form suitable for the American publication.[13] However, the response of *Physical Review*'s editor "was very discouraging," and Zweig "gave up trying to get the paper published." The result was that the first paper was never published in a journal, and the second was only published sixteen years later.[14] Copies of the "preprints," however, were widely circulated, have been referred to in the literature ever since, survive in the CERN archives, and are accessible online.

It is ironic then that within a couple of years van Hove and a Dutch colleague, J. J. J. Kokkedee, were using the quark model to understand what happens when protons collide at high energy. Kokkedee then wrote a book about the quark model, which contained reprints of seminal papers, with explanatory commentary, but did not include Zweig's.[15] While van Hove's opinion in 1964 that Zweig's idea was rubbish was a sincerely held view, it is hard to understand the omission of these seminal works from his colleague's book, which surveyed an area that Zweig had coinspired.

From the start, in 1963, Zweig had been convinced of the reality of quarks. Throughout the 1960s, the idea that quarks are genuine physical particles was hugely controversial. Most found them too exotic to accept, not least in view of overwhelming evidence that no particles with electric charges smaller than that of a proton had ever been seen.

Nonetheless, in 1965 Dick Dalitz took the idea seriously enough to imagine, like Zweig, that quarks build hadrons by rotating and orbiting around one another, following the same rules as electrons in atoms or protons and neutrons within atomic nuclei. When he spoke about these ideas at an international conference in Berkeley in 1966, Gell-Mann stood up and walked out. What created the skepticism was that although Dalitz's model was empirically successful, he was at the same time riding roughshod over the rules of quantum mechanics. If quarks are indeed real, they had to be fermions—particles with spin one-half—and quantum mechanics allows at most one fermion at a time to exist in the same quantum state. Yet for Dalitz's model to work, two, or even three, quarks had to cohabit, violating this fundamental law.

Years later we know why they can do this: Each quark carries any of three varieties of a property known as color. Color is similar in many ways to electric charge; where electric charge is the source of the electromagnetic interaction, so is color the source of the strong interaction. The threefold nature of color allows up to three quarks to occupy the same quantum state, in accord with quantum laws, so long as each carries a different color charge.[16] This triplet property became the basis for the modern theory of Quantum Chromodynamics, which explains the behavior of quarks and hadrons, as we shall see in the next chapter. However, in 1966 such ideas were far in the future; quarks carrying fractional electric charges were regarded as fictions, and for the quark model to work, you apparently had to ignore the established rules of quantum mechanics. Small wonder that many thought it rubbish.

So there I was in 1967, a student of Dalitz, working with these crazy notions, and gradually becoming depressed by it all. Research is hard work intellectually, and to succeed requires great belief in the worth of what you are doing, but in 1967 it was hard to believe in quarks.

The calculations in the model not only assumed that the quarks are objects with fractional electric charge, unlike any seen in any experiment, but also that they are very massive, yet somehow bound together so tightly that nearly all of their mass was used up in forming the relatively light cluster that we call a proton. This all seemed rather fanciful. If real, such things should have been relatively easy to find, yet experiments in laboratories, searches for them in minerals, and even eventually in rocks from the moon refused to reveal any traces. My faith in the quark religion was sorely tested.

By the summer of 1968 I had almost completed my first calculations in the model and learned that Gell-Mann himself would give two talks about particle physics at the Rutherford Laboratory, some twenty miles south of Oxford. The opportunity was too good to miss: Gell-Mann surely could verify that quarks are real. At the end of the first talk I asked him. His reply was unequivocal: For him quarks were a mnemonic and "a convenient way of keeping track of the mathematical group theory." This denial of the reality of "concrete quarks" further disillusioned me, its impact so dramatic that it has stayed with me ever since. So I was surprised when years later, by which time quarks had been established as physical entities capable of scattering electrons and other particles, Gell-Mann seemed to be singing a somewhat different tune.[17]

Gell-Mann had introduced quarks as a mathematical concept from which relations among phenomena could be abstracted, and the underlying model then dropped. He liked to compare this to French cuisine where "a piece of pheasant meat is cooked between two pieces of veal, which are then discarded."[18]

The process in cooking is known as barding and is to prevent the bird from drying out. High-quality restaurants, having no wish for extravagance, include the surrounds—be it veal or bacon—as an essential part of the dish. However, discarding veal became a common metaphor for how Gell-Mann cooked with quarks in the 1960s.

He developed a powerful program from this philosophy, known as current algebra.[19] This predicted relations among the electromagnetic and weak interactions of hadrons without requiring details of the unknown dynamics of quarks. One example, due to American Stephen Adler, used this to derive consequences for the interactions between neutrinos and nucleons.[20] A surprising result was that in some situations the chance for neutrinos to scatter from nucleons should be considerably larger than had been generally expected. There was no effective way of experimentally testing this at the time, but Bjorken then built on Adler's work, and found similar implications for the scattering not of neutrinos but electrons, which could be tested at the new accelerator of electrons at SLAC.

When I asked Gell-Mann about quarks in 1968, neither of us knew that these experiments were about to revolutionize our picture of the proton and establish quarks as a reality. For quarks, as for pheasant, throwing away the veal is not compulsory.

A month or so after this meeting with Gell-Mann, I had completed my calculations. By now I was disillusioned with the quark model, but I showed my results to Dalitz, and asked if experiment could test them. "Go and talk to Don Perkins," he advised (Perkins being the professor of experimental particle physics). I explained to Don what I had been doing, and then, unexpectedly, he suddenly asked me what I thought about quarks.

I had felt unable to say to Dalitz how I was feeling, but Perkins had asked, and suddenly I poured it all out, expressed my depression, mentioning the singular unobservability of quarks, and then told him of Gell-Mann denying them like Saint Peter. Perkins then opened a drawer and took out a graph of some data, stating in booming Yorkshire tones: "If that's not a quark, I don't know what is."

I was flabbergasted and mumbled, "Oh. Yes!" without having any understanding of why this graph was, for Perkins, like a smoking gun. Instead of asking for an explanation, I asked where it had come from. He mentioned a conference in Vienna—from which he had just returned—someone called Panofsky, experiments involving electrons scattering, something about Rutherford and the discovery of the atomic nucleus being replayed; much of it went over my head.

What was obvious was that Don was excited, and, suddenly, so was I. Something dramatic had happened in Vienna, and I had to find out what it was. After leaving his office and returning to my desk all I could with certainty recall was that electrons had scattered from quarks. I went and talked to Tony Hey, a fellow graduate student working on electron scattering, to see if he knew anything. It was news to him, too. The surprising thing was that, apart from Don Perkins, who had just returned from Vienna, no one seemed to know anything about this.

Tony and I started detective work there and then. This is what had happened.

QUARKS AT SLAC

Following the Shelter Island conference in 1947, physicists met every other year at an increasingly large conference, in a series that continues to this day. The location is peripatetic. The written proceedings of these gatherings reveal the course of particle physics throughout the decades. Those of the first conferences, in the 1950s, show that very few talks in those days were devoted to photons and electrons, for the majority of reports were about hadrons, particles such as the proton that feel the strong interaction. The next few conferences show a similar lack of interest. Then things changed, quite suddenly, and partly by accident.

Experiments in the 1950s, which bounced beams of high-energy protons from static targets, had discovered that protons have a size. They are not big, being only one-ten-thousandth the extent of an atom of hydrogen, but they do have size nonetheless.

As mentioned at the time of this discovery, there was an accelerator of electrons at Stanford University. Small by modern standards, it was cutting edge in its time, and its physics inspired the migration of 1956. The Stanford physicists had been inspired to see if they too could measure the size of the proton—by bouncing electrons off the protons that were contained in a tank of hydrogen. They succeeded, but when the results of these complementary approaches were compared, the fine details showed that the size, as measured by proton beams, differed from that as revealed by beams of electrons. The opinion around Stanford was that a proton contains some central pith, which beams of electrons could probe more cleanly than had been possible with the brute-force method of smashing protons into one another. Outside Stanford, few took much notice.

It was this faith in electrons that inspired the team at Stanford to build what would become, at that time, the world's longest particle accelerator. So, in 1962, construction began of "Project M"—*M* for Monster—at three kilometers in length. Beams of electrons were accelerated along its full length, to emerge with so much power that when directed at a target of hydrogen, they could burrow deep into the heart of a proton before being steered away by the intense electric fields that lurked within. The monster—today known as SLAC, which stands for Stanford Linear Accelerator Center—succeeded, but no one anticipated that the nature of what happened next would utterly transform the future direction of particle physics.

For SLAC was about to discover the quark.

SLAC fired beams of electrons at a target of hydrogen, whose atomic nuclei are protons. Instead of measuring what became of the protons, Bjorken urged that the team put its efforts into recording the direction in which each electron scattered, and how much energy it had lost in the process—the larger the angle through which the electron bounced, the bigger the momentum given over to the proton. His theory implied that these two pieces of information would be sufficient to map the innards of protons and that the data would be sensitive only to a ratio[21]—today known as "Bjorken-scaling"—rather than on each independently. These "deep inelastic"[22] collisions were in the experimenters' plans, and would have been measured at some point, but Bjorken's pushing for them made a significant difference to what ensued.[23]

Several worried that the experiment would not work; for example, how could you tell whether the electron that you were recording had indeed bounced off the target—which is what you wanted—rather than having been knocked out of one of the target's atoms, the electron from the beam having escaped detection? Nonetheless, in 1963 the laboratory management bravely agreed to give the experiment the go-ahead, and preparations began.[24]

The accelerator was still being built when, in 1966, Bjorken wrote his seminal paper on the subject. In this paper, he had used current algebra—the pheasant that was retained when the veal was discarded—and found that if the data were displayed in a specific manner, they would exhibit a phenomenon, which is known as "Bjorken-scaling." However, with its opaque title—"Applications of Chiral U6 × U6 Algebra of Current Densities"—and closely argued mathematics, it is hardly surprising that his paper made little impact on the experimentalists.

Finally, in 1967, SLAC was completed and experiments began. To celebrate, a conference was held at Stanford in the summer of that year, one aim being to highlight the opportunities that experiments using electrons might offer. Bjorken gave a keynote speech, using the occasion to explain the ideas that were in his 1966 paper. To make his ideas more comprehensible, for the purpose of his talk he imagined the proton to be built of quarks and explained how this was a rationale behind his theory.[25] In effect, he had both cooked the pheasant and kept the veal.

His presentation attracted huge criticism; few took the idea of quarks seriously anyway, and to make matters worse, to simplify his theory, he had had to assume that the quarks in a proton are completely independent of one another, as if feeling no forces at all. Yet if quarks were truly the basic seeds of protons and atomic nuclei, they ought to be strongly glued to one another. This contradiction worried Bjorken, and during the questions after the talk he said that he had introduced the quarks "mainly as a desperate attempt to interpret the striking phenomenon of point like behaviour [implicit in the mathematics of current algebra]." He concluded his report by saying that additional data would be needed to "destroy the model of elementary constituents."

Dick Taylor,[26] a Canadian physicist with broad shoulders like a lumberjack, was one of the leaders of the experiment at SLAC that within two years confirmed Bjorken's idea. He told me in 2009 that he believes today that Bjorken should have stuck to his guns. Taylor's experiment subsequently showed that quarks not only exist but behave exactly as Bjorken had assumed, which led Taylor to a share of the Nobel Prize in 1990.[27] This paradoxical behavior—quarks that behave independently yet are powerfully gripped together—would lead to a search for a theoretical explanation.

By spring 1968, exciting results were emerging from the experiment at SLAC. Prior to this, most people had expected that when a beam of electrons hits a target of material, most of the electrons would head onward with at most a small deviation from their path. While the primary interest had been to record the electrons scattering through small angles, huge detectors had also been set up in the laboratory to count how many electrons were deflected through larger angles. The remarkable discovery was that large numbers of electrons were being measured at large angles where most people (but not Bjorken) had expected just a few.

Along with Dick Taylor, the two other leaders of the team were Jerome Friedman and Henry Kendall, professors at MIT who shared a teaching post. MIT had arranged that each could spend six months a year on their research and the rest teaching. So while Friedman was at Stanford, Kendall would be at MIT, and vice versa. Kendall was also very active in the protests against the Vietnam War and a leading member of Scientists Against Nuclear Weapons. In addition to all these activities, he was an inveterate outdoorsman, sharing a house with Bjorken when in Stanford and climbing in the mountains with him when time allowed.

Kendall and Bjorken had swapped ideas for years. When Kendall told him that the experiment was finding big numbers, but that there was no pattern to them—that the data looked like some wild piece of abstract art—Bjorken then made his remarkable suggestion. He recommended that Kendall redraw the picture in a style that, according to Bjorken's theory, should produce a clear picture of the innards of a proton. Bjorken recalled, "I just wanted to see how things looked when expressed in my natural language."[28]

Kendall returned to his office and replotted the data as Bjorken had suggested.[29] When he saw the result, he was stunned as a picture of what appeared to be quarks suddenly emerged. Bjorken told me, "Henry freaked out."[30]

By the summer of 1968, with Bjorken's advice on how to analyze the data guiding them, the experimentalists were confident enough that they decided to announce their results at the major international conference being held in Vienna from August 28 to September 5. There Friedman made a presentation in a parallel session, which was only attended by a few people—for experiments using electrons were still a minority interest. The plenary sessions followed in the latter half of the conference, and the director of SLAC, Wolfgang "Pief" Panofsky, was scheduled to summarize the results from the laboratory.

One might imagine that the breakthrough hit the world following his talk. The reality was rather different. Panofsky was only five foot two and barely visible over the lectern. Although he had lived in the United States for many years, he hailed from Berlin and had never lost his German accent, which with a bizarre American twang made his speech almost unintelligible to those who were not used to hearing it. And on top of all this, the acoustics in the hall were so bad that only delegates in the front rows could hear him.

The incomprehension was also probably due to the audience's unfamiliarity with deep inelastic processes and the mathematical techniques for describing the phenomena. Bjorken had told the experimentalists what to do and had tried to explain the complex reasoning behind it, though none had really understood what it meant. "B. J. spoke in tongues; it was Feynman who translated it for us" is how Dick Taylor described it to me, referring to the greatest irony of all: that on the day that Feynman came to town, "B. J." was away climbing in the mountains.

FEYNMAN'S EPIPHANY

That Feynman came to SLAC, saw the data for the first time, and within twenty-four hours explained it with his celebrated "parton model" is one of the great pieces of physics lore. However, so singular was the event that it has gained a life from the many retellings over the years. Ask people who were at SLAC in 1968 for their memories, and the result is a set of mutually incompatible histories.

One memory has Feynman visiting SLAC to give a popular talk to schoolchildren and seeing the data for the first time immediately afterward.[31] With school restarting early in September this could be consistent with the memory of Elliott Bloom, then a junior assistant, and now one of Stanford's senior scientists, who told me, "All the bigwigs had gone to Vienna, so when Feynman came visiting, I was commanded to go to the Green Room [the conference room in the theory division] to show him the results."[32] It is possible that Feynman came more than once and that memories have conflated different occasions, as on Saturday, August 17, 1968—nearly a fortnight before the Vienna conference began—his own diary (Figure 12.1) records: "Visited SLAC last week. . . . Here is a summary of what I was told by M. Friedman and Tsai."[33] This chronology also coincides with the memory of Jerry Friedman, who confirmed: "We were well aware of the parton model before Vienna."[34]

Feynman seems to have confused names. There is no such person as "M." Friedman; the initial in his notes probably refers to Manny Paschos,

FIGURE 12.1
FEYNMAN'S NOTES AFTER HIS VISIT TO SLAC

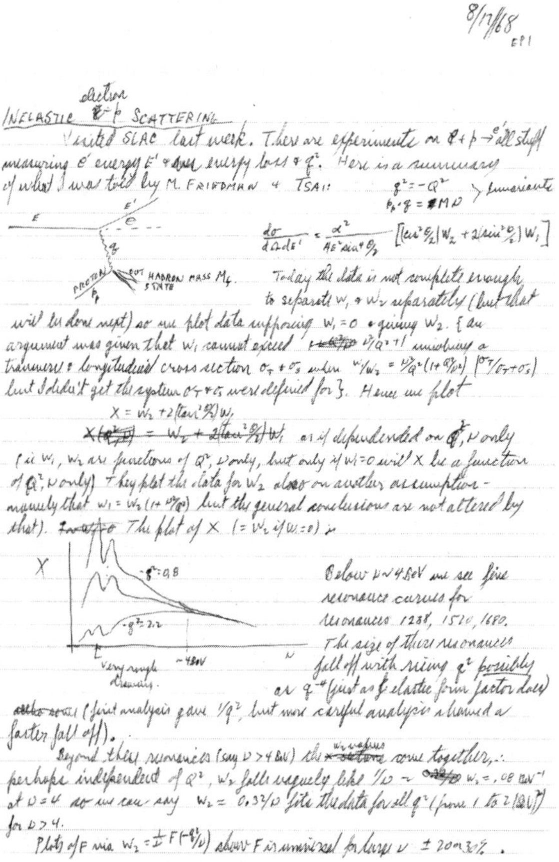
8/17/68
EP1

INELASTIC electron e-p SCATTERING
Visited SLAC last week. There are experiments on e + p → e' all stuff measuring e' energy E' & energy loss & q². Here is a summary of what I was told by M. FRIEDMAN & TSAI:
q² = −Q²
p_p · q = MD } invariants
E
E'
e
q
PROTON p_p
OUT HADRON STATE MASS M_f.
dσ/dΩdE' = α²/(4E² sin⁴ θ/2) [(cos² θ/2) W₂ + 2(sin² θ/2) W₁]
Today the data is not complete enough to separate W₁ & W₂ separately (but that will be done next) so we plot data supposing W₁ = 0 & giving W₂. {An argument was given that W₁ cannot exceed ν²/Q² + 1 involving a transverse & longitudinal cross section σ_T + σ_S when W₁/W₂ = ν²/Q² (1 + Q²/ν²) (σ_T/σ_T+σ_S) but I didn't get the system σ_T & σ_S were defined for}. Hence we plot
X = W₂ + 2(tan² θ/2) W₁
as if depended on Q², ν only (i.e. W₁, W₂ are functions of Q², ν only, but only if W₁ = 0 will X be a function of Q², ν only) They plot the data for W₂ also on another assumption – namely that W₁ = W₂ (1 + ν²/Q²) but the general conclusions are not altered by that). The plot of X (= W₂ if W₁ = 0) is
X
q² = 0.8
q² = 2.2
~4 GeV
ν
Very rough drawing.
Below ν ~ 4 GeV we see fine resonance curves for resonances 1238, 1510, 1680. The size of these resonances fall off with rising q² possibly as q⁻⁴ (just as the elastic form factor does) (first analysis gave 1/q², but more careful analysis showed a faster fall off).
Beyond these resonances (say ν > 4 GeV) the W₂ values come together, perhaps independent of Q², W₂ falls vaguely like 1/ν ~ W₂ = .08 GeV⁻¹ at ν = 4 so we can say W₂ = 0.32/ν fits the data for all q² (from 1 to 2 (GeV)²) for ν > 4.
Plots of F via W₂ = 1/ν F(−q²/ν) show F is universal for large ν ± 20 or 30%.

These notes record the first time he learned about the phenomenon of scaling in the SLAC experiments on deep inelastic electron scattering. (Credit: Courtesy of Archive, California Institute of Technology.)

who, as a young Greek postdoc in 1968, played a central role in what unfolded. Today Paschos is a senior professor at Dortmund, Germany. He recalls bumping into Feynman, who was walking along the corridor of the SLAC Theory Group one afternoon. Paschos asked if he had heard about the new results being presented at Vienna, which confirmed Bjorken's prediction of "scaling."[35] Paschos told me, "We walked to the small office of Paul Tsai"—a theorist who was aiding the experimentalists in analyzing the data. One popular version has Feynman in Tsai's office, where, upon seeing the data, he "fell to his knees" in wonder, "clasping his hands prayerfully over his head" upon realizing its significance.[36]

This dramatization has echoes of a common feature with Feynman stories, a hero for physicists around whom so many tales have grown that it is often hard to separate myth from reality. Paschos's more sober version has an aura of truth: "After seeing the data Feynman found the results very exciting and expressed the wish to stay longer. This is the first time that he saw the data." It seems likely that Feynman then visited the actual experimentalists, meeting with Friedman as his notebook records.

A room was booked at the Flamingo Motor Lodge, on El Camino Real in Palo Alto. Paschos and his wife dined with Feynman, departing after dinner "with a promise to meet him in the morning and bring him to SLAC. Up to this moment there was no mention of partons or any other interpretation for the data."

Next morning, when Paschos met him at the motel, Feynman "was all excited and mentioned that he understood the phenomena as originating from the scattering of the electron on bare constituents without structure. I drove him to the theory group where many theorists and experimentalists gathered after some time in the Green Room—the seminar room—to hear his explanation."[37]

Feynman's own memory fits with Paschos's version. Feynman recalled, "I went home [to the motel] after they showed me the data . . . and I began to think about it. I suddenly realised that I had been very stupid."[38]

Feynman was astonished for very personal reasons. Since June he had been building a model of the proton as made of pieces, parts, which he called "partons," and had been investigating what it would imply for the debris when protons smash violently into one another.[39] He imagined the two protons as clouds of partons, which are sprayed around by the collision. He was trying to find ways of analyzing the data to reveal what the true nature of these partons is.

Protons hitting protons is complexity twice over. Relative to a proton, however, an electron is a simple entity. The debris in a collision between an electron and a proton consists of the scattered electron and the spray of partons from just a single proton. This is much more straightforward to interpret, but until he visited SLAC Feynman had not realized that this was a real possibility—hence his remark about being "stupid."

In his impromptu presentation to a mix of theorists and experimentalists, Feynman explained everything, and in language that was easy to follow. Feynman's "parton model" entered the public domain. Stan Brodsky, a theorist at SLAC, recalled that Feynman admitted, "I had never thought of applying my parton model to experiments using electrons."[40]

When Bjorken returned from the mountains later that day, he found the lab in bedlam. Feynman asked Paschos to introduce him to his "famous friend."[41] Feynman and Bjorken had an hour or two together, Bjorken telling me in 2009, "Some of what Feynman was saying I already had; some was in a foreign language."[42]

After Feynman departed, Bjorken and Paschos worked out Bjorken's ideas assuming that partons are quarks and antiquarks. By early October they had it complete, but felt that they ought to wait until Feynman had published his own ideas. However, nothing happened, so Bjorken called Feynman to ask if he was writing his paper because "we are waiting." Feynman said he wasn't writing anything, and when Bjorken asked why, Feynman replied, "I'm rich."[43] Bjorken told me that he interpreted this as an act of generosity, encouraging him to go ahead and publish with Paschos.

Bjorken and Paschos duly wrote their paper.[44] By October, Feynman had worked out his ideas in detail and returned to SLAC to give a special seminar. This was a gala occasion, and there are photos of him standing in front of a board covered with equations and diagrams in chalk. Knowing Feynman's showman style, he would surely have been both physicist and raconteur. In such fevered excitement, myths germinate. Upon seeing the data, history as drama would have him falling to his knees there and then. So powerful was Feynman's persona, his Brooklyn tones making him a wonderful storyteller, the memories of the original events may well be actually memories of how Feynman subsequently told it, or of many conversations among those who had been there subsequently telling those who were not, and, like Chinese whispers, gradually adding color to the events. To me, Paschos's low-key version has the aura of reality, though the Feynman in ecstasy would be a dramatist's choice.

Three years later, Bjorken talked about developments at a conference in Cornell. Feynman spoke about his own ideas, most of which he had not formally published, and Bjorken gave a keynote review with Feynman in the audience. Bjorken told me that he didn't know how to cite Feynman and so put "Feynman's notebooks" on his display slides whenever a citation seemed merited. This became a leitmotif and started to appear on every slide, over and over again. Bjorken recalled that people began to titter and, as it continued, to laugh. "What I didn't know was that Feynman kept meticulous notebooks."[45]

Feynman kept a daily diary, being highly organized despite his public image. On the right-hand page were his thoughts about physics; on the left were cross-references to other entries.[46] Bjorken told me about an evening that he and Feynman shared later. "I said, 'I heard you had notebooks. Is the parton stuff in them?'" Feynman replied, "All but one item" but didn't say which. "The conversation got on to other things so I never found out."

News of the 1968 seminar, and of Feynman's parton model, spread rapidly around the world. Today most credit Feynman for the breakthrough that originated with Bjorken. Feynman, honest and fair, always said, "I did nothing that was not already in B. J.'s notebooks."[47] Back at Cal Tech, the home institution of both Feynman and Gell-Mann, the latter accosted Feynman: "What are these put-ons [*sic*]; are they quarks?"[48]

In the spring of 1968 I had been depressed when Gell-Mann told me that quarks were not real; by the end of that year, however, the Vienna announcement had changed everything. Jerry Friedman recalls, however, that what is established today was far from obvious at that time. The idea that the data showed pointlike structure was considered "too bizarre a concept to utter in public."[49] There were other possible explanations of the data, more compatible with the thinking of that era, so Friedman "never mentioned this possibility" in his talk. However, Panofsky did: "Theoretical speculations are focused on the possibility that these data might give evidence on the behaviour of point-like, charged structures within the nucleon."[50] In the memory of Dick Taylor, Panofsky even mentioned the word *quark*.[51] Luckily for me, Don Perkins had been one of the fortunate few who had sat near enough the front of the hall to realize the significance when Panofsky made that claim. Perkins' excitement—as in his exclamation: "If that's not a quark, I don't know what is"—offered me the chance that after all quarks were real. From that moment, I wanted to understand what it was all about.

That's why I took up a research post at SLAC in 1970 when I completed my Oxford doctorate. That is where I met Bjorken. Following the drama of the Vienna conference, and the excitement that the Stanford experiments were generating, being at SLAC felt like being at the center of the universe.

In 1990 Taylor, Kendall, and Friedman, the leaders of the experiment that had shown the reality of quarks, shared the Nobel Prize. Bjorken is still waiting. His ideas had inspired the experiment, and, even before it took place, he had predicted what the data would look like if quarks really existed. When Kendall had shown him the data, in a state of confusion, it was also Bjorken who had suggested that they try analyzing it a different way, at which point everything fell into place. But for Bjorken's intervention, it is possible that the data would have remained a morass and the history of physics radically different.

Bjorken's seminal paper, which pointed the way to proving that quarks are real, is even more remarkable. It also showed that if the *W* boson—the carrier of the weak force—existed with a large mass, a substantial fraction of its decays would be to an electron and a neutrino.[52] The discovery of the *W* in 1983 used this fact to identify the *W* and won a Nobel Prize for the experimentalists involved, as we shall see in Chapter 16. He showed too that when electrons and positrons annihilate one another, the probability for this to happen and produce various flavors of quarks and antiquarks falls steadily with energy. Any deviation from this behavior would signal new dynamics—a phenomenon that in 1974 would become the route to identifying "charmed" quarks, which won Nobels for the experimentalists involved there too. Furthermore, his exposing that protons are built of quarks has led to a whole generation of experiments whose results would have otherwise been indecipherable, if they had even happened at all. These include those that confirmed 't Hooft and Veltman's theory of the weak force and also those at the LHC, which hope to find the Higgs Boson.

In the opinion of many physicists, Bjorken is the forgotten genius behind the modern picture of particles and forces known as the "Standard Model."

CHARM

The result of Bjorken's inspiration was that experiments verified that there are quarks deep within a proton. The results from the experiments at Stanford were soon joined by data from CERN, where beams of neutrinos instead of electrons were used. These probed the protons in a complementary fashion, and from the two sets of data the electric charges of the quarks could be measured. The results were everything a quark modeler wished for: A quark of the "up" variety has positive charge in an amount that is a fraction, 2/3, that of a proton; a "down" or a "strange" quark each carry negative 1/3.

While this fitted perfectly with Gell-Mann and Zweig's original scheme, the quarks themselves were not really what the concrete quark models had anticipated. Far from being very massive and tightly bound, they appeared to be almost massless and, most radically, free.

Even after these experiments, many remained suspicious of the quark model. Weinberg himself missed a chance to complete his theory of the weak interactions: his work was titled a "model for leptons" (particles like the electron and neutrino, but not quarks). He had restricted it because if he applied his ideas to hadrons, they seemingly predicted that there should be neutral weak interactions where the strangeness of the hadron changes. The empirical absence of such "strangeness changing neutral" interactions had been a stumbling block for many, not just Weinberg. Salam and Ward, years earlier, had given up because of this.

Weinberg actually wasn't worried about that particular problem because of a much larger worry. He told me, "I called it a model of leptons to emphasize I was not getting into the morass of the strong interactions. I didn't think we understood anything about the strong interactions in 1967. I just didn't believe the quark model. That was a widespread attitude at the time. How could you take it seriously—fractionally charged particles that no one had ever seen."[53]

By 1970 the experiments from Stanford and CERN had seen fractionally charged quarks, albeit not with all the properties that the quark model would have wanted. Nonetheless, with the new confidence in quarks, Shelly Glashow, and two colleagues, Greek John Iliopoulos and Italian Luciano Maiani, discovered how to do away with these unwanted strange neutrals. The price was to invent a fourth variety of quark, which later became known as the charm quark. For charm quarks to do this trick they had to have the same electric charge as an up quark, but be distinguished from them by being more massive. Then, by the wonders of quantum mechanics, the quantum waves arising from charm quarks could interfere with those from the up, down, and strange in such a way as to destroy any strangeness-changing-neutral interaction at the source. This trick became known as the GIM (pronounced *jim*) mechanism, the acronym respecting the initials of its creators' names.

By now, if not already, you may be wondering if there is a rationale for the quarks' whimsical names. If not, just skip the next two paragraphs.

At the time of GIM's entrée, all known hadrons could be explained as built from quarks, which occur in any of three varieties, or "flavors." For decades the proton and neutron had responded to the strong interaction like two faces of a single entity—the nucleon—distinguished by their electric charges. By analogy, the names of the two varieties of quark that make them are like two sides of a coin, and their naming reflects that metaphor—hence up and down, distinguished by their electric charges, +2/3 and –1/3. Then two ups and one down are sufficient to make a proton, while two downs and one up make the neutron.

Among the many hadrons discovered in the 1940s and '50s were some that had unusual properties,[54] whereby they were dubbed "strange" particles. Today we know what distinguishes these from the familiar proton, neutron, and pion—they contain a third variety of quark, which became known as a "strange" quark. This strange flavor has the same electric charge as a down quark, and is distinguished from it by being more massive. Some of the charm of GIM's hypothesis was that by invoking a heavier version of the up quark, charge +2/3, they had created a pleasing symmetry: The basic up-down pair have two heavier siblings—charm and strange.[55] The fact that with this fourth flavor the GIM mechanism exorcised the unwanted strangeness-changing-neutral interaction further testifies to its "charm." A summary of the quarks and bosons known today is in Figure 12.2.

FIGURE 12.2
FERMIONS—QUARKS AND LEPTONS—
AND BOSONS: SYMMETRY SPOILED BY MASS

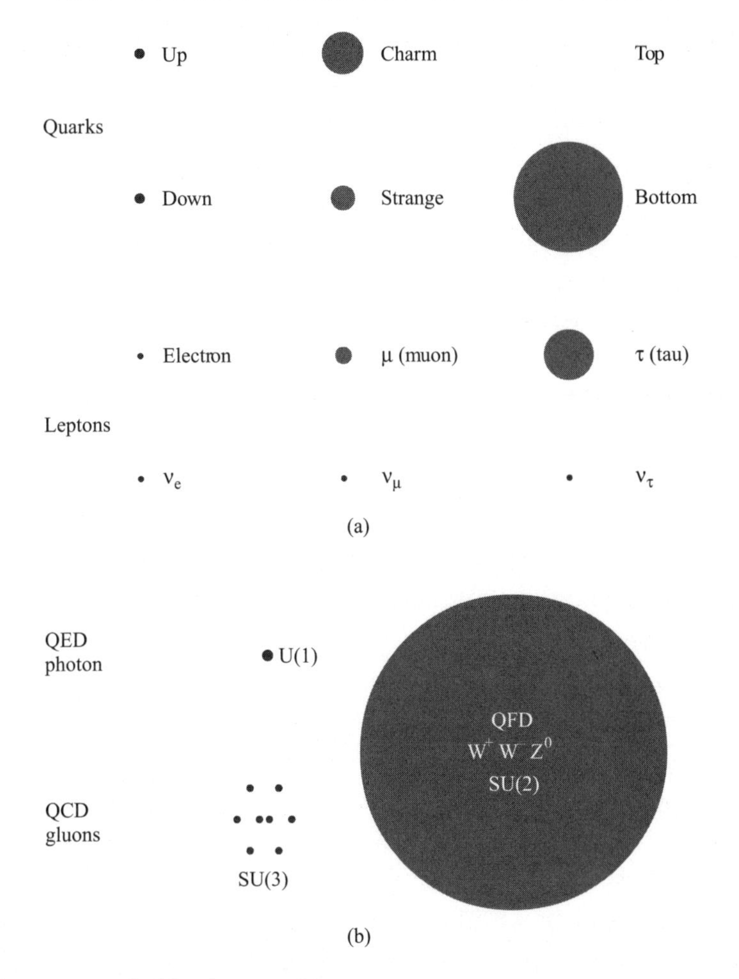

(a) Symmetry hidden by mass: Fermions

The quarks and leptons appear to have common characteristics but for their different masses. The circles give a qualitative idea of their relative masses. The top quark circle is not shown. Its diameter would fill the page.

(b) Symmetry hidden by mass: Bosons

QED and QFD (its marriage with electroweak) are U(1) and SU(2) Yang-Mills theories. QCD is an SU(3) Yang-Mills theory. The gauge bosons—the photon and gluons—are massless; the *W* and *Z* are massive. This mass, breaking the symmetry, is illustrated by the large circle surrounding the SU(2); the small circles for U(1) and SU(3) illustrate the massless nature of the respective gauge bosons. (In practice the photon and *Z* are mixtures of SU(2) and U(1), so reality is not so simple as suggested in this illustration.)

The threat, or opportunity, was that the hypothesis of a new "flavor" of quark—charm—implied that there should also be large numbers of hadrons containing one or more charm quarks. None had been seen. The excuse for their nonappearance was that they must be somewhat heavier than those hadrons without charm, but for many physicists at the time, this idea also was not taken very seriously.[56]

Weinberg's reaction is telling: "When the GIM paper came out I thought they had found a pretty solution to a problem that arises in the quark model—you might say it arises more generally, but it was certainly a problem that arises with quarks—and I just didn't take it very seriously because I didn't believe in quarks."[57] When, the following year, 1971, 't Hooft proved that the theory of weak and electromagnetic interactions is renormalizable, Weinberg was convinced that his theory was fine for leptons, but continued to doubt the reality of quarks.

While Weinberg's lack of belief in quarks at the time of his 1967 paper is easy to understand, his continued skepticism into 1971, following the observation of fractionally charged particles within the proton and neutron, is more profound. Bjorken's insight had led to a seminal breakthrough, exposing the quark layer of the cosmic onion. However, its most profound consequence was in revealing the paradoxical nature of the strong force. How can quarks act independently deep within a proton yet be so strongly glued to one another that they never escape?

The phenomenon was completely at odds to what field theories seemed to imply. In Quantum Electrodynamics, for example, which was the quintessential field theory, the electric charge grows larger at short distances—in the jargon of page 44, "The beta function is positive." The strong interaction seemingly has the opposite behavior: The source of the force becomes feeble at short distances. If the strong interaction is to be described by a field theory, the theory would have to have a beta function whose sign is *negative*. The problem was that no such theory was known, and indeed, it was believed to be impossible.

Everything seemed back to front. This was the paradox that Bjorken's insight, and the SLAC experiment, had put at center stage: Deep in the interior of the proton, quarks are almost free.

Chapter 13

A COMEDY OF ERRORS

A Yang-Mills theory—Quantum Chromodynamics—explains the paradox of the enfeebled strong force. How 't Hooft has the answer but fails to realize its significance; other physicists also have the theory in their grasp but miss the final step. Two students—Frank Wilczek and David Politzer—find the answer independently, eventually sharing a Nobel Prize. Bjorken's ideas lead to the discovery of charm, which helps to confirm the QCD theory.

By 1970 physics was in turmoil. Bjorken's idea, as promoted by Feynman, had proved that there exists a deeper level of reality than protons. Gell-Mann's taunting of Feynman with "What are these put-ons [*sic*]; are they quarks?" drew no response. Feynman was careful. He knew there could not simply be quarks inside the proton; there had to be something else responsible for gripping them together in tight bundles. In modern terminology: Feynman was aware that there had to be gluons. For example, the electromagnetic force occurs between particles with electric charge, and there is also electromagnetic radiation: photons. Feynman realized that analogous behavior should happen also for the strong force, which grips quarks inside the proton and builds atomic nuclei. The quarks must carry some novel form of charge (today known as "color"), which is the source of the strong force, and there should also be some analog of the photon, which is now known as the gluon. It would be these gluons flitting around between the quarks that would in effect glue them to one another, to make the proton. All these "parts"—quarks, gluons, and who knew what else—were what Feynman collectively referred to as "partons." What they actually consisted of was for experiment to decide, not for Feynman or Gell-Mann to second-guess.

When the news broke in 1971 that 't Hooft and Veltman had explained the weak force in a viable theory that was built along similar lines to QED—known as Quantum Flavordynamics (QFD), where the *W* plays an analogous role to the photon[1]—the idea that the strong force also might be described by a theory based on quarks and gluons began to take hold. Although for the strong force 't Hooft himself would miss what was staring him in the face, it was nonetheless his breakthrough in explaining the weak force that would lead others to the solution. The discovery that the strong force also is described by a Yang-Mills theory—Quantum Chromodynamics or QCD—is a comedy of errors.

THE ENIGMA OF BJORKEN-SCALING

Before SLAC made its discovery, the strong interactions had seemed impenetrable—the Cambridge 1967 prospectus notwithstanding. Freeman Dyson, the quiet Englishman who had helped pioneer QED but had not shared in the ensuing Nobel Prize in 1965, had asserted, "The correct theory will not be found in the next hundred years." American David Gross, who would share in the Nobel Prize in 2004 for his role in establishing QCD as the theory of the strong interaction, said, "For a young graduate student [at that time], such as myself, this was clearly the biggest challenge."[2]

By 1972 Gross had worked his way through the most well-studied field theories in search of one that fitted the properties found by the SLAC experiment: quarks that are strongly bound yet appear to free themselves when they get near one another. The mathematics involved calculating the quantity known as the "beta function," described in Chapter 2, which summarizes how the magnitude of the charge on a particle appears to change as one gets nearer to it. In QED this quantity is positive, and in all the field theories that Gross had examined it was always positive.

A positive value for beta implied that the charge concentrates to infinite density when viewed at high resolution—as in the pixel analogy earlier. However, the results of the experiments at SLAC implied that the force between quarks behaves contrary to this; at high resolution, the glue that binds the quarks appeared to lose its potency rather than become infinitely powerful. If this impotence was to be explained by a field theory, Gross needed to find one where beta was *negative.* He was beginning to suspect that this counterintuitive behavior was impossible in any field theory, and that the explanation of the strong force lay elsewhere.

Gross was not alone; a young postdoc named Tony Zee was also on the hunt. Zee had been inspired by lectures of his thesis adviser at Harvard—Sidney Coleman, a universally loved and brilliant theorist. If anyone merits the accolade "Nobel Prize runner-up," Coleman would be a contender.[3] These lectures gave Zee the inspiration that the sign of beta is key, and that the phenomena exposed by SLAC implied that a field theory of the strong interaction would need beta to be negative.

Zee had looked at established renormalizable theories and found that they all have a positive value for beta. These did not include Yang-Mills theories, which did not come into vogue until after 't Hooft's appearance in Amsterdam in 1971.

In the memory of Terry Goldman,[4] a student at Harvard in 1970, who today is an emeritus senior theorist at Los Alamos, the sign of beta was a hot topic there. Goldman proposed to Coleman that Yang-Mills theories might be a contender for the theory of the strong force, only to be told that his "excitement was misplaced because Yang-Mills theories are not renormalizable." However, after 't Hooft's talk at Amsterdam in 1971, "Tom Appelquist [a young member of the faculty who later became chair of physics at Yale] said to me that it would be good to calculate the sign of beta for Yang-Mills theories."

Goldman didn't do so because "Harvard training is that you don't do lots of hard work unless you know there's a big payoff. I asked Tom if there was any reason to think the sign was negative. He said no, but it's something that should be done." Goldman replied, "I said I'd think about it. Unfortunately, I didn't think hard enough."[5]

Zee, meanwhile, having completed his analysis of established theories, which he regarded as a "warm-up exercise" for the assault on Yang-Mills, found the latter technically demanding. He spoke freely with other theorists, with mixed results: Some "poured cold water" on his ideas.[6] Zee seems to have been one of the first to have fully appreciated the physical significance of a negative sign for beta, but he failed to complete the technical analysis. In this he was a mirror image of 't Hooft, who calculated beta but failed to realize its profound implications, as we shall see later.

At the Sixteenth International Conference on Particle Physics at Fermilab in September 1972, Gell-Mann outlined his own thoughts[7] on how quarks build hadrons. Following Bjorken's insights and the SLAC experiments, which revealed the quarks, Gell-Mann's earlier denials of their reality needed revision. In his Fermilab talk he said, "It may be possible to construct an explicit theory of hadrons based on quarks and some kind of glue," but added that these entities "are fictitious." He explained that by *fictitious* he meant that they do not appear in isolation in the laboratory and are manifested only when confined within hadrons.

Color—the threefold property that quarks carry, but that the electron, for example, does not—became possible as the source of the strong force. After all, quarks feel the strong force, whereas the electron does not. However, Gell-Mann did not make this suggestion explicitly. He focused instead on the phenomenological concept that hadrons can exist physically in the laboratory only if the color of their constituents has overall neutralized, in a mathematical sense.[8]

Gell-Mann here had almost, but not quite, invented Quantum Chromodynamics, the Yang-Mills theory of color, where quarks interact with one another by exchanging colored gluons. Austrian theorist Julius Wess seems to have been the first to raise the idea that color is carried not just by quarks but by gluons too.[9] Gell-Mann's collaborator Harald Fritzsch had worked on Yang-Mills theory in his thesis at Leipzig, and his memory is that "after Murray and I used the color quantum number, I had the idea to [build a Yang Mills theory of] color," though initially Gell-Mann was skeptical.[10] In September 1972, Gell-Mann carefully suggested that the gluons "could" be the colored gauge bosons of Yang-Mills theory, but did not pursue this idea further.[11] There was no demonstration that a Yang-Mills theory of the strong force, built on color, has the required properties to explain the data, or any mention of the significance of the sign of beta.

The cause and effect of what happened next is lost in the mists of time, and in the differing memories of those involved, along with those who were spectators.[12] What is established is that within a few months, several people at Harvard and Princeton had begun working out the sign of the beta function in Yang-Mills theories, including David Gross. Unknown to them, 't Hooft had already made the calculation.

'T HOOFT AT MARSEILLE

Not only had 't Hooft done the calculation, but he had even announced the result at a small conference in Marseille, in June 1972. However, to his great regret, he didn't write a formal paper outlining the proof. The fact that he had said anything at all was itself the result of chance.

The paradoxical behavior of quarks, which were seemingly freely moving yet impossible to shake loose, had bemused 't Hooft, like everyone else.[13] Bjorken had opened the way to the remarkable discovery with his model of the proton as made of freely moving quarks, but he had no theoretical justification for this radical assumption. Nonetheless, the experiments had proved him correct, though no one, let alone Bjorken himself, could find an explanation as to why. However, as 't Hooft admitted later, "Ironically I already had the answer in my notebooks"—"ironic" because he didn't realize the fact.

't Hooft's fame is for having proved the viability of Yang-Mills theory when the gauge bosons have mass. However, he had begun his adventure by studying those theories in a simpler case—where there are only massless gauge bosons—and in doing so had noticed something to which he had, at the time, paid little attention.

The phenomenon that struck him was the way that the amount of charge appeared to vary when looked at with a powerful microscope. In Chapter 2 we saw what happens when one looks at an image of an electron with better resolution—"smaller pixels": The electron's charge becomes concentrated more and more into the central pixel, ultimately becoming infinitely dense: the sign of the "beta slope" is positive. What 't Hooft had noticed was that in the massless Yang-Mills theory—containing just gauge bosons and no other particles—the behavior appeared to be the opposite: As the pixels became smaller, so the charge too appeared to leak away. Instead of becoming infinitely dense in the limit of perfect vision, the charge and the ensuing force seemed to have disappeared. Stated mathematically, beta is negative.

The calculations were tedious, and his purpose at the time had been to use this as a warm-up for his eventual goal—the physically interesting case of massive gauge bosons. So in 1971 he had merely noticed this mathematical curiosity and moved on.[14]

As we saw on page 213, his interest had been sparked at Cargese by lectures from Ben Lee and Kurt Symanzik. While Lee would be the one who brought 't Hooft's weak-force work to wider attention, and establish the full viability of the theory, now it was Symanzik's turn to play a lead part in the unfolding strong-force drama.

In June 1972 't Hooft had flown from Amsterdam to Marseille to attend a specialist conference on gauge theories. He told me that while disembarking, " I saw Symanzik whom I knew from Cargese. I said 'I didn't know you were on the plane,' and on the way to where we were staying we were talking about physics."[15] Symanzik told 't Hooft that he had been trying, unsuccessfully, to understand the phenomenon that Bjorken's work had exposed. He had been looking at the interaction of four spinless particles, known as "lambda phi-fourth" theory. The lambda gives the measure of the strength of their interaction. Symanzik had chosen lambda to be negative so that the theory has its beta slope negative, but the theory then becomes highly unstable.

The phenomenon of a negative value for the beta slope struck a chord for 't Hooft, who remembered that this was the very phenomenon that he had noticed occurs in massless Yang-Mills theory but had left as a curiosity. So he told Symanzik that you could achieve the same ends with Yang-Mills theory without the instability problem.[16]

Symanzik was skeptical, believing—erroneously, as it turned out—that general theorems proved this to be impossible. He suggested to 't Hooft that he must have made an error somewhere. After all, something as trivial as mistaking a plus sign for a minus in the middle of a long calculation would change everything; the whole result depended on whether the sign of beta was positive, as Symanzik thought, or negative, as 't Hooft claimed.

They continued to debate this on the journey from the airport, even up to the time that Symanzik gave his speech at the conference. Symanzik, realizing the implications if 't Hooft's claim was correct, had closed off the discussion by saying, "If you are right, you should publish this quickly."[17]

't Hooft recalls, "Much to my later regret, I did not follow this sensible advice." At the time he was very busy with other calculations for Veltman, and also he did not see how to describe his special methods in a form that would be immediately comprehensible to readers. Fortunately, he did take the opportunity to make a public remark at the conference, immediately after Symanzik's presentation.

Symanzik had concluded his talk by saying that although he had found no field theory with the right property (a negative slope for beta), "perhaps we should take another look at Yang Mills theory." 't Hooft told me what happened next: "I don't remember whether he actually looked at me, but to him it came as no surprise, he told me later, that I stood up and said that I had this remark to make. And I wrote up . . . the beta [expression] on the blackboard. I said there is one sign for . . . electrodynamics and this [other] sign for any Yang-Mills theory. So I sat down and that was that. From the audience I don't think there was much reaction."[18]

Chris Korthals-Altes, who had organized the conference, has his own memories of the events.[19] He had first met 't Hooft in Amsterdam in 1971, and on a visit to Utrecht on May 8, 1972, they talked together. During this discussion 't Hooft mentioned that in his calculations, "The sign [of beta] is funny," leaving Korthals-Altes with the impression that this was some mathematical anomaly.[20]

The year 1972 was Korthals-Altes's first on the faculty at the university in Marseille, and he felt under pressure to make the conference a success. He was sitting next to 't Hooft while Symanzik gave his talk, which Korthals-Altes told me he found "totally incomprehensible." Symanzik was infamous for his presentation style. His handwritten slides—for an overhead projector, which was the vogue in those days—looked as if he had first put lined paper underneath to guide him, and then used every line. This gave a huge amount of text on each slide, which was just about tolerable, but his habit during a talk of overlaying one slide on top of another, and then pointing to both while explaining an argument that linked the two was a challenge too far. His performance in Marseille appears to have been typical. Korthals-Altes described the moment to me: "Symanzik used a huge number of slides, and when he finished, there was silence. I felt responsible. I didn't want Symanzik left with icy silence. I was next to Gerard and said, 'For goodness' sake, say something.'"[21]

't Hooft went to the board and described his discovery. Whether it was because of urging by Korthals-Altes, or by Symanzik, is secondary; what is important is with that action, he made the first public statement that the beta slope is negative in Yang-Mills theory. The implication that a Yang-Mills theory may be a potential description of the strong interaction, and explain Bjorken's insight, was thus in the public domain in June 1972.

His comment raised little interest, however, which explains why Gross was unaware of it when his student Frank Wilczek discussed possible research projects with him four months later.[22] Tom Appelquist, who had been present at Marseille, confirmed to me that Symanzik's talk was "famously obscure with multiple overlaid transparencies," which may explain why Appelquist has "absolutely no recollection" of 't Hooft's comments afterward.[23] When Symanzik published his talk in December 1972 he made no comment about 't Hooft, but included an enigmatic remark: "It would be interesting to determine [the sign of beta] in Yang Mills theory."[24] The reason for this time delay, and the nature of this remark, will become clear later. Appelquist and a student, Jim Carrazone, tried to calculate the beta sign early in 1973 but "became confused about gauge invariance and set it aside to do other things."[25] In effect, they had confronted a similar confusion that 't Hooft had expressed to Korthals-Altes in May that year.

'T HOOFT AND OTHERS

While several in the United States were trying to find a theory to explain the phenomenon of Bjorken-scaling, 't Hooft had found the theory's key ingredient but not realized its empirical implications. He commented on this irony in his Nobel lecture in 1999: "Back in 1971 I [had already] carried out my own calculations [of the beta function] and the first I tried was Yang Mills theory." He had indeed found the negative sign[26] and "alluded to it in my [1971] paper on the massive Yang Mills theory." He then remarks, wryly, that he had failed to realize "what treasure I had here, nor that none of the experts knew that beta could be negative."

I asked 't Hooft why, in 1971, he had not applied this result to explain the SLAC experiments, whose results about the nature of the strong interaction had been around for more than two years. He replied that he had "not been sure at the time if what I was saying and doing was exactly the same thing that [people trying to understand the data] were actually interested in." He reminded me that "strong and weak interactions appeared totally different at that time, and I was mainly preoccupied with the weak interactions." More poignant perhaps was his admission that with his mathematical focus, "I didn't understand strong interactions so well. I didn't really know what the SLAC experiments were."[27]

He had a second chance, though. In February 1973 he visited CERN and had a brief discussion with Georgio Parisi, a young Italian theorist on the staff there. As we shall see, in the United States, David Politzer, David Gross, and Frank Wilczek were still some weeks away from discovering the result for themselves, leading to their Nobel Prize. On that fateful February day, 't Hooft and Parisi could have scooped the American trio, except they took a wrong turn. This is what happened.

After the Marseille conference in June 1972, Symanzik delayed writing up his talk in order that 't Hooft might formally publish the discovery of the negative beta function. However, as we have seen, 't Hooft did nothing. Symanzik was worried that if he raised the specter of Yang-Mills theory in his report, someone else would lay claim to what 't Hooft had merely mentioned. By the end of that year Symanzik must have felt that enough time had passed. Parisi told me, "At the end of December, or early in January 1973, Symanzik informed me about 't Hooft's result."[28]

Parisi was interested in field theory and was also fully aware of the SLAC data and their significance. 't Hooft at this stage was focused on trying to understand gravity. During a visit to CERN, 't Hooft met with Parisi. "It was a morning in February 1973," Parisi recalled. He told 't Hooft what Symanzik had said, and then Parisi impressed on 't Hooft: "With this you can make a theory of the strong interaction." All they had to do was decide which Yang-Mills theory to use. The electroweak theory had used the mathematics of the group SU2, which involves arrays of numbers—matrices—involving two rows or columns. 't Hooft and Parisi were aware of similar mathematics involving three, so called SU3. Parisi suggested that they construct a Yang-Mills SU3 theory for the strong interaction, based on the three flavors of quark: the up, down, and strange.

Talking together, they were initially excited. However, they quickly ran into a dead end, as 't Hooft realized that the gauge bosons required by the theory would carry electric charge and other properties in combinations that would not be renormalizable. The coup de grâce came as Parisi realized that the data from SLAC showed that the electrically charged constituents of the proton had all the characteristics of quarks: There was no room for any electrically charged gauge bosons at all. After a mere "half an hour" they dropped the idea.[29]

It turns out that there are two ways that nature has used SU3. Quarks are doubly distinguishable because each flavor—up, down, or strange—can carry any of three different "colors": the name given to describe the threefold form of charge that, we now know, is the source of the strong force. Parisi and 't Hooft had fixated on the three flavors, which were well established by 1972, whereas the concept of color was at that time not generally accepted. Parisi admitted to me, "I didn't think to try color. I don't know if Gerard 't Hooft knew of color. My mistake was to think only for half an hour; it deserved more than half an hour."[30]

Unfortunate indeed: it is the three *colors* that nature uses to produce the forces that grip the quarks. It is this "three" that is the seed for the SU3 Yang-Mills theory describing the strong interaction (today known as Quantum Chromodynamics, or QCD), for which a few weeks later Gross, Wilczek, and Politzer would produce their papers and eventually share the Nobel Prize.

But so many others could have done so if only they had put two and two together, and made the correct three. In 1965, eight years before Gross, Wilczek, and Politzer, in Russia, Vanyashin and Terentyev had discovered that beta is negative in such a theory and concluded that the result "seems extremely undesirable."[31] Not only this, but in 1969 another Russian, I. B. Khriplovich, performed a similar calculation, found beta to be negative, and left it without comment, failing to make any connection with the strong interactions, and this a full year after SLAC had made their discovery. Then in 1971 't Hooft had the result in his famous paper about electroweak interactions, and not only he himself, but anyone reading his paper carefully, could have made the connection. Astonishingly, no one did.

In the spring of 1973, a few weeks after 't Hooft and Parisi's wrong turn, Gross, Wilczek, and Politzer found that beta is negative and advertised this as the basis for a theory of the strong force. Three decades later, in 2004, the trio shared a Nobel Prize. Quite how these theorists independently settled on the correct—negative—sign for the beta slope has, however, been "fraught with ambiguity" ever since.[32]

THE WINTER'S TALE

Princeton I

Frank Wilczek's undergraduate specialty had been mathematics. When he began his doctoral studies at Princeton in 1972, gauge theories were having a renaissance following 't Hooft and Veltman's breakthrough, and Wilczek saw that this field was ideal for him. First, it was a new field, so he didn't have to know much of its physics history, and—most significant—it was mathematical, based in the group theory techniques with which he was familiar. He became immersed in it right away, giving a series of lectures about Yang-Mills theories to his fellow graduate students in the fall of 1972.[33]

Wilczek's initial interest was abstract mathematics, to see if the new insights in quantum field theory could be applied to understand what occurs in the limit of perfect resolution.[34] The experiments at SLAC had resolved distances nearly 1/100 the diameter of a proton, and Wilczek's supervisor, David Gross, meanwhile, wanted to see if the empirical phenomenon of Bjorken-scaling, which the experiments had revealed, could arise in quantum field theory in such a case. Gross's intuition at the time was that it could not. This would have implications later, as we shall see.

Wilczek told me that using the mathematical techniques of that time, the analysis was very complicated, with a lot of algebra. He did the calculations over and over, gradually building up the edifice. First, he did it roughly, calculating Feynman diagrams without worrying about numerical factors or whether the overall sign was positive or negative. Gradually, he began to understand what was involved and was able to map the way through the labyrinth. Finally, he was confident that he both knew the sign of beta and that his result was gauge invariant.

At some point, Wilczek learned that a student of Sidney Coleman's, at Harvard, was also working on this problem. It was not until later, Wilczek recalls, that he met the student—David Politzer—"when he came to Princeton to compare results to ensure that we weren't making contradictory claims."[35]

Harvard I

David Politzer first became interested in Yang-Mills theories during a drive from Harvard to Hoboken, New Jersey, en route to a conference in 1970. He was with Erick Weinberg (no relation to Steven Weinberg), who was a year ahead of him in graduate school and "something of a mentor." The journey took several hours, during which Politzer asked Weinberg to tell him about Yang-Mills theories as he had heard the name but "was otherwise ignorant." Weinberg explained the basic ideas, adding that nothing was known about their consequences and that many leading theorists had become very confused about them.

In 1972 Politzer visited SLAC during the summer. At that time SLAC—where I was working—was at the center of particle physics, and visitors came and went all the time. So the visit of a graduate student from Harvard made no impact on me—it would only be later, in 1978, when I met and interacted with him. I did recall the presence of Sidney Coleman, his supervisor, a remarkable scientist, whom I had already met, and of Tony Zee, who was well into his own search for a theory of the strong interaction.

Excitement about the strong force was intense that summer. The corridor of the SLAC Theory Group is famous for its chalkboards, at which physicists freely discuss, enabling spectators to join in. This freedom stimulates ideas, sometimes consciously, leading to formal collaborations, at other times subconsciously. Zee talked with Coleman, telling him how none of the theories that he had examined were able to explain the phenomenon of Bjorken-scaling because they all had "beta positive." Zee told me that Coleman was encouraging, saying, "Keep looking."[36]

Whether this inspired Politzer, even subconsciously, or not at all is long forgotten. In any event, six months later, when Coleman was absent on sabbatical, Politzer decided he needed a research program "that might not meet Coleman's high standards but on which I might have some chance of success."[37] As we shall see, he was certainly successful, because his research led to his very first paper, which in turn led him to a share of the Nobel Prize. However, Coleman's choice of venue for his sabbatical—Princeton, where David Gross was thesis supervisor to Frank Wilczek—unwittingly led to a controversy that has haunted this episode for years.

Early in 1973 Politzer visited Coleman "a couple of times in Princeton" and asked him if he knew whether anyone had calculated the beta function for Yang-Mills theories. Coleman thought not and suggested that they "should ask David Gross down the hall. David said no, and we [Gross and Politzer] discussed briefly [how to do the calculation]."[38]

Politzer's memory has himself physically present at this meeting with Gross, with no hint that anyone at Princeton was already actively pursuing this goal. Gross's memory differs, both about the circumstances and the time scale. In his Nobel Lecture, Gross recalled that when Coleman asked: "I told him we were working on this," Coleman then "expressed interest" because his student Politzer was pursuing a similar course.[39] He also recalled that the calculation was well under way, as he and Wilczek "both kept on making mistakes," the pace picking up "in February," and they "completed the calculation in a spurt of activity."[40]

Politzer confirmed to me that when he first talked to Coleman about this, he received no indication that Coleman was aware of any activity at Princeton.[41] He recalls that he and Coleman went together to David Gross, which was the first time that Politzer's interest became known. Politzer had not at that time actually started calculating the sign of beta; he had merely described to Coleman his thoughts that it might be a good thing to do.

His colleague Erick Weinberg told me that he remembered Politzer returning from a visit to Coleman at Princeton and saying that he planned to see if Yang-Mills theory has beta negative. Weinberg told Politzer that he also was calculating the quantity, beta, in his thesis, and later they compared notes. Weinberg confirmed to me that, in his memory, there had been no mention that any similar activity was taking place at Princeton at that time.[42]

Princeton II

I asked Frank Wilczek how he finally reached the goal of determining the sign of beta. He recalled that he had made many calculations in different gauges and checked his algebra by comparing the expressions for different Feynman diagrams, which had common features.[43] After a lot of filling up pages and getting things straight, he almost had everything agreeing, but not quite. One hiccup was that he had taken on trust one of the standard references, which turned out to have a sign error in one of its rules for calculating Feynman diagrams.[44] Finally, he had everything consistent and complete. It was just a question of the overall sign. At this point he went to David Gross and said, "It looks like it's negative."[45]

Wilczek laughed as he admitted that he had no prior intuition as to what the result would be. His discovery that the answer for beta was the negative sign was, at the time, "no more or less likely than if it had been the other result." However, after several years chasing for an understanding of Bjorken-scaling, and a field theory that would explain it, David Gross had begun to suspect that the counterintuitive behavior (of beta negative) was impossible in any field theory and that the explanation for the strong force lay elsewhere.

When Wilczek presented him with his negative-beta result, Gross therefore needed to be convinced. Wilczek continued with his reminiscence: "David pointed to this and pointed to that. . . . 'Are you sure of this, are you sure of that?' I said okay, I'm not actually totally sure."[46]

Somehow at this juncture Gross decided that the sign of beta is positive—in line with what had been found in other field theories, and, unfortunately, opposite to what was needed for a viable field theory of the strong interaction. Whether this was because the sign calculated by Wilczek was indeed positive, as in Gross's memory of events recorded in his talk at the Nobel ceremony,[47] or due to misunderstanding between the two of them, as in Wilczek's memory as told to me, there is no doubt that at this juncture Gross believed beta to be positive, began to draft a paper to that effect, and Sidney Coleman was informed.

Harvard II

At Harvard, as we have seen, Erick Weinberg was also investigating an aspect of this problem in his own thesis, but in a different context. Politzer, with help from Weinberg,[48] succeeded in calculating beta and found it to be negative.

Politzer was "excited over the possibilities" about its being a theory of the strong interactions, and phoned Coleman. Coleman said it was interesting but must be in error, as "David Gross and his student had completed the same calculation, and they found [the sign to be positive]."

Weinberg never had any doubt that the sign was negative. Politzer had calculated a negative sign, and Weinberg in his own thesis work had found an independent example with beta negative.[49] Politzer, however, was somewhat taken aback. He felt that Coleman had more faith in a team of two, involving a seasoned theorist, rather than in a single student, so "I said I'd check it yet once more."[50]

Princeton III

Meanwhile, Wilczek had given Gross a written summary of all the calculations. Gross started checking through Wilczek's report, in the course of which he realized that the sign was indeed negative, namely, the phenomenological property for which he had been looking but had doubted was possible. From this point Gross and Wilczek were agreed. Wilczek summarized this confusion: "There was an interval of perhaps a couple of days when it was up in the air and I wasn't confident to stand up and defend it."[51]

David Gross now told Sidney Coleman the new conclusion.

Harvard III

Politzer called Coleman "about a week later,"[52] and said that he could find nothing wrong with his calculation. Weinberg recalls that it was during the Harvard spring break[53] when Politzer checked his calculations, while visiting a cabin in Maine. This is at least consistent with Politzer's memory[54]—both that it rained a lot and of the amount of time that had elapsed.

If correct, it may also provide a time frame within the confusing memories of what happened next. The spring break was, until recently, the last week of March.[55] If that was the case in 1973, this was March 24 to April 1, and consistent with Weinberg's memory that early in April the issue was settled.

The problem is that when Politzer told Coleman that he had checked and was certain that the sign of beta was negative, Coleman said that he agreed, adding that he knew because in the interim "the Princeton team had found a mistake, corrected it, and already submitted a paper to *Physical Review Letters*."[56]

In trying to understand the chronology and time scales of the events, I asked David Politzer whether Coleman had said "already" submitted or that Gross and Wilczek "were [i.e., in the process of] submitting" a paper. After nearly four decades his response shows the traumatic shock that he must have felt: "At that point I would not have noticed the difference between 'were submitting' and 'had submitted.' [Today] I couldn't swear which it was, although the 'had submitted' better carries the sense of what it felt like. I was being scooped!"[57] When Politzer heard that the Princeton team had "already" submitted a paper, he rapidly produced one himself.

Writing a paper is an art developed over years of experience. We saw in Chapter 4 how Abdus Salam, who later became the author of fluent articles, was advised to redraft one of his first papers, as it was "badly written." Politzer, with his supervisor absent, had to go it alone, under pressure:

"It was my first paper and I was proud," he told me. The result was that it was too long for *Physical Review Letters*, the leading American journal with major international impact. Politzer recalled sending it elsewhere—possibly to *Physics Letters*,[58] a European journal, which had the advantage of a less severe page limitation than *Physical Review Letters*, though with the disadvantage of reduced impact for his work. He handed out photocopies. He was dismayed and discouraged that he was being scooped and in danger of being overlooked, but his colleague Roman Jackiw, with his experience as a professor, proposed that Politzer simply cut down the long version into a short one and submit it to *Physical Review Letters*.[59] This he did, where it was received on May 3, a week after the paper by Gross and Wilczek.

ALL'S WELL THAT ENDS WELL

I referred to the saga of the sign of beta as a comedy of errors. The events involving Gross, Wilczek, and Politzer during the winter of 1972–1973, where I have focused on who said what and when, might, to continue with Shakespearean metaphor, seem to be much ado about nothing. However, for years until they shared the Nobel Prize, and since, there has been conjecture about the history, including suggestions of skullduggery. Politzer's Nobel address hints at tensions beneath the surface and differs in several respects from the version in Gross's talk. I have therefore tried to assemble the evidence in order to suggest a possible chronology, in an attempt to deconstruct some of the conflicting versions.

That Politzer had the negative value of the sign right, and that within the Gross and Wilczek duo it was believed to have the opposite—positive—value, at least at some point along the way, is not disputed. What has remained unclear are the circumstances: whether David Gross changed his opinion on the sign as a result of intelligence from Sidney Coleman's innocent mention of Politzer's result, or whether the correct negative sign was found at Princeton independently. Even Gross and Wilczek have conflicting versions on how they settled on the negative sign, Wilczek insisting that he found beta negative but was unable to satisfy

Gross's questions, whereas Gross has claimed that Wilczek had its value as positive, Gross subsequently finding the error and correcting it, while drafting a paper.[60] The difference between Gross and Wilczek's accounts is one of detail, and owes nothing to Politzer. Unfortunately, there is no independent documentation of this.[61]

So what can we establish?

There seems little doubt that both Erick Weinberg and Politzer already knew the sign of beta in mid-April. By April 1973, Weinberg had completed his thesis, and in passing displayed a negative value for beta in a similar, though different, example.[62] The acknowledgments in Weinberg's thesis make an enigmatic remark that Politzer "helped [him] learn that two people working together can calculate Feynman diagrams which neither can calculate alone," Politzer's own paper in turn making a reciprocal acknowledgment. Weinberg's thesis defense was on Friday, May 11, 1973.[63] He was not sure when the thesis was submitted to the department, although he did "recall that it was a Monday, and that the department secretary would not schedule a defense until the thesis was in her hand, so it was presumably mid- to late-April." Thus, the possibility that beta can be negative in Yang-Mills theory was known in Harvard several weeks before the end of April.[64] This is consistent also with Politzer's account, supported by Erick Weinberg, which has him checking his negative sign during the spring break, *after* having been told by Coleman that the Princeton duo had beta positive.

Politzer calling Coleman "about a week later" brings us to mid-April, at which Coleman apparently tells him that Princeton had "found a mistake" and submitted to *Physical Review Letters*. In fact, the manuscript for Gross and Wilczek's published paper was received by the editor on April 27, Politzer's arriving on May 3. The gap between these two is understood: Politzer is agonizingly drafting a paper, aided by Roman Jackiw, after learning about his rivals' submission. However, if this chronology is correct, there appears to be an interval around the third week of April, which is as yet unexplained. This is my interpretation.

The chronology seems consistent with the possibility that the Princeton pair were vacillating about the sign of beta at the start of April, around the time that Politzer was checking his result during the spring break. Whether Sidney Coleman had already made anyone at Princeton aware of Politzer's work at this stage, even subliminally, I don't know. But it seems that it was at this time when Princeton settled on the correct—negative—result.

This matches what Wilczek recalled. He said to me that after telling David Gross his result (that beta is negative) there was an interval of "perhaps a couple of days when it was up in the air when David doubted it and I wasn't confident enough to defend it. During that period he talked to Sidney Coleman and said that we had done the calculations. The stories that we had got the wrong sign are simply that David had told Sidney [that the sign was positive]."[65] But when the reality became clear—that beta is negative, and hence here is a solution to the strong interaction—Gross told Coleman the conclusion.

When on returning from spring break Politzer contacted Coleman, he learned of Princeton's favoring of beta being negative, in agreement with what he has believed all along. Then, possibly during the third week of April the Princeton write-up is put on hold, or at least an insurance policy is developed, with Gross, Wilczek, Politzer, and Coleman all meeting to compare notes.

Wilczek's memory is that "a short time" after he and Gross agreed on the final result, they all communicated before sending off any papers, and that he first met David Politzer "when he came to Princeton to compare results to ensure we weren't making contradictory claims. This I absolutely remember."

During this, Wilczek discovered that although Politzer had the overall sign correct—negative—he had an error in one of the contributions, known as the "fermion loop." This did not affect the overall conclusion.[66]

Wilczek told me, "Sidney Coleman, the two Davids, and I were in David Gross's office comparing the calculations. David Politzer had the opposite sign for the fermion loop. Sidney said, 'These two guys who actually did the calculations should go off in another room and figure out what the right answer is,' and that's what we did. It was just about ten minutes. It was before the papers were written but not long before. All the central results were there."[67]

This seems to me to be utterly sensible. Sidney Coleman was in an embarrassing position: His student back at Harvard had made a calculation during Coleman's absence, which was also being done by two colleagues at Princeton, the institution where Coleman was then spending his sabbatical. There had been doubt, at least in Gross's and Coleman's minds during the previous days, about what the correct answer was. Ensuring that everyone agreed was paramount—for settling their scientific curiosity no less than for pragmatic reasons: The stakes were high and the potential for egg on the face huge.[68]

The two papers were submitted as April turned to May and published back-to-back in the same issue of the journal.[69] As one student, Gerard 't Hooft, had made a Nobel Prize–winning discovery by solving the enigma of the weak interaction in his first papers and thesis, so now had two young American students done the same for the strong.

B. J.'S CHARM

In his inspirational 1966 paper, which had led to the discovery of quarks in the proton, Bjorken had made another startling prediction. When an electron meets its antimatter doppelganger, a positron, there is a chance that they mutually annihilate, creating a momentary flash of energy out of which new forms of matter and antimatter can emerge. As the latter have no memory of the electron and positron that created the conditions for their birth, "electron-positron annihilation" proved to be a novel way of creating forms of exotic matter, hitherto unknown on Earth.[70] Bjorken predicted that the chance of electron and positron annihilating one another is inversely proportional to the square of their energy: Double the total energy of the pair, and the chance of annihilation falls fourfold. This simple relation was another example of what has become known as "Bjorken-scaling,"[71] and by 1972 this prediction had been verified also. Furthermore, the data agreed best with Bjorken's theory if each flavor of quark—up, down, or strange—carried three varieties of color. So the annihilation of electrons and positrons, combined with Bjorken's ideas, had produced some of the best evidence for this enigmatic property, which is the seed of the strong interaction. Bjorken's prediction, which had already led to the discovery of quarks within the proton, now helped to establish their colorful nature.

In 1973, SLAC built a machine, called SPEAR, which could collide electrons and positrons with energies larger than before. This immediately produced a surprise: Bjorken-scaling—the smooth variation of annihilation with energy—failed. Instead of continuing to fall as the energy increased, the probability appeared to stay the same for a small range of energies.[72]

By this stage, Bjorken-scaling had achieved such a pedigree that its apparent overthrow demanded explanation. Several months passed before the puzzle was solved: SPEAR had just enough energy to create a hitherto unknown fourth variety of quark—charm—whose unrecognized appearance was masking Bjorken-scaling.

The revolution began on November 10, 1974, with the discovery of the first example of the charmed landscape: a particle—called the J-psi—formed from a charm quark and a charm antiquark.[73] After this needle had been separated from the haystack, and production of other novel charmed particles also allowed for, Bjorken-scaling was seen to hold true once again.

Bjorken's prediction applied in situations where the energy in the experiment is much greater than the rest mass—the "$E=mc^2$"—of any particle involved. This had been the case when SPEAR was creating particles made from the lightweight up, down, and strange flavors of quark, but the massive charm quark, previously unseen, was being produced almost motionless, which ran counter to the conditions required for Bjorken-scaling. Thus, it was the resulting failure of Bjorken-scaling that led, indirectly, to the discovery of charm.

Charm had been proposed by Glashow, Iliopoulos, and Maiani in 1970 to expunge the strangeness-changing neutral weak processes (see page 254). Within the two years following the discovery of the J-psi the whole landscape of particle physics came into focus. Several examples of charmed particles—where a charm quark combines with an antiquark of the up, down, or strange variety—were isolated at SPEAR, and their properties turned out to be exactly as required by GIM. This breakthrough showed that electroweak unification works for hadrons: There was no longer any need for Steven Weinberg to restrict his model to leptons. With charm confirming the GIM mechanism, even the most ardent skeptics had to admit that the electroweak theory agreed with the phenomena.

The discovery of the J-psi, furthermore, turned out to be seminal for understanding the strong interaction. Known as the first example of "charmonium,"[74] the J-psi would have been little more than a curiosity had it not been for an astonishing property of the beast: It lived thousands of times longer than most would have expected.

Being made from a quark and an antiquark, the J-psi survives only so long as these two do not meet and annihilate. By 1974 many examples of such hadrons—known as mesons—had been found, built from up, down, or strange quarks and antiquarks. In all cases they are unstable: The strong force attracts quark to antiquark, whereupon they annihilate one another. Relative to the attraction of up, down, or strange flavors, the mutual affinity of the more massive charm quark and antiquark within the J-psi is less, much less. This enfeebling of the strong interaction for the massive charmed flavor enables the J-psi to survive for longer. The phenomenon turns out to be a remarkable triumph for QCD.

According to the theory of QCD, the strong force is enfeebled at very short distances—quarks appear free inside the proton. The equations imply that the same impotence should occur at high energy, and as Einstein tells us that energy can be exchanged for mass, this should occur when particles with large masses are involved. The charm quark is much heavier than the up, down, or strange; the force binding it is thus less potent, and charmonium survives longer than would otherwise have been the case.

So in the space of a few years, the appearance of charm had profound implications for understanding both the electroweak and the strong forces. First, by confirming the GIM mechanism, it showed that the electroweak theory applies to hadrons. And the anomalous longevity of the J-psi, along with other examples of charmonium states, also provided convincing evidence that QCD is the correct theory of the strong interaction. The fact that charm was exposed in the debris of electron and positron annihilation was due to Bjorken's prediction of scaling.

Thus, in addition to Bjorken's insights having led to the discovery of the quark layer of reality, they had now inspired crucial discoveries about the electroweak and strong forces. There would be a further consequence of Bjorken's genius: the discovery of the *W* and *Z* bosons, which were the smoking guns of electroweak theory. They had still to be found.

Intermission: 1975

It is 1975. The strong interaction has been solved.

Quarks have been verified as the seeds of the proton and other strongly interacting particles. This has led to a viable theory of the strong interaction built on the fundamental quarks and gluons: QCD—Quantum Chromodynamics—a Yang-Mills theory like that describing the electromagnetic and weak forces.

The existence of charm quarks has been established, which provides a major piece in confirming the theory of the weak interaction.

The strong interactions of charm also help to verify the predictions of QCD.

The theory of electromagnetic and weak forces, marrying them into a single "electroweak" force, remains beautiful mathematics, awaiting full experimental confirmation.

Chapter 14

HEAVY LIGHT

The electroweak theory is tested experimentally. The prediction of a neutral weak interaction is confirmed. The Higgs Mechanism and Higgs Boson enter the lexicon, along with the Weinberg-Salam model. The neutral weak interaction is found to violate parity. In 1979, Glashow, Salam, and Weinberg share the Nobel Prize.

By 1973 viable theories of the strong, electromagnetic, and weak interactions were to hand thanks not least to the maiden papers of three students: Frank Wilczek and David Politzer in the case of the strong interaction and Gerard 't Hooft in the case of the weak. 't Hooft's discovery of a viable theory of the weak force had inspired Ben Lee, who in 1972 reformulated the arguments in ways that were easier to understand, completed some technical details himself, and became hugely influential in making the world aware of the profound implications of the breakthroughs.[1]

At the Sixteenth Rochester Conference in Chicago, in 1972, Lee's talk set the scene for much that followed. He mentioned the Higgs Mechanism by name, so much so that a colleague of Higgs came back to Edinburgh from the conference and told him "Peter, you're famous."[2] Whatever impact Lee's mention of Higgs at the Berkeley conference in 1966 may have had, from 1972 certainly Higgs's name became coupled with both *mechanism* and *boson* in the world of particle physics.

There was a powerful belief in that community that the mathematical breakthroughs had exposed the way that nature actually functions. Nonetheless, the marriage of the electromagnetic and weak forces into a single "electroweak" force awaited experimental confirmation. Attention now turned toward testing the theory.

NEUTRAL CURRENTS

Although the *W* was predicted to be about ninety times more massive than an atom of hydrogen, and thus out of the reach of experiment for some while, the theory made another prediction, which would not prove unification of weak and electromagnetic interactions directly, but nonetheless would be an early indicator of whether nature really worked this way. The implication of the "electroweak" marriage was that in addition to the electrically neutral and massless photon, there should also exist an electrically neutral but massive *Z*, like a photon in many respects but for two critical differences. One, of course, is that the *Z* too is massive—I described the *Z* in a popular article, twenty-five years ago, as "Heavy Light." The other was that the *Z* would give rise to a new form of the weak force.[3]

Its sibling, the electrically charged *W*, not only alters the motion of particles but also moves electrical charge around, changing one variety of particle to another as a consequence. The *Z*, however, will transmit energy from one particle to another but cares naught for electric charge. The consequence of this is that the *Z* can affect the motion of electrically neutral particles, such as neutrinos, by what is known in the jargon as the "neutral current."

Neutrinos are like electrons but without electric charge. Highly elusive, they pass through the universe almost as if they are mere spectators, so shy that it is remarkable that we know of them at all. If the *Z* did not exist, neutrinos would be unable to bounce off anything.[4] So the challenge was obvious: See a neutrino scatter from something, and the force of the *Z* would be revealed. This was easier said than done, however, stretching technology and ingenuity to their limits. Just imagine, for a moment, what you are trying to observe: An invisible neutrino comes in, bounces off a proton, say, and an invisible neutrino flies out.

The clue that this had happened would be analogous to the way that the hero in H. G. Wells's tale *The Invisible Man* was revealed—by jostling the crowd. Fire a beam of neutrinos at some target, and look for an electron or a proton in that material recoiling as if by magic. A team at CERN in 1973 had evidence for this, and hence that the *Z* was at work, and announced this at a conference during the summer. A rival group in the United States also reported the same phenomenon at that conference, and so the evidence looked good.

However, later that year the American team modified their apparatus, and to their horror, the signal disappeared. They backed off from their claim and announced instead that they had no evidence for the neutral current. This made the CERN group worried, and for several months many were convinced that the CERN physicists were wrong. Nonetheless, they answered all criticisms, and subsequent experiments have confirmed the phenomenon without any doubt. It is now generally agreed that the team at CERN had indeed been the first to see the neutral current.

A problem was how to be certain that any recoiling electron was due to its being hit by a neutrino, as against having had a chance encounter with some other neutral particle, such as a neutron, which had contaminated the beam. The contamination was very small, but as the neutrino interaction is itself very rare, there was the chance that the handful of examples might have had nothing to do with neutrinos at all. This was part of the reason for uncertainty in the United States and CERN.

A way of filtering these is to use the fact that neutrons don't travel very far before interacting with something, whereas neutrinos pass through easily. Events due to neutrons therefore tend to be relatively rare toward the back of the detector, downstream from the beam, whereas neutrinos would have the same chance everywhere. The large bubble chamber "Gargamelle" at CERN found examples of neutral events spread throughout the detector, in accord with what was expected for neutrinos and unlike that for neutrons.[5]

So the first piece of evidence for the marriage of weak and electromagnetic interactions was now in place, and speculation began about an eventual Nobel Prize. Four people—Glashow, Salam and Ward, and Weinberg—could lay claim to the SU2 × U1 theory, while 't Hooft and Veltman had combined to show that it is mathematically viable, leading to the modern theory of Quantum Flavordynamics. At most three people can share a Nobel Prize; maneuvering had already begun.[6]

When 't Hooft and Veltman proved in 1971 that the SU2 × U1 model is renormalizable, soon followed by the discovery in 1973 of the neutral weak interaction, the physics community began to take notice. At that stage I was a young postdoctoral fellow at CERN, and although my specialty was quarks and the strong interaction, I became aware of a new term describing a unified electromagnetic and weak force: *electroweak force*. At seminars, the conjunction *Weinberg-Salam* and the label *Weinberg angle* were the new terms in the lexicon.[7]

For the model to succeed, one had to invoke the GIM mechanism (see page 254), which predicted the fourth variety of quark, known as "charm." John Iliopoulos, one of the creators of the GIM mechanism, was so confident that in his talk at the seventeenth conference of the Rochester series—held in London in 1974—he made a wager. In order to encourage speakers to keep to the allotted times, the conference organizers had presented a bottle of wine to each who did so. Iliopoulos, who happily received one, declared, "I am prepared to bet a whole case of wine that the next conference in this series will be dominated by the discovery of [charm]."

Two years later, in 1976, I spoke at that very conference, in Tbilisi, USSR. The organizers did not offer any bottles of wine; the highlights included sight of Andrei Sakharov sitting in the back, as an unofficial delegate whom the Soviet authorities would have preferred remained hidden. But in physics, that conference was remarkable. On my return, I wrote my first popular article about particle physics for *Nature*, headlined "Iliopoulos Wins His Bet."[8] Charm had been discovered, and the implications of this had indeed dominated the conference.

With the discovery of charm, the "electroweak" theory was consistent with everything known: Neutrinos had been seen to bounce—as expected if the Z^0 was at work—and strangeness-changing-neutral processes were killed by the destructive presence of charm. The actual production of the still hypothetical *W* and *Z* lay outside the reach of any accelerator of the time, but there was a growing awareness that here was a breakthrough of Nobel quality.

In my report in *Nature* of that conference I wrote about the remarkable amount of data that had emerged in the previous two years, in particular the discoveries of the neutral weak interaction and charm. Then, echoing the mantra that was typical of the time, I wrote that these meshed with the "original unification idea of Weinberg . . . and Salam," citing Weinberg's 1967 paper and a talk by Salam in 1968 titled "Elementary Particle Physics," which had been published in the proceedings of the Nobel symposium.[9] I further added that almost all of the data on weak interactions reported at the conference agreed with the "Weinberg-Salam" theory. This included some hints from an uncompleted experiment that the predicted weak neutral force indeed violated parity—mirror symmetry— in a manner predicted by the eponymous theory.

I have long remembered this, not just because it was my first article in *Nature*, famous for the breadth and depth of its scientific readership, but because of its reception: I received a note from Abdus Salam, saying how much he had enjoyed it.

THE WEAK LEFT HAND

By this stage the one accessible missing link was a clear demonstration that the neutral force violated parity in accord with what the theory predicted. The source of this parity violation is as follows.

In electromagnetic theory, an electron and a proton interact by exchanging photons. This process conserves parity, and hence is equally probable for electrons corkscrewing left or right-handed. In weak-interaction theory, neutrinos can turn into electrons and interact with protons by exchanging electrically charged *W* bosons. However, in the combined "electroweak" theory, two additional types of effect are predicted, see Figure 14.1.

FIGURE 14.1

THE PHOTON, *Z* AND *W* AND THE NEUTRAL WEAK FORCE

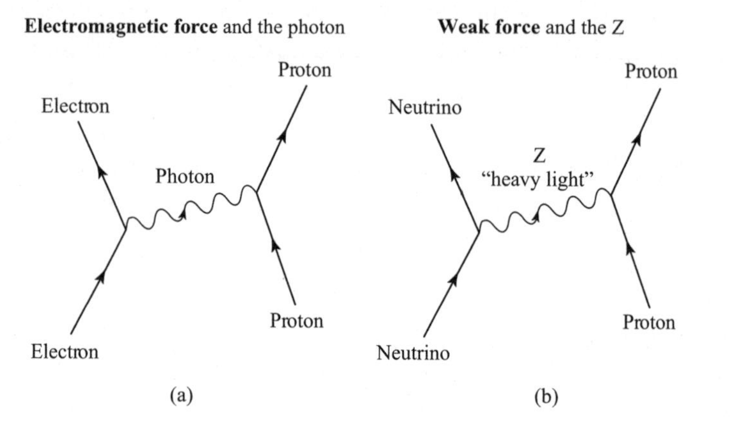

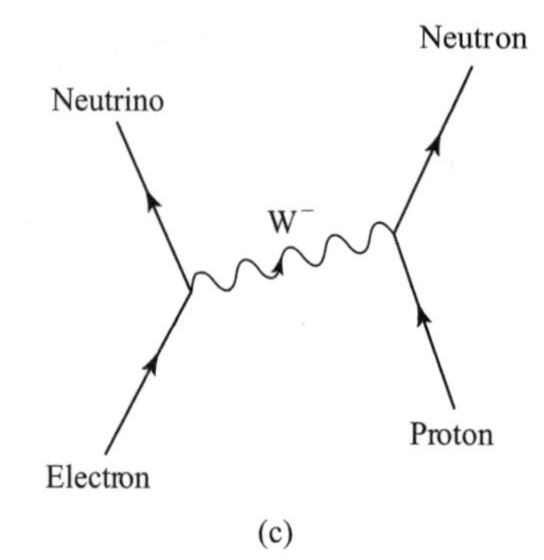

(a) Electron scatters from proton by exchanging a photon-gauge boson of the electromagnetic force. (b) and (c) show examples of the weak force. In (b), a neutrino scatters from a proton by exchanging a Z^0—an example of the weak neutral force, and in (c) the exchange of an electrically charged *W* gives the historically familiar force, as manifested in beta decay. In (b), the neutrinos may be replaced by electrons. In such a case the Z^0 can also be exchanged between an electron and a proton, giving a novel contribution to the behavior of electrons in atoms, or other examples that are traditionally thought of as purely electromagnetic.

In the first of these, as we have seen, neutrinos can remain neutrinos when they interact with other particles by exchanging the Z^0. This interaction violates parity because neutrinos are left-handed. That this reaction occurs had been first demonstrated in 1973, but was such a delicate phenomenon that proving that it occurred at all had been at the limits of practicality. Measuring whether parity was violated was beyond capability.

Just as the theory predicted that a Z^0 can couple to the (left-handed) neutrino, so it predicted that the Z^0 couples to electrons. This is the second effect referred to above. You may be anticipating that the Z^0 couples only to left-handed electrons; this is not quite correct. It is the case, however, that it does not couple to left- and right-handed particles the same. The actual preference for the one or the other is expressed in the parameter that by 1973 was being increasingly referred to as the Weinberg angle.

If the neutral weak force indeed violated parity, it would affect the properties of atoms. Electrical forces, which bind electrons in atoms, do not distinguish left from right. However, the neutral weak force also disturbs the motion of electrons in such a way that subtle effects of parity violation should appear in atoms: in such a situation the behavior of atoms viewed in a mirror would differ from what can happen in reality. Several experiments looked for this asymmetry, with conflicting results. For some time there was no clear support for what the theory predicted, and there was a possibility that the theory would be overthrown.[10] This led to my second article in *Nature*, on December 9, 1976. Titled "Parity Violation in Atoms?" it again mentioned the Weinberg-Salam model, concluding, "The clear blue sky of summer now has a cloud in it. We wait to see if it heralds a storm."[11]

The editor of *Nature* expressed his excitement and said he looked forward to further news. I don't recall receiving any other reactions to that article. However, the storm never came, though a couple of years were to pass before the all clear.

As parity is violated in the weak interaction when an electrically charged W is involved, it is not surprising that a marriage of weak and electromagnetic would lead one to expect that interesting parity violations should be found in processes involving the Z^0 but that are superficially electromagnetic. The issue was finally settled by an experiment at SLAC.

The electron accelerator at SLAC was able to produce beams of electrons that corkscrew left-handed (like a neutrino) or right-handed. According to the electroweak theory, there should be a marginal preference for left-handed electrons to interact.[12]

When this experiment was completed in 1978 it confirmed the phenomenon.[13] The left-handed excess was only one event in every 10,000 or so, but this was a large effect within the sensitivity of the experiment. Consequently, here was a new way of distinguishing the real from the mirror world: There are both left-handed and right-handed electrons, but it is the former that marginally prefer to interact in reality.

I concluded that this observation of the left-handed excess, and with this very size, pointed "clearly to the Weinberg-Salam model as being the correct model for unifying the weak and electromagnetic interactions."[14] By this stage—July 1978—the twinning of "Weinberg-Salam" had become imprinted throughout the field, and I had adopted this mantra. By now almost everyone was happy.

GLASHOW AND NOBELITIS

With the SLAC experiment confirming the existence of a neutral weak force, which violated parity precisely as the electroweak theory predicted, all pieces of the SU2 × U1 construction had been established. There was as yet no sign of either the *W* or *Z* themselves, nor any test of renormalization or the hypothesis of spontaneous symmetry breaking. But with the successful prediction of the neutral weak force, many suspected that the 1978 Nobel Prize would celebrate the creation of the electroweak theory. That summer the successor to the 1976 conference, which had been dominated by charm and set the circus moving, took place in Tokyo.

I was responsible for organizing discussion on the Quantum Chromodynamics theory, which by 1978 had rapidly become established as an essential piece of the new paradigm. In the week before the conference, the speakers were hosted in Kyoto while preparing their contributions. I shared an office with David Politzer, who had already prepared his and advised me to spend my time touring Kyoto's many temples rather than spending time in the office. It was during those days that I first began to appreciate that the history of the discovery of asymptotic freedom—the crucial negative sign of the beta function in QCD—had not been without controversy. In 1978, any thought of a Nobel Prize for this idea was on few people's radar screens; the possibility that the electroweak theory would soon be honored was, however, very much on the agenda.

Having followed the developing history closely, and reported on it in my articles for *Nature* during the previous two years, I was interested in how the conference would assess the landscape. However, I was then naively unaware of the politics concerning the summary talk. This, the closing talk of the leading conferences in the field of particle physics, is traditionally given by one of the leading physicists of the day. The electroweak theory was at center stage: Aware of the delicate symmetry, the organizers scheduled what amounted to two summary speeches, one on the subject of unification by Abdus Salam, and another on the weak interaction by Steven Weinberg.

This led to my third successive report for *Nature* on these conferences.[15] A third of a century later, I am surprised at how little of the landscape, as it then appeared and was described in my account, has changed. Nonetheless, in the sensitive climate of the time, some found my article "disturbing."[16]

By this stage, several were becoming nervous about their places in history. At the time I knew nothing of Glashow's early paper, which had been the first to present the SU2 × U1 structure, with its prediction of the weak neutral force and the Z^0, and already contained what corresponded to the Weinberg angle. Glashow himself was not at the conference, having chosen to spend the summer with his family in Aspen.[17] In his memoir, written in 1988, he recalled that he was "amazed and disconcerted" when he saw a transcript of Weinberg's talk. Glashow described it as a "revisionist history" of a field in which "[I] thought I had played something of a role."

While it is true that Glashow was not cited, neither was there specific mention of "Weinberg-Salam." Weinberg's talk was less a summary of the implications of the latest results—other talks in the conference covered that—than a discussion of ideas on where to go next. Glashow's reaction probably reveals as much about the nervousness that can develop when the Nobel Prize is at stake as it does about the talk itself.

Scientists who have made major discoveries covet the prize while attempting to suppress the feeling. Whereas scientists may be unaware who among their colleagues are in the national academy of their home country, or be uncertain of who has won various international awards, such as the Wolf, Sakurai, or similar prizes, everyone knows who the Nobel laureates are, such is the unique prestige of that award. While most awards are readily described in the media as "prestigious," you never see or hear reference to the "prestigious Nobel Prize": The adjective is redundant.

The concept of "Nobel Prize Runner-Up" adorns student T-shirts on university campuses. Among the faculty may be found those who feel that they qualify to wear the shirt by right, and could have been a winner—"if only." "If only" they hadn't delayed publishing their big idea, or had joined in a successful collaboration, or had publicized their own ideas more vigorously. At any time there are ideas, which are widely accepted of being of Nobel quality, but have not yet been chosen for the prize itself. Announcements are made in October. As the date approaches there is speculation among the crowd and tension for the contenders. Winners' stories of their behavior on the night of the award are publicized; the reactions of those whose hopes are dashed are surely little different, except that their secret tends to stay with them. What Glashow called the disease of "Nobelitis"—the psychological condition that takes over the minds of scientists who see the finishing line of the Nobel race coming into their view—is a highly selective epidemic in the months between the major summer conferences and the announcement of the Nobel Prize.[18]

In September 1978, Glashow was invited to give a talk at a small meeting organized by the Swedish Academy of Sciences in Stockholm. Glashow recalls that the night before he was due to speak, he received a transatlantic call from Weinberg, who was in Harvard and spent an hour debating the content of Glashow's impending speech. In Glashow's assessment, "Steve [was] presumably also a victim of Nobelitis," and Glashow's visit to the "Nobel heartland" had made Weinberg seem "a bit distraught."[19]

Glashow was keeping an open mind and wanted to consider alternatives to their electroweak theory that might still be logically viable. He had told Weinberg, a lifetime friend and colleague, as well as rival, about his plans for the talk. Glashow sensed that Weinberg disapproved of this aspect of his talk, but Glashow felt that as the *W* and *Z* still remained to be discovered, there might still remain the possibility of surprises.

The phone conversation ended without satisfaction. "Steve was adamant. I was stubborn," Glashow recalled, also admitting that he was jet-lagged and rather drunk. After an hour, Glashow "hung up" on what he described as Weinberg's "interminable tirade."[20] (Weinberg's recollection of this differs.)

By late 1978 Glashow's work was already being noted by those who would make the decisions. During the meeting in Stockholm that had so concerned Weinberg, Glashow was approached by an eminent elderly Swedish physicist Ivar Waller.

Waller had been a member of the Nobel Physics Committee for more than twenty-five years, and up to 1970 he and Abdus Salam had interacted a lot.[21] By 1978 Waller had retired, but he remained active in physics and retained much influence. During what superficially appeared to be an innocent chat, it became clear to Glashow that Waller had Glashow's old papers more at his fingertips than did Glashow himself. Critically, Waller asked whether a parameter that Glashow had introduced in his 1961 paper was the same as what had become known as the Weinberg angle.

Glashow replied that he thought that the Weinberg angle, and the angle that he had introduced six years earlier, were perhaps different. Waller, however, assured him that they were, in fact, identical.[22] Suspecting that his name was now in the minds of the Nobel committee, Glashow, overcome with Nobelitis, noted the date when the announcement was due.

How do you learn that you've won the Nobel Prize? Many winners recall receiving a phone call from the king of Sweden, having been alerted by a secretary to expect it. This is not always 100 percent efficient, as in 't Hooft's case in 1999. The day that the award was announced that year, he was in Bologna, where Italian experimentalist Antonino Zichichi had invited him to give a talk and hold discussions. 't Hooft told me how "Zichichi had become more and more convinced of the standard model and kept pointing at me saying that I was one of the originators. So he was more certain than I was that a Nobel Prize would come my way. I was giving a talk about QCD and during it some students sneaked away to look at the Internet—at the Web page of the Nobel Committee." (This was before WiFi was the norm, enabling people to multitask during conferences.) "The Web, sure enough, showed the decision. So the students quickly made a transparency, and when I had finished my talk, Zichichi came in, put it on the viewgraph [projector], and announced that this was the last talk by a non–Nobel Prize winner." 't Hooft, however, was looking at the audience, not the transparency, and was unaware of what was on the screen. The audience started to applaud, which he assumed to be the standard statement of appreciation for the speaker, but "the applause lasted much longer than usual, and I began to think there was something going on. So I looked around and saw the transparency."[23]

The Swedish Academy had wanted to call him, but as he wasn't in Utrecht, they had failed to locate him immediately. He had said that he was going to Bologna, but not exactly where. "It took them about half an hour to locate me, but by that time journalists had managed to find me earlier, and so they [the academy] couldn't get through for quite some while."

Back in 1978, Glashow took a sleeping pill the night before the announcement, as the news from Stockholm breaks at dawn in Harvard, and Glashow feared that otherwise he would be awake all night. However, there was no phone call: The prize that year went elsewhere.[24] In 1979, however, Glashow and Weinberg duly shared the award, together with Abdus Salam. The citation read, "for contributions to the theory of the unified weak and electromagnetic interaction including the prediction of the weak neutral [force]."

Given the number of places that one sees Weinberg and Salam's names joined for incorporating the ideas of spontaneous symmetry breaking into the SU2 × U1 model that Glashow had originally invented, you would hardly expect that there is any doubt about the history. Yet when I delved into this while researching this book, the questions began to multiply. The contributions by Glashow and Weinberg are both outstanding and stand out; but the symmetry that had united Salam and Ward had been broken—the mechanism, truly hidden. Ward's omission has been a cause célèbre ever since.

Chapter 15

WARMLY ADMIRED, RICHLY DESERVED

Glashow, Salam, and Weinberg shared the Nobel Prize, but Ward missed out. This chapter assesses what separated Abdus Salam from his erstwhile collaborator. Salam's original nominations for the prize cite his work on neutrino and parity violation. The critical role of spontaneous symmetry breaking in Weinberg and Salam's work in 1967, the role of Tom Kibble and its significance for the Nobel Prize. What are the criteria for priority in discovery?

Whereas Glashow experienced an acute bout of Nobelitis toward the end of the 1970s, Abdus Salam had a chronic version spanning many years, which was assuaged in 1979 when he shared the award with Glashow and Steven Weinberg. This chapter focuses on Salam, how his name became linked with that of Weinberg and separated from Ward. Which leads to a bigger question: What separates Nobel winners from losers? More specifically, which category did Ward deserve to be in? Was Ward, as some believe, robbed?

When Lee and Yang won the Nobel Prize in 1957 for their work on the failure of mirror symmetry in weak interactions, Abdus Salam's contribution was bypassed. For nearly two decades unsuccessful efforts were made on Salam's behalf to have his work similarly recognized. The experience may help explain some of the issues surrounding Salam's eventual prize, which was for his part in the electroweak theory.

The story began in the early 1950s when Rudolf Peierls, who had already played a singular role in the Ph.D. examination of Salam's collaborator J. C. Ward, made a critical intervention in Salam's own Ph.D. viva.

Salam frequently told how Peierls had asked him if there was a fundamental reason that the neutrino was massless. Salam knew of none, at which Peierls responded that neither did he. That an examiner would ask a question to which he himself had no answer was the minor punch line in this story; more dramatic is what then ensued. The question nagged Salam for a long time, until one night on a transatlantic flight he had an insight: The neutrino would be massless if the equations describing its dynamics satisfied "gamma-5 symmetry." Furthermore, such symmetry would make a neutrino spinning like a left-handed screw act independent from one that was right-handed, and in general would spoil mirror symmetry—violate "parity."

This story became part of the Salam folklore. His 1957 paper[1] setting out this idea was cited in nominations that he be awarded a Nobel Prize[2]—on the grounds that his theory had preceded the definitive experiments proving parity violation, and that Lee and Yang "had not proposed a theory of parity violation, nor attempted to link it to any deep principle of physics."[3] Peierls, however was more sanguine. In a letter to Salam in 1982 he remarked that their discussion had focused on a question that had interested him for some time, and although Salam had been "the only person to show a serious interest in it," in Peierls' opinion Salam "took it to what seemed to me an unwarranted conclusion." Peierls declined to coauthor Salam's paper, as he felt the argument was not compelling. He pointed out that there was no reason to suppose that the neutrino is massless (and today we know that it is not), and as for parity violation, Dalitz had already advocated this as a possibility, which in turn had inspired the work of Lee and Yang.[4]

Whatever Peierls's opinion of the gamma-5 symmetry idea may have been, Salam believed in it passionately. In 1969 Salam even wrote at length about it to Ivar Waller, who had been a member of the Nobel Committee for Physics since 1947.[5] In a letter to his Imperial College colleague Paul Matthews in 1971, Salam wrote about himself in the third person: "Salam's greatest contribution is no doubt the modern theory of the neutrino," adding a covering letter with the instruction, "Paul, I think the same letter could go to Hulthen[6] [a former chair of the Nobel Physics Committee] and I now feel you should send it. In your letter you can enclose . . . my letter to Waller which for Hulthen must be changed to read as my letter to you (like last time)."

Notable is that Salam here described this to be his "greatest" contribution, and this in 1971,[7] three years after his Göteborg talk, which dealt with different issues and would form the nub of his eventual Nobel award. The nomination that Salam had prepared for Matthews to submit also cited his work with gravity as "giving perhaps the definitive resolution of the 70 year old infinity problem."

Whatever significance the "perhaps" may have, Abdus Salam clearly believed his gravity work to be primary when he spoke at Amsterdam in 1971, alongside 't Hooft. Nowhere does Salam appear to set any great store by his work on the weak-electromagnetic unification until after 't Hooft's grand entrance. From this juncture, repetition of the phrase "Weinberg-Salam model" raised this work to everyone's attention.

REPRISE: 1967

Weinberg's 1967 paper,[8] which led to his share of the Nobel Prize in 1979, has been cited more than 8,000 times. However, as Sidney Coleman had noticed,[9] this explosion of interest arose only after 't Hooft's appearance in 1971. Coleman looked at the number of citations as recorded by *Science Citation Index*, and found that it had received only one mention prior to 't Hooft.

Salam had referred to Weinberg in a talk published in the *Proceedings of the Nobel Symposium*.[10] These occasional gatherings of specialists in a small field have a standing similar to the Solvay conferences—useful in the history of science as a test bed for ideas, but not the normal archive for announcing major advances.

In 1968 Salam had given a talk there, in which he reviewed his work with Ward on the SU2 × U1 model, and added some comments that spontaneous symmetry breaking could provide the link to complete a viable theory. His paper had but three references, two of which were to his earlier work with Ward and the third to Weinberg's 1967 paper about these ideas. Later this talk at Göteborg became a springboard for Salam's share in the Nobel Prize.

There has been a history of controversy about this. At the actual proceedings, Salam's presentation made limited impact.[11] Furthermore, it was hardly new: Weinberg's creation was already on the record. There are few prizes for coming in second. Questions asked ever since include: Why is this given credit when Weinberg's work had been published several months earlier? Why is Salam's paper in such an obscure place? And how did he become separated from Ward and linked with Weinberg in general awareness after this juncture?

Salam's priority for discovering this golden path, independent of Weinberg, who published his theory in November 1967, is that he had already given some lectures at Imperial College during 1967, in which he outlined his ideas on spontaneous symmetry breaking. These lectures, which, as we shall see, probably occurred in October 1967, collectively form the germ of what became known in the 1970s as the Weinberg-Salam model. However, they made little impact on the audience, and there is no surviving record of their content.[12]

Tom Kibble, who was based at Imperial College in those days, was at the epicenter of the research about symmetry breaking. Following his work with Guralnik and Hagen on this topic in 1964, Kibble had completed his proof that the "Higgs" mechanism could be extended, in particular to Yang-Mills theories with an SU2 × U1 mathematical structure, and had told Salam about this early in 1967—probably in March.[13] The ideas of spontaneous symmetry breaking, as generalized by Kibble, formed a kernel of Weinberg's paper, and of Salam's claim.

In July 1967 Kibble left for a sabbatical, spending the summer at Brookhaven—as we saw in Chapter 10—and the autumn of that year in Rochester, New York.[14] Consequently, he missed Salam's lectures.

Chris Isham, who became one of Salam's research collaborators, recalled having heard "some lectures on electroweak theory just after the end of my first year," which is consistent with the autumn of 1967.[15] The only detailed memory comes from Bob Delbourgo, who was a research collaborator of Salam and is today emeritus professor in Tasmania.

During Kibble's absence, Delbourgo took responsibility for organizing the seminars. Abdus Salam liked to talk about his pet projects, and Delbourgo told me that Salam wanted to give a series of lectures about his recent work on the spontaneous breakdown of symmetry in gauge theories. Delbourgo recalls, "To the best of my recollection I organized some three talks on Tuesday afternoons where he explained his now famous work." Unfortunately, Delbourgo took no notes and knows of no one else who did: "It all seemed so esoteric at the time."[16]

"Soon after this," Delbourgo told me, he was in the physics departmental library and saw Weinberg's article in *Physical Review Letters*, which Delbourgo thought "looked suspiciously like what Salam had recounted to us." He mentioned this to Salam, who "looked really chagrined and worried. So I urged him to write up his work as soon as possible, as it was done independently and at roughly the same time as Weinberg's. Salam mentioned that there was to be a Nobel Symposium, and that this would provide a vehicle for rapid publication of his own work."

The symposium was held in Göteborg, in the spring of 1968. Six months after Weinberg's paper had appeared in *Physical Review Letters*, the content of Salam's talk was based on the old ideas of Salam and Ward amplified with inclusion of spontaneous symmetry breaking.

As details of these events are so sparse, I sought further confirmation. Salam, apparently, created his masterpiece out of thin air. As van Gogh's *Potato Eaters* is the culmination of years of cartoons and paintings, developing the characters and the design for the final creation, one might expect that Salam also would have left some trace in his oeuvre, or in his notes, of his own seminal work. Weinberg's epiphany, which revealed the secrets of the weak force, was like a butterfly emerging from the chrysalis of spontaneous symmetry breaking that he had been applying to strong interactions. What clues do Salam's papers give?

The answer is: little or none. Throughout 1966, '67, and '68, papers with Salam's name appeared at the rate of one every six weeks, a huge production rate for a theorist.[17] These works, in a variety of collaborations, were principally on the strong interactions and gravity. There was nothing about electromagnetic and weak unification, or even spontaneous symmetry breaking, other than the talk in Göteborg in May 1968.

He seemed bent on writing up every idea, with the exception of the one that would later be his Nobel ticket, separating him from John Ward—with whom his electroweak papers had hitherto been united.[18] The only written record of his electroweak interest—his Göteborg talk—hardly screams that here is a major stride forward, nor of being in tune with nature in the way that Weinberg's paper has turned out to be. To a considerable extent, Salam's talk revisited the papers that he had written years before with Ward.[19] There is no prediction of the masses of the *W* and *Z*, a singular feature that sets Weinberg's paper apart.

Having given this talk, Salam seems to have dropped the thread and focused on gravity. That remained his chosen theme until 't Hooft's grand entrance in 1971. As soon as 't Hooft had announced that the SU2 × U1 model was renormalizable, and instantly elevated it to the level of a fully fledged theory, Salam advertised some historical priority by citing his work with Ward, together with mention of Weinberg.[20] But he then added remarks about gravity being an essential requirement for the 't Hooft work to be fully complete.

It is hard to understand why Salam, if he really understood the profound significance of the role that hidden symmetry played for the electroweak unification, did not formally publish it. Much else in his oeuvre would suggest that he would hardly have held back. Later, when awareness grew that here was a viable theory, he promoted it vigorously.

Abdus Salam is no longer here to explain how his ideas emerged and crystallized. So this is what seems to me to be consistent with the written record—surprisingly sparse, considering its subsequent perceived importance—and with the lack of detailed memory of a singular moment in history for those who were around then.

There is no doubt that some lectures by Salam took place, though it is less clear whether more than a handful of people were present. Delbourgo's testimony is confident even after four decades; he played a seminal role in the events, which are burned in his memory. The only written record that I have found is a letter, which was sent in 1976 to Ivar Waller, of the Nobel Committee for Physics. Salam's friend and collaborator Paul Matthews had long played a role in lobbying for his recognition, as we have seen. He wrote to Waller, confirming that he had attended "the course of postgraduate lectures . . . during the autumn term of 1967 at which [Salam] described the unified gauge theory of weak and electromagnetic interactions using the recent work of Kibble. . . . At the time these lectures were delivered, Weinberg's work had not appeared."

Matthews is no longer alive to confirm what stimulated this letter, nor to the accuracy of dates remembered nearly a decade after the events in question.[21] Terse and to the point—a mere ninety-six words—it provides no insight into the actual content of the lectures, other than to say they took place, and to give a tantalizing hint of when.

If Matthews and Delbourgo are correct, the timing of Salam's lectures paralleled Weinberg's creation in uncanny fashion. Autumn term at Imperial College in 1967 began on October 2,[22] and Salam was in London from October 5 to October 21, in Trieste from October 22 to October 31, and then back in London until November 13, returning for the first two weeks of December.[23] Delbourgo recalls that there were "three talks on Tuesday afternoons," and felt it unlikely that he would have organized a talk the very first Tuesday.[24] That at least fits, as Salam was absent on Tuesday, October 3, but October 10 and 17 are possible. Whether there were in fact three lectures, all given on successive Tuesdays, is unresolved. Chris Isham's memory[25] is that there was "only a handful of people [present] but whether it was 3, 4 or 5 I do not recall"; if this is correct, the lectures would seem to have been more of a customized teach-in, which would allow more flexibility in their timing, enabling them to fit in with Salam's hectic schedule. In any event, it seems that the period October 5 to 21, 1967, is when some, at least, of the lectures were given. If so, there was a remarkable coincidence: It was on Monday, October 2, that Weinberg made his comments at the Solvay conference in Brussels (see page 190), and his paper—already written during that conference—was received by the editors of *Physical Review Letters* on the seventeenth.

The similar time frames probably are due to the common stimulus that Tom Kibble's insight had provided. Published in March 1967, Kibble's paper played a role for Weinberg through the summer, leading to Weinberg's eventual application of the ideas to the electroweak interaction in September, and his paper in October. As for Salam, the tutorials from Kibble—which Salam's diary, reviewing the year, records as March 1967[26]—stimulated him; the lectures during the autumn term then provided an opportunity to air these speculations. It was this introduction to the opportunities of hidden symmetry, from Kibble, that may explain why Salam always referred idiosyncratically to the "Higgs-Kibble" mechanism.

So how do this letter from Matthews and Delbourgo's version mesh with the singular lack of recall of many others, or with the significant absence of anything on Salam's oeuvre in the years surrounding the episode that relates even remotely to an interest in the electroweak work?

The difficulty in finding people who remember the discussions may be understandable if only a handful were present. The lack of recall may also reflect the general attitudes of the time—Weinberg made little impact on the audience in Brussels. In Salam's case, the focus at Imperial College then was rather far from electroweak interactions, but the low level of reaction was probably amplified by Salam's personality and style.

For Salam, ideas sprouted like mushrooms—Matthews's role was to filter the wild from the delicacies. Salam would talk about what he had currently in his mind, and the next month something quite different would have overwritten it, only to be replaced a month later by something else. I recall an occasion when he was due to speak at Rutherford Lab at the annual meeting for the UK community of theorists. His train was delayed, and when he eventually arrived Salam announced that during the extended journey he had made a calculation, literally on the back of an envelope, which had convinced him that the proton—the basic constituent of hydrogen and all atomic matter—was inherently unstable. This is what he wanted to tell everyone about—and did.

Like some of his other ideas, such as the ferment in 1965 when the *Sunday Times* announced that he and Delbourgo had solved the great problem of uniting relativity and the internal dynamics of particles, this was like a supernova—blinding for a while all who experienced it, only to dim and fade from view with the passage of time. The 1965 sensation had been the mathematical scheme that was known formally as "U-twiddle-12," which formed Salam's primary interest in the mid- to late 1960s; it was far removed from the electroweak theory. When "U-twiddle-12" came to nothing, there was retribution following the earlier hyperbole.

In 1965 Salam ran a research conference at the new International Centre in Trieste.[27] Murray Gell-Mann had given one of the talks, and in the questions afterward was asked about "U-twiddle-12." The episode may have become embellished in memories over the years, but supposedly Gell-Mann took on a puzzled expression, as if searching his Olympian memory to decide what the questioner could possibly be referring to, and then, as if suddenly enlightened, exclaimed, "You mean twiddle-twaddle."

The remark was made in jest, but was taken up by the community. Abdus Salam had been in the audience. In the opinion of John Polkinghorne, who was also present and a longtime colleague of Salam, reactions to this episode may help explain his later reticence to pursue the electroweak ideas. Polkinghorne explained that behind Salam's bubbly exterior was a private personality: "If he was hurt by this episode, he wouldn't show it, but he might have felt that he got his fingers burned." Years earlier, when Polkinghorne had been a starting on his career, Salam had advised him, "Publish everything you do. They'll remember the good ones, and forget the bad." Recalling this, he commented to me, "I always found it odd [that Salam did not publish his 1967 idea immediately, given his propensity to generate papers]. It was most un-Abdus-like."[28]

While this fear of being burned might have played a role, there may also be a more mundane explanation. Salam's response to Kibble's tutorials on spontaneous symmetry breaking in 1967 was an example of the fast-flowing stream and transient optimism for which he was famous. At that time, the application of spontaneous symmetry breaking in his and Ward's model was nothing special in Salam's mind, nor of those who heard it. I put this to Delbourgo, who agreed: "I don't think he realized the import of what he had talked to us about. It was just another flurry as far as he was concerned. Then when [I showed him] Weinberg's paper he realized he was on the same track as Weinberg."

For Salam, as for his audience, it was the latest "flurry." But as events developed, years later, it would take on significance and meaning much greater than it had in 1967. Kibble had given Salam an idea; Weinberg, courtesy of Delbourgo's accidental intervention, elevated its significance in Salam's mind; 't Hooft's breakthrough gave it urgency.

FROM 'T HOOFT TO NOBEL

't Hooft and Veltman's work gave physics something to reckon with. That Nobel Prizes would follow became a certainty; who would be winners and losers were the main unknowns. That Weinberg and 't Hooft would be among them was generally agreed; however, with a maximum of three for a single award, assessing the apportioning of credits began to be a parlor game. The publicity machines began to operate.

Ideas, which had previously been ignored, were born again following 't Hooft's debut. More than two hundred citations to Weinberg's 1967 paper appeared in 1972 alone, whereas, at this stage, Salam's 1968 talk in Göteborg was almost unknown.

In 1971 even Salam himself appears not to have rated his Göteborg paper,[29] but he soon became active in ensuring that his name be paired with Weinberg in the general consciousness. At conferences Salam was noticeable at the front of the hall, within earshot of the speaker. If anyone referring to the electroweak model mentioned only the name of "Weinberg," a voice would be heard with a familiar bubbly humor, though serious nonetheless: "And Salam."

His personality and high profile ensured a steady flux of visitors at the International Centre for Theoretical Physics in Trieste, and invitations to conferences. Research collaborations with visiting university academics during the summer vacation are the lifeblood of science, and distinguished scientists are needed for advisory committees; members of the Nobel Committee for Physics were prominent among these at ICTP.[30] Regular exchanges of letters kept them up-to-date on progress with what was becoming universally known as the Weinberg-Salam model.

The first awakening for Salam seems to have come in November 1971 with sight of Weinberg's own entry into the new age, post–'t Hooft.

Weinberg had written a paper updating his 1967 theory and discussing its experimental implications in light of both 't Hooft's proof of its viability and of the GIM mechanism for including hadrons.[31] This was received by the editor of *Physical Review Letters* on October 20, 1971, and appeared in print on December 13. The references cite Higgs first, followed by Brout and Englert, for their use of hidden symmetry, but there is no mention of Salam.

A preview of the paper had been sent to Salam, however, as on November 18 he wrote to Weinberg, drawing attention to his lecture in Göteborg. On the twenty-fourth Weinberg replied that he had already alerted the editor, as he had meanwhile recalled that Salam in Göteborg "did then discuss what I consider to be the essentially new element in my 1967 letter, the use of the Higgs phenomenon to build renormalizable theories of massive vector mesons." He promised Salam that in a further paper, to go to *Physical Review*, he would cite Salam's "earlier work on renormalizability and gauge fields, as well as the Salam-Ward article." This he duly did, in the paper received by the editor on December 6, 1971.[32] Weinberg recognized Salam's influence by adding in his letter to Salam that he was "very much aware that all theorists who work in this area are greatly in your debt for what we have learned from you."

Following this, letters were exchanged between Salam and those who failed to match his name alongside that of Weinberg. The most significant of these exchanges is probably that with Ben Lee in 1972.

Lee, recall, had been seminal in creating the new paradigm. It had been his paper in 1964[33] and his 1966 talk at Berkeley[34] that had led to Higgs's entry into the lexicon of physics; his lectures at Cargese had inspired 't Hooft to create a viable theory of the weak interaction, and it had been Lee's own work that had helped complete 't Hooft's proofs, bringing the results to general attention. By 1972 Lee, no less than 't Hooft, was widely regarded as the authority on applying the ideas of field theory to the weak interaction. The Sixteenth International Conference on Particle Physics was scheduled for Chicago that summer, and Lee was the natural person to review the field.

In advance of that occasion, Salam wrote him a "painful" letter on June 26.[35] Salam wrote several drafts (which I saw in the ICTP Archives in Trieste), and even then the final version had both a p.s. and a p.p.s., the latter requesting that after reading it, Lee should "kindly tear this letter up and destroy it." It would seem that Lee did so, as no trace exists in his collected papers.[36]

The letter dealt with the ethics of references concerning the renormalizable theory. Whereas Lee in his earlier papers had bracketed Salam's name with Weinberg's, more recently he appeared to be referring to Weinberg alone. Salam then raised the point, which is at the nub of what constitutes credit for discovery, suggesting to Lee that the omission was because "even though I claimed in my Nobel Symposium lecture that I invented the theory independently of Weinberg, and lectured on it to an audience of postgraduate students and physics staff during the fall of 1967, my date of publication being later I deserve no credit."

Salam then argued: "I believe, in our subject, there is a general convention that, in a case of independent invention, credit is shared, even though some of the authors may have seen the work of the others, while their own was in preparation for publication." Here Salam has addressed head-on the issue that has haunted the field ever since.[37] In reply, Lee agreed, pointing out that the omission was inadvertent and apologized "most profusely."[38]

An example of a different response came in 1978, when Salam had an exchange with Norman Dombey, today a senior theorist at the University of Sussex in England. In the 1960s Dombey had been a colleague of Glashow's and was very conscious that Glashow had made his entrance in the electroweak unification saga in his doctoral thesis, his insight recognized by Murray Gell-Mann at the fateful lunch in France in 1960. Glashow had then developed and completed his publications during his time in Cal Tech, which is where Dombey, then a student, had befriended him. Years later, Dombey, who was working actively in this area himself, closely followed the mushrooming emergence of the electroweak theory.

In January 1978, with his colleague David Bailin, he had written an article in *Nature*, which referred to the "Weinberg model."[39] They cited Weinberg's seminal 1967 paper, Glashow's 1961 paper, and also the 1964 paper by Salam and Ward, but included no mention of Salam's solo 1968 talk in Göteborg. Salam objected and sent Dombey a copy of the correspondence that he had exchanged with Ben Lee. Having read Salam's 1968 article, Bailin and Dombey defended their citation on the grounds that the SU2 × U1 model had been introduced by Glashow, only to be rediscovered later by Salam and Ward jointly; furthermore, Weinberg alone had shown how the critical prediction of the relative masses of *W* and *Z* emerge.[40] In Dombey's opinion, on the one hand Salam's Göteborg talk contained nothing fundamental that was not already in the literature, and on the other there were already important contributions in the literature that were not in Salam's solo talk.[41] In response, Salam claimed that his 1968 paper was essentially accepted as the "standard model" but that he had not set down the "precise values for the masses of the particles."[42]

If the maximum of three Nobel laureates were to include 't Hooft and Weinberg, then by the middle of the 1970s Salam's name was, by repetition, becoming entwined with that of Weinberg. Veltman, who eventually shared a Nobel Prize with 't Hooft, and had some self-interest in the outcome, sensed that the momentum favored that trio:[43] Weinberg had speculated that spontaneous symmetry breaking would lead to a complete theory of the weak interaction, 't Hooft had proved this, and Salam's 1968 talk was the basis for his own claim. In Veltman's opinion Weinberg freely cited 't Hooft's work and tended to overlook Veltman's contribution, and this combined with the continual repetition of the mantra "Weinberg-Salam" in the literature.[44]

Science apart, politically the award of a Nobel Prize to Abdus Salam was like pushing on an open door. He had a high profile as founder of the ICTP and, through this, his role in UNESCO; he was well connected throughout the developing world, where he was both a hero and an icon. Momentum grew behind the twinning of Weinberg and Salam. Ward's name, meanwhile, was hardly mentioned, if at all.

One of those who nominated Salam for the 1979 prize was Tom Kibble.[45] In eight pages his nomination provides an elegant summary of the state of physics, recalling Salam's breadth of contributions and identifying various aspects of Salam's work on the weak interaction—spontaneous symmetry breaking, electroweak unity, gauge invariance—where although several had contributed, only Salam "played a leading role in every stage in this success story."

Sometimes the quality of the reference, as much as that of the candidate, determines who is chosen. Kibble's document showed remarkable modesty, nowhere mentioning his own singular role in having developed in his 1967 paper the mathematics that brings group theory into the physics of spontaneous symmetry breaking, and then explaining it to Salam.

W. A. R. D.

Salam's biographer, Gordon Fraser, had been a student at Imperial College during the 1960s and later, in his capacity as editor of the *CERN Courier*—effectively the monthly record of the ongoing history of particle physics—had watched the developments closely. He commented to me, "While people have always spoken of the Weinberg-Salam model, its mixing parameter is just as invariably called the 'Weinberg angle.' Salam did not like this, and [to justify his own claim] used to point people back to his work with Ward."[46] Which brings us to the question: Was Ward cut out of a rightful share in the electroweak prize?

In attempting to answer that, I shall first set out what I feel are the three essential steps in uniting weak and electromagnetic interactions and solving the Infinity Puzzle for the weak force. First was the recognition that SU2 × U1 is the empirical matchmaker to the marriage, which led to the prediction of Z^0 and neutral currents. Second was incorporating ideas of spontaneous symmetry breaking along the lines that Kibble outlined, leading to the electroweak theory, which in its simplest form relates the masses of *W* and *Z*. Third was the proof that the combination of SU2 × U1 and spontaneous symmetry breaking produces a renormalizable viable quantum field theory—Quantum Flavordynamics—which successfully describes the data.

We have already seen the first step's prediction of neutral currents, which violate parity, confirmed; in the next chapter we will see the discovery of *Z* and *W* with masses as predicted by the second step, followed by precision tests of Quantum Flavordynamics, which confirm the third. Direct experimental demonstration of the dynamics of electroweak symmetry breaking, such as through the production of the Higgs Boson, remains to be achieved.

As to the history of the theoretical ideas, the first step—SU2 × U1 and the Z^0—was the essence of Glashow's 1961 paper. Ward and Salam were attempting to marry the weak and electromagnetic forces at that time, but did not find the correct route until three years after Glashow's work had been published.

The third stage—the birth of QFD—leads to 't Hooft and Veltman's Nobel Prize in 1999. Weinberg and Salam had each speculated that incorporating spontaneous symmetry breaking might lead to a renormalizable theory, but never demonstrated this. Weinberg's share of the 1979 prize was due to his singular role in the second stage—creating the electroweak theory incorporating spontaneous symmetry breaking, with its prediction for the masses of *W* and *Z*.

Inspired by Kibble, Salam had grafted similar ideas onto his 1964 work with Ward, lectured about them at Imperial College in 1967, and written them up following his Göteborg talk in 1968, but never put them to the test of a peer-reviewed journal. If the 1979 Nobel Prize recognized the priorities for what I have described as the first two stages, then rewarding Glashow with Weinberg, and Salam but not Ward, has a clear rationale. One might argue that Salam was fortunate to be included, but there is no evidence that Ward has any claim to priority in the first stage, nor to involvement in the second.

Much hangs on how singular one regards the incorporation of spontaneous symmetry breaking into the mix relative to having identified the importance of SU2 × U1. This has been the subject of much debate. In particular, Murray Gell-Mann said that he had "argued a little bit" with the Nobel authorities: "the only time I ever have."[47] While he judged Salam and Ward together to have made many very important contributions to particle physics, worthy of consideration for a Nobel Prize, in the specific case of the electroweak theory Gell-Mann "didn't think that Abdus Salam *without* John Ward had played such an enormous role." However, he felt that "Shelly [Glashow] had played a *tremendously* important role" in being the first to bring the weak and electromagnetic interactions together in line with nature's scheme.

He added that if he had been awarding a prize to Salam, he would have done so for a set of things, which *included* the weak interaction rather than just for that alone, and "I might also have coupled it with John [Ward] as [Salam and Ward] did so many wonderful things together." Gell-Mann made no negative arguments against Abdus Salam but "did argue positively on behalf of Shelly [Glashow]."[48] Gell-Mann's assessment mirrored Kibble's nomination of Salam: A prize for Salam could be awarded for a set of things.

Ward felt marginalized and continued to do so throughout his life. He aired some of his opinions following an article by Weinberg in *Scientific American* in 1974. In a letter to that journal, Ward objected to Weinberg's having described Salam's 1968 talk as occurring "later" than Weinberg's own contribution. In Ward's perspective, Salam's 1968 talk was reporting on work done with him, which had occurred "several years earlier" in 1964, and any additional remarks about spontaneous symmetry breaking were secondary.[49] Whereas Weinberg—and eventually the Nobel committee—regarded spontaneous symmetry breaking as the "essentially new element,"[50] Ward regarded the concept as a "commonplace" in physics, going back several years.

As we have seen, the idea of spontaneous symmetry breaking does permeate physics, and has a long history. However, it was not mentioned in Salam and Ward's 1964 paper. If Ward and Guralnik had not aborted their conversation in 1964 (see page 161), history could have turned out very different; however, they did. Furthermore, what Weinberg had done, and no one else, notably not Ward and Salam in any of their papers, was to predict the *W* and *Z* masses.[51] However, there is an irony here; this prediction of *W* and *Z* properties could not have been regarded as primary in the decision to award the prize, because in 1979 neither of these particles had yet been discovered.

Ward was a complex person. Not only had he worked with Salam and not received the same recognition, but as we saw earlier he always claimed to have been the "father of the British H-bomb" and, in his opinion, had not been recognized for that either.[52]

Whatever the facts, it is ironic that Ward, who worked on the British H-bomb, was also engaged in open science with Abdus Salam, whose advice on atomic affairs was called upon by the Pakistani government. But there the parallel ends. Salam regarded nuclear weapons as an abomination and advised on the peaceful uses of nuclear energy. Salam was lauded in his own land;[53] Ward's government work received no formal recognition in the United Kingdom. Then, when the mathematical structures that he had forged with Salam were drowned in the hyperbole of the 1970s, Ward missed out again.

Several colleagues had memories that bitterness formed a leitmotif for Ward's later years.[54] Some felt that Ward was convenient when required and ignored when he became inconvenient. This may have been so, but Ward himself was sanguine in print, his enigmatic summary being "what you gain at the swings, you lose at the roundabouts."[55]

The last words are Ward's. Among the congratulatory telegrams received by Salam upon the award of the Nobel Prize was one that consisted of four words: "**W**armly **A**dmired **R**ichly **D**eserved." Friends and adversaries were split as to whether or not this was irony.

Chapter 16

THE BIG MACHINE

The Big Machine—literally in the form of particle accelerators of increasingly Brobdingnagian scale, and metaphorically in the size of the research teams, and the international planning and political maneuvering required to realize the goals of particle physics. America, Europe, LEP, and the LHC. What is required to complete the solution to the Infinity Puzzle.

The Nobel committee had awarded the 1979 Physics Prize to Glashow, Salam, and Weinberg, even though the central players, Schwinger's "invisible instruments"—the W^+, W^-, and Z^0—remained like yeti: their footprints had been seen, in the form of the neutral weak interaction and a range of other phenomena, but there was no sighting of the actual animals. By the end of the 1970s people knew what they were looking for—giants heavier than a nucleus of iron, ten times as massive as any other particle then known. The challenge was clear; the question was how to reach out so much further than before in the hope of capturing them.

The standard route, well traveled for more than twenty years, had been to use "atom smashers," the colloquial phrase symbolizing beams of protons that smash into solid targets, releasing energy, which can materialize as new varieties of matter. The experiments are like entering a lottery: Most of the collisions produce familiar particles, but once in a while you get lucky and something unexpected shows up. If the amounts of energy invested are large enough, you might win a major prize: the W^+, W^-, or Z^0. That was the dream, but unrealizable with the existing accelerators in the 1970s—until a novel approach[1] gave the breakthrough.

Imagine that you are in a car, waiting at a red light, in neutral and without its brakes on. Another car, traveling at 30 mph, suddenly hits you in the rear. Your machine will suffer serious damage, to be sure, but you will probably survive, as the recoil of your vehicle takes up much of the energy of the invader. However, if you had collided head-on, each traveling at 15 mph, hence with a relative speed of 30 mph, the results would have been more serious. A similar phenomenon occurs when particles collide. A beam of protons hitting a static target wastes a lot of the energy, but if two opposed beams hit one another, results are more dramatic.

The problem is that it is hard to get two particles, each smaller than an atom, to make a hit rather than a miss. Beams of particles are like diffuse swarms of flying ants, passing through one another as if neither was there. However, if those beams are recycled, traveling around and around two intersecting circles, crossing each other on each successive orbit, now and again two particles will collide. Like Robert the Bruce: If at first you don't succeed, try, try again.

Particles in a high-energy accelerator move at very near to speed of light, traveling approximately 300,000 kilometers every second. Use magnets to guide them around a loop, so that if they miss one another on the first circuit, they may try again—hundreds of thousands of times every second.

In 1971 CERN built the ISR (Intersecting Storage Rings). At 300 meters in diameter, this prototype collider of protons was not big by today's standards, and not powerful enough to produce the *W* or *Z*. In the United States, the idea of colliding counterrotating beams of protons was taken up at Brookhaven, the laboratory on Long Island, where in 1978 they started building a larger machine, known as Isabelle.[2]

Isabelle was a conservative approach, safe, guaranteed to produce the *W* or *Z*—assuming of course that they exist. However, in a sense it was looking backward to an age that was fast disappearing, where "atom smasher" symbolized bits of standard matter—such as protons—smashing into one another. The discovery of charmed particles, beginning with the "November Revolution" in 1974, inspired a new vision for high-energy physics.

The startling first sight of particles containing charm quarks had come at Brookhaven's old "atom smasher," the so-called Alternating Gradient Synchrotron (AGS). Protons hitting a solid target had produced vast numbers of particles, among which were a handful that seemed to be the debris from a hitherto unknown beast—the J-psi, the signal for charm. This was like finding a needle in a haystack, and it took so long to extract it with certainty that by the time the team had succeeded, another group, at Stanford in California, had discovered the J-psi produced in electron-positron annihilation, as we saw on page 277. What's more, far from being a needle in a haystack, this technique produced the J-psi standing proudly in full display.

The codiscoverers—Sam Ting at Brookhaven and Burt Richter at Stanford—shared the Nobel Prize in 1976. This would be, at least for several decades, the last hurrah for traditional experiments like Ting's at the AGS where protons smashed into a static target. Richter's approach, of head-on collisions between electrons and positrons, offered more tempting possibilities.

To create a Z^0 in the annihilation of an electron and a positron would need beams with about ten times more energy than those in Richter's small machine. With twenty times the energy, you might even produce W^+ and W^-.[3]

To make higher energy requires a bigger machine. However, the size required grows much faster than the increase in energy. Richter's machine had been about one hundred meters in circumference, but to achieve the extreme goal of W^+ and W^- would require a ring more than twenty-five kilometers in circumference. If one was prepared to settle for the Z^0, a smaller machine would suffice, but even this would be several kilometers long. The discovery of the J-psi at SPEAR had stunned everyone and inspired new ideas. In Europe, at CERN, almost immediately people began to wonder if such a monster could be built as a Europe-wide collaboration. The idea of LEP—the Large European Project, later known as the Large Electron Positron collider—was born.

During the period 1976–1981 when the big decisions were made, the director-general at CERN responsible for accelerators and general management of the laboratory was British engineer John Adams. Born in 1920, he never went to college. The reason was that his father had been shell-shocked in the First World War, lived as an invalid thereafter, spent long periods on the dole, and the family could not afford the cost of a university education for their son. It was while Adams was working on microwave radar during the Second World War that his talent and natural aptitude with machinery became recognized. Following the war, he began to build the first particle accelerator at Harwell, and then joined the infant CERN in 1953.[4]

The idea of CERN had been born in 1950, following discussions at UNESCO, which recommended building a proton accelerator laboratory as a pan-European collaboration. CERN was a totally new kind of international laboratory. Adams played a major role in designing the original proton accelerator and in creating the organizational structure.

Physics is the end-product, but accelerators are the tools, and the choice of which device to use can be the difference between success and failure. Adams believed in protons. Although in 1976 he went along with the desire to build LEP, in private he expressed reservations,[5] adding the cogent advice: "If you want to build LEP, make sure that you keep the tunnel large enough to put a proton accelerator in there one day."[6] The concept that would eventually become the LHC—Large Hadron Collider—was already in John Adams's mind in 1976, long before LEP had begun.

Although the idea of LEP had been born, the route to the discovery of the *W* and *Z* particles, and confirmation of the electroweak mystery, proved to be quite different. This came about largely due to a force of nature called Carlo Rubbia.

MAKING HEAVY LIGHT

Bjorken's work, as developed by Feynman, and the SLAC experiments described in Chapter 12, had exposed that a proton is a swarm of quarks. As a proton is made of quarks, so is its antimatter doppelganger—an antiproton—made of antiquarks. When quark and antiquark meet, they annihilate one another, and out of the flash of energy emerges a shower of new particles. Usually these are familiar varieties, such as electrons, protons, and pions, but according to Bjorken and Feynman's model, about once in every million collisions a *W* or *Z* could be produced. Furthermore, Bjorken's 1966 paper had even predicted how to find the *W*. As the concept of the *Z* grew during the 1970s, Bjorken's method for finding the *W* was easily adapted to finding that as well.

All that was needed was an accelerator of protons and antiprotons powerful enough to do the task. In 1976 the brilliant Italian experimentalist Carlo Rubbia had realized that one of John Adams's creations, the Super Proton Synchrotron (SPS), which was an accelerator of protons at CERN, could be converted to do the job. Rubbia proselytized for the SPS to have counterrotating beams of protons and antiprotons capable of producing the *W* and *Z*.

A bulky man, with immense energy and single-minded will, sweeping his hair back from his forehead, he would speak so fast that it was hard to keep up with the flow of his ideas. Rubbia harangued anyone willing to listen, along with many who weren't. During the latter part of the 1970s and early '80s he held a professorial position at Harvard as well as being a staff scientist at CERN, crisscrossing the Atlantic weekly, seemingly oblivious of jet lag. With missionary zeal, he lectured around the world about the requirements for the accelerator, the design of an experiment, and the means for detecting and identifying the *W* and *Z*.

He had so much to say that his talks invariably went overtime, and even once the lecture was over, he didn't stop. My first memory of him was in 1978. While waiting in a slow-moving line at one of CERN's cafeterias, I became aware of agitation near the back of the line where someone was getting impatient. Full of nervous energy, this was Carlo. With his tray balanced precariously, he kept turning from side to side while explaining to those around him why the proton-antiproton collider was the way forward for high-energy physics and bemoaning that the management was unable to appreciate the fact.

The proton-antiproton collider had been initially approved in 1976, but support for LEP was also growing. Physicists at large were confident that LEP could certainly find the *Z*, and possibly also the *W*, but there was considerable doubt whether Rubbia's idea to use the SPS could reveal the beasts. The *W* and *Z* would leave only faint traces, which many people feared could be missed, if particles escaped from the production site without being detected. Chris Llewellyn Smith—a British theorist who became director-general of CERN and a leading player in realizing the LHC—told me that even as late as 1978, when the first major plans for LEP were under way, he believed that Rubbia should be able to discover the *Z* at the SPS, but the *W* would be harder. He continued, "Many people thought we couldn't find either of them. We didn't know then that it would be possible to build hermetic detectors—[where nothing escaped]." In the CERN policy committee, Tini Veltman was also skeptical, fearing that proceeding with Rubbia's machine might jeopardize LEP; several worried that if LEP wasn't started at once, it might never happen.

In competition with Rubbia, there could only ever be one winner. Leon van Hove, who was scientific director at CERN, also argued powerfully that a discovery of *W* and *Z* at CERN would be a feather in their cap. The CERN management duly agreed to convert the SPS, and it began its new life in 1981. Making the required intense beams of antiprotons was a remarkable feat of engineering, achieved thanks to the Dutchman Simon Van der Meer.

Two detectors were built, code-named UA1 and UA2.[7] They were indeed sufficiently leak-proof that telltale traces of the transitory presence of *Z* and *W* showed up. Bjorken's idea of 1966 had worked: The *Z* decayed into an electron and a positron, each easily identified as they flew away back-to-back from the site of the *Z*, while the *W* decayed into an electron and a neutrino. The electron and neutrino also flew off back-to-back, the neutrino escaping from the detector without a trace, while the electron's trail was revealed in glorious isolation, hinting that something invisible must be balancing it on the far side.

The sight of the lone electron was like the sound of one hand clapping, whereby the presence of the unseen partner could be inferred. When the sums were done, and enough examples seen, everything fitted with these trails being the death masks of the short-lived *W*. Both W^-, decaying into electron and neutrino, and W^+, which decayed to a positron and neutrino, were identified in January 1983.[8] These were followed in May by the sighting of the *Z*, their masses agreeing with Weinberg's prediction.[9] A half century after Enrico Fermi had made his first tentative model of beta radioactivity, another Italian had uncovered the "invisible instruments" that are responsible.

The Nobel Foundation awarded their 1984 prize to Rubbia and Van der Meer. Glashow, Salam, and Weinberg had been invited back as guests; perhaps the Swedish Academy was grateful for the proof that their 1979 award had at last been confirmed.

I was in Stockholm that December to invigilate a Ph.D. exam[10] and by a series of chances was invited to the Nobel proceedings. I heard Rubbia summarize the history in a speech that went at his usual frenetic pace, and still managed to run overtime. While that was normal for a Rubbia speech, it was not universally greeted with amusement at such a special event.

John Adams, whose commitment to designing accelerators had helped bring the enterprise to fruition, had died in March 1984. Although he survived to see the successful discoveries of *Z* and *W*, he never witnessed the award of the Nobel Prize for the work in his laboratory. His widow, Renie, was present, and at the last minute I was invited to sit at the CERN table in order to balance the numbers.

Male attire is traditionally white tie and tails, but there are exceptional alternatives. Abdus Salam was resplendent in his national costume, which attracted the attention of the television cameras at the ceremony. I too was captured on television, though not in a clip that I would want to see on YouTube. Because of the doctoral examination, I had brought my academic gown and was thrilled to discover that full academic dress was one of the options allowed for the celebratory banquet. This was a wonderful occasion, with the tradition that immediately after the meal, the laureates go to a private party with the Swedish royal family, and the remaining guests, numbering more than a thousand, retire for a formal dance. I learned that the first dance would be a waltz, and protocol required that I partner the lady who had sat to my right during the banquet. The position of our table was such that the two of us were among the first to reach the magnificent ballroom. Wearing my scarlet doctoral gown, and my partner—the wife of a Swedish academician—being dressed in green, we were highly visible. This part of the evening was shown live on Swedish television, and we were a favorite of the cameras. Whether this was due to our color coordination, or because I cannot waltz, my hosts were too polite to say.

THE END OF HIGH-ENERGY PHYSICS?

By 1984 the *W* and *Z* had been found, electroweak theory verified, and the Nobel banquet digested. Four years earlier, in the summer of 1980, I had been lecturing at a summer school in Akaslompolo, above the Arctic Circle in Finland—the farthest north that I have ever been. Previously I had only seen Gerard 't Hooft giving talks about his work; it was in Akaslompolo that we met for the first time.

We discussed physics, and the history of his ideas, under the midnight sun while being eaten by mosquitoes. I was impressed at his certainty that the *W* and *Z* would be found—remember that the actual discoveries were still three years in the future—and that Rubbia's upgraded SPS would be just the beginning. The SPS could confirm the ideas that Glashow, Salam, and Weinberg had proposed, but to establish the mathematical foundations, which 't Hooft and Veltman had built, would require a much larger and hugely more sophisticated machine.

't Hooft explained that, according to their theory of Quantum Flavordynamics, careful measurements of the masses, lifetimes, and other properties of the *W* and *Z* should show subtle deviations than what Glashow, Salam, and Weinberg's relatively simple model had predicted. The precision required would be far beyond the capabilities of the SPS. A dedicated factory for making millions of *Z* and *W* would be needed to measure such fine details.

That was when I first began to appreciate how singular LEP—the Large Electron Positron collider—a twenty-seven-kilometer ring underground at CERN, would be. The project was approved by the CERN Council in December 1981, and construction began in 1983. Nothing like this had ever been built. Such a Brobdingnagian accelerator seemed fantasy. A sequence of political events on both sides of the Atlantic almost made it so.

In Europe, CERN had the SPS proton-antiproton collider and plans for LEP. In the United States, Stanford had beams of electrons and positrons; the national accelerator at Fermilab, near Chicago, had protons, comparable to or even exceeding in energy those at CERN; and as we have seen, the original hadron laboratory, Brookhaven on Long Island, was looking to its future. They had plans for Isabelle, a proton accelerator with enough energy to make the *W* and *Z*.

By 1983 the W^+, W^-, and Z^0 had all been discovered by CERN, and much of the rationale for Isabelle had evaporated. There was certainly a feeling that the center of power had shifted to Europe, CERN's discoveries of W^+, W^-, and *Z0* having been greeted by an editorial in the *New York Times* headlined "Europe 3, U.S. Not Even Z-Zero." There was a call from George Keyworth, the science adviser to President Reagan, that the United States should go for broke and "regain leadership" in high-energy physics.[11]

The five years of construction of Isabelle had cost $200 million when it was canceled. The result was a proposal to build the "Superconducting Supercollider," or SSC, with energy of 40 TeV.[12]

The idea that one day CERN would build the LHC in the tunnel then being built to house LEP had been common knowledge since 1978. Many European physicists believed the American choice to go for 40 TeV was in order to trump anything that CERN could do in the LEP tunnel.[13] American physicists, however, insist that the decision to design for 40 TeV was driven by the need to ensure the optimal physics reach, which depends on both the energy and the intensity or "luminosity" of the beams. The paramount need was to make a machine that had a very good chance of discovering the origin of electroweak symmetry breaking. The SSC planned to reach that goal with a luminosity that was certainly achievable; it was this pragmatic approach that led to the 40 TeV energy device.[14] To achieve the same reach with lower energy would require higher luminosity—in the case of the LHC, a tenfold increase, which was daunting.

In any event, the challenge had been set. Particle physics was by then a global enterprise, its future plans discussed at an annual summit of the International Committee for Future Accelerators (ICFA). The next meeting was planned for Japan in May 1984, and it was clear that the United States would present the SSC project. So, in the autumn of 1983, CERN started serious planning for the LHC.

The LHC idea had been in the wind for some years, but there had been no serious work on it. Over the winter of 1983 people came up with draft designs. A meeting was organized in Lausanne—to make the community aware of the possibility of the LHC and to prepare a presentation for the ICFA summit.

In the middle of the Lausanne meeting Chris Llewellyn Smith received an urgent phone call from Sir John Kendrew, the Nobel Prize–winning biologist, with startling news: Kendrew had been asked by the British authorities to chair an inquiry into whether the UK should stay in CERN. As his first act, Kendrew wanted Llewellyn Smith to be the committee's scientific adviser. Llewellyn Smith immediately refused, explaining that Kendrew really needed an experimentalist rather than a theorist for this task.

Kendrew replied that would be a great pity, as the news was about to be announced in the House of Commons that very afternoon, and he wanted to be able to say that there is a scientific adviser: "If you don't agree then there may not be one." So Llewellyn Smith had to say yes.

If Britain pulled out of CERN, the funding of the whole organization would have been severely diminished and the continued construction of LEP threatened. As particle physics was rapidly becoming a global enterprise, some even feared that the future of the subject itself could have been at risk. Kendrew's committee included scientists from other disciplines, engineers, and industrialists, with Llewellyn Smith as their adviser on particle physics. To my astonishment I then discovered that a book I had written—*The Cosmic Onion*, which is the story of how the structure of matter was uncovered—had been prescribed as "bedtime reading" for the committee to get up to speed with particle physics. My worry was that the book might not have convinced them, but although opinion was split, their report was favorable. If there had been any doubts, the discoveries of the *W* and the *Z* were like trump cards that could not have come at a better time for both CERN and the entire field of particle physics.

David Saxon, an experimental physicist, never doubted that the outcome would be successful. Saxon, who enjoyed wordplay no less than

Nick Kemmer, had realized that an anagram of Christopher Hubert Llewellyn Smith spelled out his unique ability to convince the government of British prime minister Margaret Thatcher: "Tory PM huh. Bet his wit'll sell her CERN." As, indeed, it eventually proved. Britain remained a member of CERN, and the laboratory has gone from strength to strength.

In January 1987 President Ronald Reagan endorsed the SSC as an American machine, with a price tag of $4.4 billion. However, by May 1990, the cost had risen to nearly $8 billion. The House of Representatives limited the federal contribution to $5 billion, with the host, Texas, putting in $1 billion. The remaining $2 billion would have to come from international partners.[15]

Once the decision had been made to site the SSC in Texas, it became harder to maintain support from senators and congressional representatives from other states, especially those that had bid to host the facility and lost. In June 1992 the budget for the SSC was defeated in the U.S. House of Representatives but resurrected in the Senate. By 1993, with George Bush Sr.'s presidency having ended, there was no longer a Texan in the White House to bat for the SSC in his home state. The new president, Bill Clinton, was not prepared to make auto concessions to Japan, and financial support from outside America never materialized. In June 1993 the budget was again defeated by which time the General Accounting Office estimated that the costs had escalated to $11 billion, and the SSC was under threat of extinction.

The cost was such that the SSC appeared as a line item on the budget, and Congress could see its demise as a means to save money. With about a quarter of the eighty-seven-kilometer tunnel already dug beneath the Texas soil near Dallas, in October 1993 funds were withdrawn and the project was terminated. By this time, at CERN, the LEP—with beams of electrons and positrons instead of protons—was already at work.

LEP

Compared to electrons and positrons, protons and antiprotons are monsters. Two thousand times heavier, they pack a much bigger punch. The power in a collision between protons or antiprotons moving at speed exceeds by far what you get from the tiny electrons and positrons at that same speed, which is why the former have been the first choice when pushing into previously uncharted areas. If protons are like archaeologists' pickaxes, used to uncover a site, then electrons are like trowels or brushes, for careful examination of any precious ornaments that are revealed.

That is how it was in the 1950s, when experiments with protons had hinted that these protons themselves possessed some inner structure, followed in the 1960s by SLAC's experiments with electrons, which were able to resolve the quarks and gluons that form the proton. That is how it would turn out also at the end of the century, as the SPS, having discovered the *W* and *Z*, gave way to LEP, which would examine these agents of the weak force in minute detail.

The ability to make forensic examination with electrons or positrons comes at a price. To give the same amount of power to these puny particles as to bulky protons requires machines that are considerably larger. Thus, in the 1950s, the so-called atom smashers, which used beams of protons, were typically tens of meters in size; contrast SLAC's electron accelerator, which needed to be three kilometers long to achieve comparable power. By the 1980s, the SPS, using protons and antiprotons in the discovery of the *W* and *Z*, was about seven kilometers in circumference; the forensic examination of this pair, using electrons and positrons as the tools, would require a twenty-seven-kilometer circuit. That was the challenge that led to LEP.

LEP—the Large Electron Positron collider—was indeed Large. Fifty meters below the surface of Switzerland and France, in a tunnel as long as the Circle Line on the London Underground, magnets steered beams of electrons and positrons to their goal. The machine was as large as could be fitted into the stable geology surrounding CERN, between Lake Geneva and the Jura Mountains. It was also at the limit of what could be afforded, and even then it would take the combined efforts of more than a dozen nations, and of the CERN management focusing its resources on this single enterprise for several years, to bring LEP to fruition. And all because theoretical physicists, by scribbling equations on pieces of paper, had found that the hieroglyphs of the mathematics revealed remarkable messages about the universe.

The raw statistics give an impression of the engineering marvel. The enormous ring consisted of eight curved sections, each nearly three kilometers long, with five-hundred-meter straight sections between them. Thirty-five hundred separate magnets bent the beams around the curves, and another thousand were specially constructed so as to focus the beams into intense concentrations of electric charge. The tubes, within which the beams of electrons and positrons traveled, ran through the center of the magnets, and formed the longest ultra–high vacuum system ever built. The insides of the tubes were pumped down to a pressure lower than on the moon.

The reason for doing so was because antimatter, such as positrons, is destroyed the instant that it touches even a single atom of matter. Consequently, having gone to so much trouble to make, store, and focus intense beams of particles of antimatter, that was the quality of vacuum needed in order to stop them from being destroyed by stray atoms in the air before they reached their intended goal.

The positrons sped around the twenty-seven-kilometer ring beneath Swiss vineyards, crossing the international border into France 11,000 times each second, scurrying under the Chinese restaurant near the statue of Voltaire in the French suburb of Geneva where he spent his final years, rushing beneath fields, forests, and villages in the foothills of the Jura Mountains. For electrons there was a similar story, as the magnetic fields steered electrons and positrons on the same circular paths but in opposite directions. Keep those paths slightly apart, and all is well. At four points around the circuit small pulses of electric and magnetic forces deflected the beams slightly, so that their paths crossed. Even here these beams were so diffuse that almost all of their individual electrons and positrons missed one another and carried on circulating. However, occasionally a positron and an electron made a direct hit, leading to their mutual annihilation in a flash of energy.

That was the key moment. The ability of antimatter to destroy matter and release all its energy was being used here by science to create, momentarily in a small region of space, a miniature representation of what the universe as a whole was like in the first moments after the Big Bang.

It is the aftermath that interested the scientists. By seeing what forms of matter and antimatter emerged from this simulated "mini bang," they learned how energy was first converted into substance in the real Big Bang of the early universe—for matter is energy that has congealed and taken on substantial form.[16] Highly complex pieces of electronics encircled the collision sites, capturing and recording the emergence of these primeval particles, as LEP repeated over and over the long-ago act of the Creation.

There was a moment in the early universe when the conditions were ripe for making the Z^0. LEP was especially tuned to replicate this. During the final decade of the twentieth century, some ten million Z^0 were made this way. This large sample enabled that particle's properties to be measured with forensic precision, which were then compared with the predictions flowing from Quantum Flavordynamics.

Hundreds of scientists from around the world collaborated in these experiments.

Needing up-to-date access to the latest data, and having to communicate the results of their analyses to one another easily, were great challenges when preparations first began in the 1980s. The World Wide Web was invented at CERN in 1989 as a result.[17] So while antimatter may destroy matter, it also indirectly created the World Wide Web.

When LEP tested Quantum Flavordynamics, it showed that there was no longer any problem with infinity. 't Hooft and Veltman's construction of that theory made predictions, which the experiments confirmed. Within a few months, theory and experiment agreed to an accuracy of one in a thousand. Over the years, as data accumulated, the precision improved yet further. But in so doing, the first signs began to emerge of a mismatch, a slight difference between the theoretical predictions and what the experiment was finding.

Far from QFD being wrong, this turned out to be the first hint of something novel in the vacuum, previously unknown, and unaccounted for. This was a hitherto undiscovered variety of quark—the "top quark." The top quark was too heavy to be produced at LEP, but, according to quantum theory, its presence in the bubbling froth of the vacuum could influence the values of quantities that the experiment was measuring.[18]

Theorists calculated how heavy the top quark would have to be if it was indeed responsible for the discrepancy. The answer turned out to be some 180 times the mass of a hydrogen atom, far bigger than CERN could produce. However, since 1985 Fermilab in Chicago possessed the world's most powerful collider of protons and antiprotons, at which in 1995 they succeeded in producing the top quark. Its mass agreed with what the theory had predicted. This was the crowning success of 't Hooft and Veltman's work, which had banished infinity, and from the finite numbers that ensued had enabled the missing top quark to be discovered.

With the top quark established, theorists included this in the accounts. The result was that the theory matched with data to better than one part in ten thousand. With 't Hooft and Veltman's solution to the Infinity Puzzle fully vindicated, a Nobel Prize was ensured. The pair were duly honored in 1999. However, by this stage they had drifted apart. Even though 't Hooft's work had been made with tools handed him by Veltman, attention focused on the younger man's meteoric rise as "the biggest genius to hit physics for years," and Veltman "withdrew from the limelight."[19]

THE LHC

When the idea of LEP had first been mooted, John Adams had presciently advised that the tunnel be made large enough to include a hadron machine one day. As we have seen, when the Americans came up with the idea of the SSC, CERN had quickly responded by preparing a first outline for the LHC.

Although the Kendrew committee had agreed that the UK should remain a member of CERN, which meant that the future of CERN was ensured in the short term, much concern remained about its plans to follow LEP with the LHC, because until 1993 the SSC promised to be an even more powerful machine. In the United Kingdom the mantra was value for money. For them the decision was clear: Why build the LHC if the SSC is going to exist? On the other hand, for the French and Italians, among others in Europe, this was a matter of European pride: "If the Americans can do this, then so must we."[20] Carlo Rubbia, with the stature by then of Nobel laureate, became director-general of CERN in 1989 and a powerful advocate of the LHC.

In December 1991 the CERN Council formally recognized LHC to be "the right machine for the advance of the subject and the future of CERN."[21] They set out the road map, asking Rubbia to return with a complete proposal before the end of 1993, when his tenure as director-general expired.

In September 1992 Llewellyn Smith was appointed as Rubbia's successor. By this time, the SSC was entering terminal decline, and the long-term future of the field increasingly depended on a successful approval and construction of the LHC at CERN. In May 1993, during the fifteen-month changeover period, there was a hiatus with the plans. Llewellyn Smith had a meeting in Rubbia's office. He told me, "I remember it clearly. It was a Saturday." The outcome was that Rubbia, having realized that the projected costs of the LHC had become untenable, effectively washed his hands of the problem and handed the task over to Llewellyn Smith to present the LHC and a new long-term plan for CERN to the CERN Council.[22]

When Llewellyn Smith responded: "OK. Let me have what you've done on the long-term plan," he learned there was nothing. "I found a completely empty cupboard, apart from unrealizable requests for extra manpower and a costing that was too big to be acceptable."

So from May 1993, he worked flat-out with the help of the German Horst Wenninger, who knew the CERN manpower inside out, and Welsh engineer Lyn Evans, who became responsible for the actual design and construction of the LHC. They put together a complete long-term plan, some of Evans's redesigns saving 300 million Swiss Francs in all.

Even after this, the delegates representing the UK and German governments said that the proposal was too expensive and that CERN must come back with something cheaper. British physicists saw this as the latest battle in the ongoing government skepticism toward European physics collaboration. I recall all too clearly our suspicion that the government would have been happy to kill CERN and get all the money back.

Then suddenly we found an unexpected ally, in a science minister—William Waldegrave—who showed a genuine intellectual interest and wonder about science. The idea that the LHC might discover the Higgs Boson, leading to Nobel Prizes, and that Higgs is a British physicist at Edinburgh University, was unashamedly used to gain media attention. The trump card was that Waldegrave's parliamentary constituency was Bristol West, in the same town where Higgs had grown up and attended high school. Waldegrave became interested, so much so that in 1993 he set us a challenge: In order that he would be able to present the case for the LHC in the best way during discussions with other cabinet ministers, including the chancellor of the exchequer in an upcoming budget, he suggested that we give him a concise explanation of the LHC's purpose. As incentive, he offered a bottle of vintage champagne as prize for the best effort: The catch was that to maximize attention, the description must fit on a single side of A4 paper.

At the time, I said to the minister that his challenge was like being asked to describe the Maastricht Treaty on the European Union in a similar amount of space: Could he do so?

Each of us failed the other's challenge (Figure 16.1). The winning entry in Waldegrave's competition came from Professor David Miller, of University College, London. He described the ideas of Peter Higgs by making a political analogy. He imagined the prime minister, Margaret Thatcher, passing through a crowd of supporters and being slowed down by their persistent attempts to get her attention. These admirers were the analog of the Higgs Field, which fills the vacuum. The inertia that Mrs. Thatcher experienced was analogous to the way that particles gain their masses through their own interactions with the Higgs Field.

In Miller's analogy, the Higgs Boson was like the effects of a rumor passing through the gathering. Small clusters of supporters gathered in intimate groups to hear the news and then dissolved and re-formed as the message spread. I once saw an example of this in a film of several thousand people gathered in a huge hall in Dallas, on November 22, 1963, waiting for President John F. Kennedy to arrive for lunch. He was assassinated en route, and as the news spread through the hall, waves of people could be seen moving toward the speaker and then away to pass the message onward. Those mobile clumps were like Higgs Bosons.

FIGURE 16.1
AUTHOR'S LETTER TO WILLIAM WALDEGRAVE
ABOUT THE HIGGS BOSON

Rt Hon William Waldegrave
Chancellor of the Duchy of Lancaster
Office of Public Service and Science
Cabinet Office
70 Whitehall
London SW1 2AS

Dear Mr Waldegrave,

The Higgs Boson

I understand that you have read "Genius", Gleick's biography of physicist Richard Feynman. You may recall Gleick's story that when Feynman was asked by a reporter to explain his Nobel Prize in under 30 seconds he replied that if that were possible, then his work would not have merited The Prize! Similar thoughts perturb me concerning your request to explain the most profound ideas at the frontiers of scientific understanding on a single sheet of A4 (though you did not specify the print size). My attempt is enclosed; in return could I issue a similar challenge relating to the Maastricht treaty?

Yours sincerely

Frank Close

Prof Frank Close

Miller's metaphor did not use that example, but an imaginary one of news of some political event being spread among the crowd at a political conference. The political angle added to the positive reaction among Waldegrave's colleagues. To what extent this played a significant role is hard to gauge, but the fact is that Britain maintained its support for the project, and the media discovered CERN and turned Higgs into a local celebrity.

After much more political maneuvering,[23] in 1994 the LHC project was approved, the final plan relying on borrowed money and huge risks, which involved a cash flow where every assumption was pushed to be as optimistic as possible. If the slightest thing had gone wrong, it would have been disaster. The whole calculation was made assuming that two large numbers—the borrowing and the costs—would cancel precisely.

For a half century, theorists had been canceling positive and negative amounts of infinity with another, in the renormalization of quantum field theory. The only risk was to their reputations, and the cost was limited to pencil and paper. No one anticipated that to test these theories would involve playing such games with real money. Llewellyn Smith told the council that it was a crazy way to build the LHC, reminding them that it would involve a totally new technology, and that if anything went wrong, the whole project would probably fail.

The Germans were the first to say that they understood the risk and were prepared to share it. As Llewellyn Smith remembers it today, the message was, "If the worst happens, come back and tell us and we'll discuss, but for the moment that's the only thing we can do." And his reaction: "Bingo; we'd approved the LHC."[24]

Intermission: End of the Twentieth Century

By 1980 electroweak theory has been established enough that Glashow, Salam, and Weinberg share a Nobel Prize. Ward gets no award. The *W* and *Z*, however, remain to be discovered.

By 1990, big machines have been designed. This led to the discovery of *W* and *Z*, which were identified using Bjorken's ideas, and to Nobel Prizes for Rubbia and Van der Meer. Then LEP makes precision tests of the quantum flavordynamics, which reveal hints of the top quark, which is subsequently discovered at Fermilab. The 1999 Nobel Prize is awarded to 't Hooft and Veltman. Likewise, QCD, following Bjorken's insights, is established, leading to Nobel awards in 2004 for Gross, Wilczek, and Politzer.

By 2000 the precision data are good enough to give tantalizing glimpses of phenomena that might be caused by the Higgs Boson. Direct observation of the Higgs remains to be achieved. The LHC is approved, designed to explore the region of energy where the breaking of electroweak symmetry is expected to be revealed.

Chapter 17

TO INFINITY AND BEYOND

One hundred years from Rutherford's discovery of the atomic nucleus to the new frontiers of particle physics. Why is particle physics today so big, whereas Rutherford was a genius with "string and sealing wax"? Looking for the Higgs Boson, or whatever else nature has in store. Why is there something rather than nothing, and why are things as they are? Questions for the twenty-first century.

The proof that the theories of the strong force, QCD, and of the united electromagnetic and weak forces, which have become known as Quantum Flavordynamics, or QFD, are all Yang-Mills theories, gave the first hint that nature reads a single set of rules. As these new ideas began to sink in, physicists realized that the mathematics revealed something very profound, namely, that if experiments could resolve matter on a scale that is a million billion times smaller than the nucleus of hydrogen, nature would be found to exhibit a unity that is completely hidden to our gross senses. It was only after having peeled away the layers of the cosmic onion—from atoms to the nucleus, then protons and neutrons, then quarks—that this first hint of a far-off Shangri-la had been glimpsed.

To get there would require microscopes a million billion times more powerful than anything before. To achieve such a goal with a particle accelerator would require beams with huge energies and machines so vast that they are likely forever to remain beyond practical possibility. However, nature itself performed such feats long ago—within the first instant of the Big Bang. In the intense heat of that epoch, particles crashed into one another with energies far greater than any yet achieved in our most powerful accelerators. So the early universe has been likened to a state of extreme energy where the underlying symmetry of the fundamental laws was manifested.

This has inspired an intriguing thought: Our present material universe, filled with structures—atoms, crystals, galaxies, and even life—may itself be an example of hidden symmetry. In the beginning, goes the mantra, the universe was created in a state of perfection, where a single superforce ruled. Then, according to the mathematics, at a temperature of a billion-billion degrees, all changed: As beautiful patterns appear when mist freezes on a frosty windowpane, so order and structure emerged from the primeval uniform hot soup of the Big Bang as the cosmos cooled. Disparate forces—which today we perceive as the strong, electromagnetic, and weak forces (and gravity)—and distinct varieties of particle, such as electrons and quarks, were born. It is because the original symmetry became hidden that the variety of forms required for an evolving universe, leading to life itself, emerged.

That is what the mathematics of QCD and QFD seem to imply. QED on its own had logical inconsistencies, which may be assuaged if QED is actually but one piece of a grander theory. Recall the problem: The charge in QED is increasingly focused in the smallest pixel, heading toward infinity, whereas the color charges of QCD become ever more dilute. At some scale the concentrations of the two become equal, at which point it is possible that they merge into a single conglomerate of charge. That they do so is the hypothesis underpinning unified theories of the forces. The promise is that when the pixel scale is small enough—or, correspondingly, the energy in an experiment high enough—the forces are unified: At high energy, there is a symmetry among the forces, which becomes hidden in the cool conditions on Earth.

Which is all very exciting—*if* you trust the mathematics. Theory can show what might lie beyond current knowledge, but only experiment can verify how nature is. These ideas emerged during the late 1970s. To show whether they are visions of reality or just dreams first required experimental proof that the electromagnetic and weak forces themselves are unified. According to the mathematics, that would be easier to test. To reveal their original united self—the "electroweak force"—required microscopes capable of resolving distances just a thousand times smaller than an atomic nucleus.

That is what was achieved in 1983 with the discovery of the *W* and *Z* and confirmed with the experiments at LEP during the 1990s. Electromagnetic and weak interactions are merged into a single "electroweak" force at high energies, this symmetry being hidden at low energies by the high mass of the *W* and *Z*. The challenge now is to expose the dynamics that broke the electroweak symmetry by having given mass to this pair of bosons while leaving the photon without mass.

The idea that spontaneous symmetry breaking is the source of mass for the fundamental particles, not just for the *W* and *Z* but also the quarks and leptons, has been the received wisdom among particle physicists ever since the work of Higgs and the rest of the Gang of Six. However, these pioneers uniformly sidestepped the question of how nature actually performs the feat. Guralnik, Hagen, and Kibble did not consider a potential field but assumed that the symmetry breaking nonetheless occurs; Brout and Englert recognized that there is a field but made no discussion of its form; Goldstone had a shape for the field—the wine-bottle analogy—as an example, which Higgs adopted in 1966. In 1964, Higgs merely assumed its general features. None of these seminal papers discusses the dynamics of what spawns the field responsible.

In 1964 the challenge had been to demonstrate that such a mechanism for generating mass is possible, even in principle. Thus, it is no surprise that these authors chose the simplest mathematical examples to achieve this end. It is possible that nature itself follows such a simple route, whereby the field spreads uniformly in all directions and the massive particle is simply a scalar boson, as in Higgs's original example. However, in other examples of hidden symmetry, such as those described in Chapter 8, the dynamics has been richer. In superconductivity, for example, fermions pair off to form composite structures that have the overall properties of a boson, in particular, undergoing Bose condensation.

It is possible that a similar dynamic occurs in particle physics, with novel forces gripping pairs of fermions such as the top quark, which then undergo Bose condensation. The top quark is intriguing in its own right—so much more massive than any other known particle, a pair of them would weigh in at hundreds of times the mass of a proton. This is the scale of energy where the breaking of electroweak symmetry is predicted to occur, so some theorists suspect that this coincidence is not an accident and that top quarks may somehow be linked to the hiding of electroweak symmetry. It is even possible that totally new fermions, bound to one another by hitherto unknown forces—as in a theory known as "technicolor"—might be the answer.[1]

The exciting feature is that, until the experiments are done, we do not know which if any of these will be revealed as the source of electroweak symmetry breaking. Whatever awaits us a century after Ernest Rutherford first discovered the atomic nucleus, which revealed atomic structure and led to the modern science of particle physics, the modern conceit is that the heat of the Big Bang congealed into matter and antimatter in perfect symmetry, the symmetry becoming hidden as the universe cooled, thereby providing the structures that Rutherford explored. After thousands of years of speculation and searches for the basic pieces of matter, the outcome of the revolution that pursuit of the Infinity Puzzle helped inspire is that physics has for the first time a testable theory about the origins of the material universe.

THE ROAD TO UNITY

A hundred years ago, the structure of matter was tantalizingly simple: Just two varieties of particle—electron and proton—were enough to build the periodic table of elements. The attraction of opposite electric charges held these constituents within atoms, their individual negative and positive values being counterbalanced at atom's length, leaving matter in bulk electrically neutral.

By the time that QED was born in the 1930s, two further particles had entered the stage. The electrically uncharged neutron, sibling of the proton, and the neutral neutrino, which partners the electron, maintained a hint of symmetry: The proton and neutron are two faces of the "nucleon," the electron and neutrino forming a dual pair of "leptons."

The beautiful alignment of the electric charges in each of these separate families itself hinted at some further symmetry connecting them all. However, what this might be is even today deeply hidden. Electric charge apart, much else is asymmetric. The neutrino is utterly neutral to electromagnetic forces, whereas the neutron feels intense magnetism: The magnetic poles of a neutron are swapped relative to those of a proton, and some two-thirds as strong. In size and mass also they differ: A nucleon is at least 1,000 times broader and nearly 2,000 times more massive than an electron, while it would take at least 100,000 neutrinos to balance the mass of a single electron.

Bjorken's intervention in 1967, which led to the discovery of lightweight tiny quarks inside the nucleon, also exposed symmetry at this deeper layer of the cosmic onion. The proton and neutron were revealed as compounds, built from "up" and "down" quarks. With electric charges that are respectively +2/3 or −1/3 relative to a proton, these form a pair, whose properties appear much more akin to those of the leptons. Lightweight, with no measurable size, the up and down quarks take on characteristics of leptons, which in the proton and neutron as a whole are hidden. The similarities are so marked that it is as if a quark is a lepton that has been "painted" with any of three colors, electrical charge also being fractionated three ways. That this is the case seems clear; how it is done is not.

The mass of bulk matter is dominantly due to the nuclei in its atoms and arises from the energy involved in trapping quarks within the breadth of each nucleon; it is not due to the Higgs Mechanism. By contrast, the mass of the fundamental quarks and leptons may be. The source of their masses is an enigma, but the Higgs Mechanism was the prime suspect as far as theorists were concerned; whether it is indeed responsible would be for experiment, and the LHC, to determine.

A clue that mass hides some underlying fundamental symmetry is that nature has tripled these various "flavors" of leptons and quarks, differentiating one "generation" from another by their masses. As the up and down quarks form a pair, so do charm and strange, and top and bottom; the electric charges and other properties of these latter pairs are identical to those of up and down, but they are heavier—in some cases much heavier (recall Figure 12.2). The leptons are also tripled. The electron has two heavier, but otherwise similar, counterparts—the muon and tau; there are also three varieties of neutrinos, their masses differing, although their individual magnitudes are not yet known.

Forty years of experiments since the theoretical breakthroughs that first set physics on this path hint that there is a fundamental symmetry at work among the particles and forces, but that it is hidden by their masses. It was because this symmetry is empirically hidden—in the case of the (massive) *W* and *Z*—that the Infinity Puzzle for the weak interaction was so resistant to being solved.

As we have seen, the proof in 1971 that we have a viable theory, built on the assumption that the *W* and *Z* acquire their masses due to hidden symmetry, inspires the general belief that this mechanism of hidden symmetry is indeed at work. While the empirical success of 't Hooft and Veltman's construction gives indirect support for the belief that the *W* and *Z* indeed acquire masses this way, this does not necessarily imply that leptons and quarks do also. To establish if spontaneous symmetry breaking is the source of their masses required production of the Higgs Boson,[2] and then by examination of the debris when it decays, we may determine if its affinity for various members of lepton or quark families is in proportion to their masses. This is why discovery of the Higgs, which as we shall see has now been achieved beyond reasonable doubt, became *the* central question of particle physics at the start of the twenty-first century.

If it should turn out that we have indeed exposed nature's fundamental symmetry, and The Mechanism that hides it, we will have found the source of structure in nature. A question that might be troubling you, though, is: "Why?" Why, given a perfect symmetry, with the energy of the Big Bang spawning particles without mass, did nature spoil it? Why are we the result of "nearly symmetry"?

To be honest, I don't know. However, had it not happened, I am sure that we would not be here having this discussion. This is not simply the trivial comment, "If it had been different than it is, then things would have been different." The reasons are more profound: A universe without asymmetry would not create structures on which life depends.

The pathways of life depend on the flow of electric current, the ability of atoms to gain and lose electrons as ions, and for energy to transfer over large distances from one source of power to a receiver, such as from the sun to the biosphere. If there were only the electromagnetic force, with the massless photon, all this could be achieved. However, to create the smorgasbord of elements from hydrogen to plutonium and beyond, transmutation of elements in beta radioactivity—the most noticeable manifestation of the weak interaction—appears essential. The fact that the *W* has mass is critical here. Had the gauge bosons remained without mass, the "weak" force would have had as much strength as the electromagnetic. In such a circumstance, instead of atoms being built by the attraction of opposite electric charge, with the potential to form ions, they would be bound by the exchange not only of photons but also of *W* and *Z*, democratically. This would radically change the nature of the forces and could even cause electrons to be permanently confined within atoms, as quarks are inside nucleons. It is thus possible that, at a fundamental level, chemistry could not happen were the fundamental symmetry not hidden.

Thus, the phenomenon of hidden symmetry, and its possible manifestation in the fundamental particles and forces, could be the key to sentient existence. Far from referring to the Higgs Boson as "the God Particle," one might suppose that a mathematical designer would have created perfect symmetry; in such a case, our existence is the legacy of whatever hid the original creation. Establishing how and what is the inspiration that has led to the LHC.

When William Waldegrave, in 1993, set the British physics community the challenge of describing the Higgs idea, Miller's clever use of a political analogy in what was, ultimately, a political exercise deserved to win the bubbly. However, opponents responded that it did not explain why the LHC had to be so big and expensive. After all, had not Rutherford discovered the atomic nucleus, in 1911, using little more than "string and sealing wax"? Why, by 2011, has science so utterly changed—in size, in the numbers of people in an experiment, and in the vast sums of money needed to achieve their ambitions?

I shall attempt to address these questions in the context of the LHC, but also as a preview of what may lie beyond, twenty or more years from now.

ONE HUNDRED YEARS

When Rutherford discovered the atomic nucleus in 1911, he did so by means of a small piece of apparatus on a tabletop, working mostly on his own, aided by a couple of assistants and technicians. Today, teams of physicists from more than a hundred institutions collaborate together. A single experiment at CERN or Fermilab may involve more than a thousand people. For scientists in a bygone age who, like Rutherford, were trying to discern what matter is made of, nature had done the hard work billions of years ago by locking energy into tiny particles, which eventually built up atoms and matter as we know it. Billions of years later it was relatively easy for scientists to examine this stuff, and discover how matter is constructed.

First atoms and then the atomic nucleus were revealed with tabletop experiments. Then protons, neutrons, and hordes of ephemeral particles were revealed with the advent of bigger facilities. Finally, as a result of Bjorken's insight, the quark and gluon seeds of hadrons were exposed. With the fundamental particles identified, and with the forces that weld them into large-scale structures also understood, by 1990 ambitions had become much grander. The breakthroughs that have become associated with the names Higgs, Bjorken, and 't Hooft have caused us to become less interested in what matter consists of and more in wanting to know how it came to be. The new paradigm, unleashed by the insights of the theorists, gave the prospect that experiment could show how the material universe emerged from the cauldron of energy that we call the Big Bang. To do so, we would need to re-create the conditions of the early universe in the laboratory and then use special cameras to record the event.

There are no mass-produced test tubes that can make an experiment of such magnitude. There is no "Big Bang Apparatus" for sale in the scientific catalogs, whereby we can experience the first moments of the universe in our living rooms, nor even in the laboratory of a single university. This is not mere hype; to voyage to the start of time you have to design and build all the pieces from scratch, transforming the earth, rocks, and gases of our planet into tools that extend our senses. Sand produces the raw materials for the nervous system of the ubiquitous computer chips that orchestrate the enterprise; hydrogen gas, from which protons can be stripped, provides the beams of the LHC; ores, dug from the ground, are melted, transformed, and turned into magnets able to guide beams of protons at 99.9999 percent of the speed of light. A myriad of other tools—the results of centuries of invention—have to be assembled. When all is done, these tools of the millennium can reveal the universe, not as it is now, but as it was at Creation.

Even as LEP began its experiments in 1989, plans for the next step were already afoot. The vision even then was that if the electroweak unification were confirmed at LEP, the final piece in the puzzle—the Higgs Boson—would move into the frame. 't Hooft's work had relied on the phenomenon of hidden symmetry. The mathematics in Quantum Flavordynamics, and the properties of the W and Z as measured in experiment, were already enough to imply that the Higgs Boson, or whatever nature really uses to break electroweak symmetry, would be revealed by experiments that lay beyond LEP's reach.

Why the rush? In 1983, when serious planning for the LHC began, LEP had not yet started operation; by 1990, as the LHC moved toward final approval, LEP had yet to produce its major results, so to contemplate building an even more powerful machine may appear profligate. The reason for such time scales reflects the nature of Big Science today. Projects like LEP and the LHC are so huge that they take years to build, following on from years of design, prior to which there have been years of dreaming and scheming to convince others to come on board. And not least, it is necessary to build political support in order that governments sign the checks for constructing and then operating these behemoths.

One result of this qualitative escalation of ambition is that it can take twenty years to think up, build, and perform an experiment. The useful lifetime of a facility tends to be similar to the time it has taken to design and build it in the first place. SLAC entered the scene as the mammoth electron accelerator in the 1960s; it did its major work during its first ten years.[3] The SPS at CERN using protons and antiprotons then came to fruition around 1980 after ten years of development; its discovery of the *W* and *Z* was its main accomplishment, and by the end of that decade it had been overtaken by the greater sensitivity that LEP offered. The inspiration for LEP developed in the late 1970s, as a result of the confidence born from 't Hooft and Veltman's breakthrough with Quantum Flavordynamics. After much political maneuvering—for nothing of such a magnitude in physics had been contemplated before—the project was approved in 1981 and construction began in 1983. It took six years to build and began experiments in August 1989. LEP, as the electron successor to the SPS, operated for eleven years. The LHC, which at a cost equivalent to $10 billion has also been an exercise in politics and high finance, has taken twenty years to design and build. After an equipment malfunction in 2008, it began operation in November 2009 and is likely to produce discoveries for twenty years.

When an accelerator ends its operation, the physics output does not stop. The vast amounts of data that its experiments have produced are pored over for years. Ten years after LEP ended, some research papers are still emerging. Analyzing the results, and redesigning the capabilities of the accelerators and the detectors in light of the unexpected, can take many years. A substantial part of a scientist's professional life may be taken up with just two or three projects. At any moment you may be analyzing the results of one experiment, taking data at another, testing equipment for the next experiment, and sitting on committees discussing the future facilities that are most likely to be pushing the frontiers twenty or more years in the future. So the LHC was not born in a rush; rather, the infancy of LEP was the natural time for serious thinking about the next step to begin.

A SIGHT OF HIGGS

With the SSC's demise in 1993, support for the LHC project grew, aided by the fact that LEP was proving fortuitously successful. It was around this time that LEP's data were beginning to confirm that the theory of the weak force, as demonstrated by the work of 't Hooft and Veltman, is correct.

As we saw on page 30, while energy is conserved over ordinary time scales, for brief moments the account can be overdrawn. This facility allows the Z to transform momentarily into a "virtual" top quark and antiquark, which then coalesce back to a Z during its brief lifetime. Such quantum fluctuations leave an imprint, which can be manifested in some of the experimental measurements. That is what happened at LEP (see page 326). When the precision data were compared with calculations in 't Hooft and Veltman's QFD theory, the virtual presence of top was revealed.

As we have seen, the subsequent production of the top quark, at Fermilab in 1995, in line with those predictions added to the general confidence that physics had found the Golden Path. Then, as the precision of experiments at LEP continued to get better, exceeding even one part in ten thousand, and with the top quark in the bag, further more subtle discrepancies begin to show up between LEP's data and the theoretical predictions.

As the dawn heralds the sunrise, so these quantum effects preview phenomena that lie beyond the current energy horizon. Excitement mounted when it was realized that here, possibly, were the first signs of the Higgs Boson. It began to look as if the requirements for producing the Higgs Boson in the laboratory were probably tantalizingly just out of the reach of LEP but achievable at the LHC.

There is one further feature of these results, which touches on the Infinity Puzzle of electroweak interaction and the role of the Higgs Boson. This concerns the mass of the electron.

The Higgs Boson made its entrance as The Mechanism for giving masses to the W and Z. The idea that leptons and quarks also acquire their masses through this mechanism was suggested in Weinberg's seminal 1967 paper, but in principle the mass mechanism for fermions is independent of what happens with gauge bosons. When LEP was operated at its maximum energy, it was able to produce large numbers of W^+ and W^-. This is where an intriguing thing happens.

The quantum processes that are involved include those illustrated by the three Feynman diagrams of Figure 17.1: The electron and positron can annihilate into a photon, or into a Z^0, or can exchange a neutrino. However, when the fact that electron and positron have mass is included in the equations, nonsense erupts. For example, if you performed the experiment at high energy—above 1,000 GeV say—the calculations imply that an electron and positron produce W^+ and W^-, in contrast to some other possibility, with a chance that exceeds 100 percent.

The Higgs Boson comes to our rescue, at least in theory. As the peak and trough of two waves can cancel, so can the wavelike nature of particles cause cancellations among the contributions of various Feynman diagrams. That is what happens here. If the electron's mass comes from its interaction with a Higgs Boson, then there is a fourth diagram to include: where an electron and positron annihilate, producing a Higgs that then decays into the W^+ and W^- (Figure 17.1d). The quantum waves associated with the Higgs Boson cancel those from the other three processes, giving sensible answers (chance never exceeds 100 percent) consistent with the empirical data at the large, but finite, energies of LEP.

When these theoretical calculations were compared with the data, everything worked best if the Higgs Boson—or, at least, something that gives mass to particles and can be simulated in calculations as if it is a simple spinless boson—exists within the reach of the LHC.

SUPERSYMMETRY

While the inspiration for the LHC was to discover how nature hides electroweak symmetry, there are many other exciting possibilities waiting. Some of these ideas—that there are extra dimensions of space into which gravity leaks, hence its apparent feebleness for the individual particles in the universe of which we are currently aware, or that miniature black holes might be produced and evaporate—make it hard to distinguish science fact from science fiction. However, there is one further aspect of the Infinity Puzzle that the LHC might resolve. That is the idea that the particles and forces so far discerned are part of a more complete "supersymmetry," which itself became hidden at energies within the reach of the LHC. If this is confirmed, it could provide a remarkable symbiosis between particle physics and cosmology with profound implications for our place in the universe.

The relation between the present Standard Model and the more profound theory of reality, which eventually will subsume it, can be compared

FIGURE 17.1
THE NEED FOR HIGGS

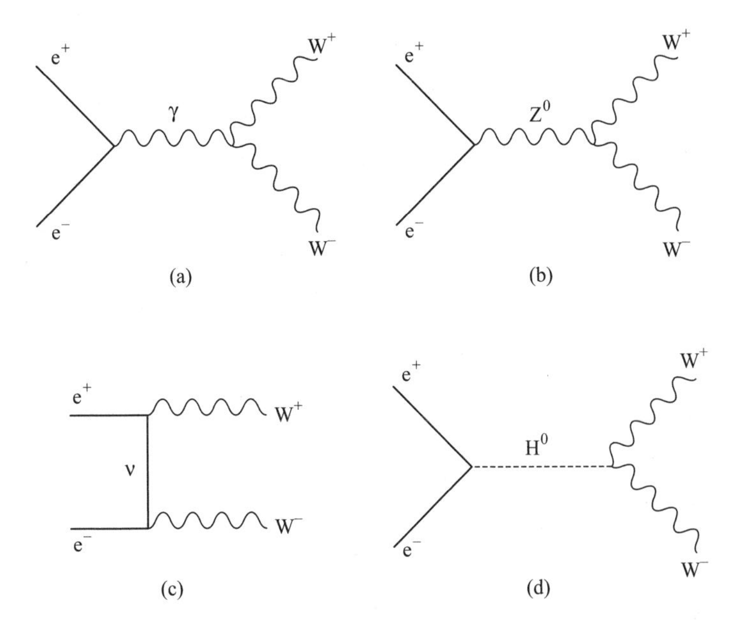

The most important Feynman diagrams contributing to electron-positron annihilation, which produces W^+ and W^-, according to the standard model of the electroweak interaction. These include possible intermediate virtual states of (a) photon, (b) *Z*, and (c) the exchange of a neutrino. To complete the computation, and to agree with experiment, it is necessary to include a further contribution, such as (d), containing a virtual Higgs Boson.

to Newton's Theory of Gravity and Einstein's Theory of General Relativity. Newton's theory described all phenomena within its realm of application but is subsumed within Einstein's grander edifice, which reaches parts that Newton's theory cannot. So it is with the Standard Model. It describes phenomena from energies on the atomic scale all the way to those at which the LHC is beginning to explore: unimpeachable success over twelve orders of magnitude. However, while spontaneous symmetry breaking is believed to be the source of mass for the fundamental particles, there is currently no explanation for the specific magnitudes of those masses or of other parameters—such as the strength of the forces—which appear so critical for our existence. At present the empirical magnitudes of these quantities are inserted into the equations of QFD and QCD by hand. The origin of their values lies in some richer theory of which the Standard Model will one day be seen to be only a part.

Finding this richer theory occupies the dreams of physicists.[4] It will need to combine Einstein's General Relativity with Quantum Field Theory—a challenge that has yet to be resolved. Many theorists suspect that the answer will involve a radical concept: that one can never force a lump of charge to a point because particles themselves are not points but ultimately vibrating strings with shape and form in dimensions beyond the three dimensions of space and one of time accessible to our senses.[5]

Although it is believed that the dynamics of such strings will be revealed only at energies far beyond those of the LHC, nonetheless, one aspect underlying the theory could be revealed: the concept of supersymmetry, affectionately known as "SUSY."

All known laws of physics remain true when particles are moved from one point to another (invariance under "translation"), rotated in space, or "boosted" (have their velocity changed). These invariances are known as symmetries. In the 1970s a further symmetry was discovered lurking in the mathematics of relativistic field theory. This changes the spin of a particle by one-half—thus converting a boson into a fermion or vice-versa while leaving other properties such as mass and electric charge unchanged. Deep mathematics, beyond the scope of this book, have proved that this is indeed the one remaining possible symmetry in the entire formulation of relativistic quantum field theory. Its discovery would therefore close the book on the symmetries of space and time.

If SUSY is fundamental in nature, then for every variety of boson there exists a fermion with the same mass and electric charge, and to every fermion there exists such a boson. Clearly, SUSY cannot be exact: There is no boson with electric charge and mass equal to that of the electron, for example. Such a superelectron (or "selectron") would not only be easy to see but probably also destroy all structure in the universe: Recall that structure occurs because fermions are like cuckoos, whereas bosons are like penguins. Electrons are fermions, but selectrons are bosons. If lightweight bosons with positive and negative electrical charge were to exist, their mutual electrical attraction would create condensates that would destabilize everything that we know. Thankfully here, as before, the hiding of the symmetry is necessary for the menu of established particles and forces to do their work.

It is suspected that, once again, spontaneous symmetry breaking has hidden a fundamental symmetry—SUSY—making some particles light enough to create familiar stuff, while their superpartners, such as the selectron, are so massive that they have remained hidden so far. If this is true, then hidden symmetry seems to be essential for all structure. There are indirect hints that in the case of supersymmetry, selectrons and other supersiblings of known particles could be revealed at the energies of the LHC. The hope that the known forces might merge into a single superforce at high energies seems mathematically to work best if the energy threshold for exposing supersymmetry is that of the LHC. Along with the possible discovery of new families of superparticles, supersymmetry would also imply that the Higgs Boson is itself but one of a family—its supersymmetric sibling being known as the Higgsino.

How much of this is true, and how many of these bizarre names will one day become part of the lexicon of science, is for the future. However, there is one aspect of supersymmetry that could prove most profound: SUSY might be the source of the dark matter that appears to dominate the material universe.

In cosmologists' simulations of the large-scale universe, dark matter is the glue that binds the galaxies, preventing them from spiraling apart and probably having helped to form them in the first place. From the motion of galaxies we can infer that 90 percent of the universe consists of stuff that is dark—in the sense that it does not shine in any wavelength of the electromagnetic spectrum. This implies that it is impervious to the electromagnetic force. Lightweight neutrinos have these properties, but the models of galactic dynamics work best if the dark matter consists of massive particles. However, no such particles are known in the standard model.

In SUSY, such beasts are expected. Moreover, if the lightest superparticles are electrically neutral siblings of the photon or gluon—known as "photino" and "gluino"—they could be stable and form large-scale clusters through their mutual gravitational attraction. This is analogous to the way that the stars form, but whereas stars made of conventional particles feel strong and electromagnetic forces, which enable them to shine, the conglomerates of neutral "-inos" have no such ability. Inert lumps of dark matter populating space accompanied by occasional "conventional" particles, such as make us, could be the nature of the universe.

If SUSY particles are discovered, and turn out to have these properties, it would prove a beautiful convergence between high-energy particle physics and cosmology. Having once believed that we were at the center of the universe, and then seen our location shifted into the outer reaches of one out of billions of galaxies, the stuff that makes us might turn out to be no more than flotsam in a sea of dark matter.

BIG BANG DAY

When the Large Hadron Collider was approved in 1993, some wondered if it would ever become reality. Huge technical challenges had to be overcome; nearly every particle physicist on the planet would have to join in a collective effort; new technology would have to be invented.

Although the LHC was to use the tunnel where LEP had been, the magnets that had steered the electrons and positrons would be useless for bulky protons. The whole infrastructure of LEP had to be removed, and then 10,000 new magnets designed, built, tested, delivered, and installed. The enterprise became truly global. An engineer in Novosibirsk might have designed a special bolt, say, while a technician in Japan constructed the piece of apparatus into which that bolt would go. It would only be after the various pieces were delivered to CERN in Switzerland that they could be fitted to one another. Thankfully, modern communication via the World Wide Web is so powerful that this jigsaw of collaborations came successfully together.

The energy in the beams of protons is so vast that any resistance to the electric current in the power supplies generates large amounts of waste heat. So the magnets had to be made of superconducting material, free of electrical resistance, which in turn required them to be cooled to within 1 degree of absolute zero of temperature—colder than outer space. The entire twenty-seven kilometers had to be like this. As a result, the LHC has become the largest refrigeration plant on the planet. Superconductivity, which inspired the theoretical ideas leading to the Higgs Boson, is the technological key to the LHC's operation.

After fifteen years of planning, design, and construction, "Big Bang Day" was scheduled for September 10, 2008, with the first current of particles ready to enter the LHC. After years in the background, Peter Higgs suddenly became a celebrity, the subject of articles and interviews in the media. His visit to CERN earlier that year—to see the experiments that hope to find his eponymous boson—was like that of an aging rock star.

As the scope of the LHC became a world news sensation, unusual bedfellows emerged. Scaremongering began, with blogs, newspaper articles, and even the BBC advertising claims that the LHC—in colliding particles at energies beyond anything previously achieved in experiments—might make black holes and destroy the planet. The director-general of CERN was so concerned by these scare stories that he even set up a committee of scientists to evaluate the likelihood of this. They duly concluded that there was no such possibility and wrote a report in the hope of calming people. However, for conspiracy theorists this merely added fuel to the fire. Some of the media whipped up this alarm, such that on Big Bang Day, a news headline read: "If You're Reading This at 10 A.M. on 10 September 2008, the World Has Avoided Destruction."

In the special edition of the CERN house journal celebrating the start of the LHC, I made the analogy of nature revealing itself through a picture sewn on a tapestry, writing that "we have just devoted 15 years to making a needle [the LHC] and today for the first time [by injecting beams of protons] we will attempt to put thread through its eye."[6] Only after many pieces of thread had been prepared could the actual sewing begin, and then it would take many months, or even years, before even the outline of the image began to emerge. Patience is needed; discovering the Higgs Boson, supersymmetry, or whatever else nature has to surprise us is a program that will take many years. On Big Bang Day no camels passed through the metaphorical eye of my needle, merely the first pieces of thread. But twenty years after the dream began, at last beams of protons successfully sped around the ring. The machine worked!

And then, like the straw that broke the camel's back, the failure of a minor piece of equipment rapidly escalated to disaster. A single spark caused a small hole, through which ultracold liquid helium began to escape. This vaporized and froze everything for meters around. Wonderful stalagmites of ice filled the tunnel and ruined sensitive apparatus. It took a year to repair before the LHC was ready to start its first exploration. (You can see how it is doing today by following http://public.web.cern.ch/public/.)

The quest for the Higgs Boson has been likened to searching for a needle in a million haystacks. The beams—protons—are themselves swarms of quarks and gluons. Quantum Flavordynamics predicts how the Higgs Boson is created by collisions between quarks or gluons, and, thanks to Bjorken's insight into how protons transport these constituents, one can calculate under what circumstances collisions between protons will create a Higgs Boson. As Bjorken's work led to the discovery of the *W* and *Z*, and has enabled Quantum Flavordynamics and the theoretical ideas of 't Hooft to be confirmed, so it has now enabled the Higgs Boson to be found—beyond reasonable doubt. Our ability to filter the signals from the noise, and eventually decode the message, is because Bjorken and Feynman showed how protons are the vehicle for transporting swarms of fundamental partons.

Epilogue

THE BONFIRE OF THE INFINITIES

EARLY LHC DEVELOPMENTS

On March 30, 2010, the Large Hadron Collider finally started to show what it was capable of, smashing together two beams of protons with energies of 3.5 TeV. The LHC's planned research program could at last begin. The occasion was marked by breathless "revelations" from some quarters that a discovery would come soon, as if they alone were privy to what in reality was common knowledge: barring accidents, the status of the Higgs Boson would be settled, one way or the other, within a few years.

In theory, looking for the Higgs Boson is analogous to trying to detect a pair of loaded dice by rolling them inside a closed box. If sixes happen every time, you can be certain that the dice are special—in our analogy, the boson exists. However, if sixes occur on average once every thirty-six throws, the results are no more than chance and there is no Higgs Boson.

One major problem is opening the box: at the LHC this is done by smashing two protons head-on and detecting the debris. Recording the characteristics of the by-products—the varieties, energies, and flight paths of the constituent particles—is like determining the numbers on our dice. Correlations among the particles may show them to be the debris from the decay of an ephemeral boson—each die being a six.

In practice, however, the task is much more complicated. The sixes don't always turn up; instead, they turn up slightly more often than mere chance predicts, and you have to decide whether this is significant—evidence for the "Higgson" or "higgsons," the name depending, as we will see, on what else the world might hold—or mere vagaries of luck, where there is no real boson at all.[1] To be sure, you need to do more tests, checking on the demon time after time, until either the excess of sixes becomes convincing, or instead gradually dies away toward randomness. In the physics jargon, one is trying to decide between signal and noise.

The beams at the LHC circulate 11,000 times a second. Two independent teams of physicists—known as ATLAS (A Toroidal LHC Apparatus) and CMS (Compact Muon Solenoid)—have designed their detectors to be sensitive to collisions where higgsons are produced. The longer that the LHC takes data, the more observations are possible, and the more confident one becomes in discriminating signal and noise.

That's the good news. The bad news is that most collisions reveal nothing for or against higgsons. It's as if our dice-filled boxes don't hold dice at all, but instead a roulette wheel, or coins, or the tools for some other game of chance. These may be interesting to other physicists, but the precious dice, the ones relevant to the Higgs search, are very rarely on view.

Fortunately, if nature does use the mass mechanism in the same way as originally proposed in 1964, it is possible to predict not just how the dice should fall, but how often we will see them at all. The latter is in essence the chance that a Higgs Boson—or things that might be confused with it—will be produced, and the former is whether you can be sure to have identified the real Higgs from imposters.

In the early months of the LHC, luck was in: dice came up more often than expected, and by the middle of 2011 the accumulation of a slight excess of sixes began to create excitement.

The first public presentation of results was scheduled for a major conference in Grenoble in July 2011. Rumors began to circulate. The parents of a leading scientist in ATLAS are neighbors of Peter Higgs in Edinburgh. They received an enigmatic message that they should tell Higgs to "watch for an announcement in Grenoble."[2]

Among the data were thousands of examples where the debris in collisions included a pair of photons. It is possible that these arose because a Higgson had been produced and then decayed into two photons. By measuring their energies and momenta it is possible to calculate the mass of their parent. There are many other ways that photons can be made, but in such cases their energies and momenta combine to "random" values. So a graph of the masses corresponding to these pairs will show a random distribution—the "background"—together with peaks at values where the photons are the decay products of genuine short-lived particles. Looking for such a peak on top of an underlying background was the goal.

In practice, with a limited amount of data, the background is not smooth but is itself lumpy. So distinguishing a genuine bump, signifying a particle such as a Higgs, from random fluctuations of the background is difficult. The more data one accumulates, the smoother the background becomes and the clearer genuine signals will be.

The size of signal to background is measured by a statistical quantity known as "sigma." Sigma is a measure of the chance that random events have conspired to make an apparent signal; the larger the value of sigma, the more certain you are that a signal is real. A sigma of 2 is the first hint of possible interest, indicating that we have about a one in twenty chance of being fooled; 3 becomes more serious, with odds against being about one in three hundred. Particle physicists regard 5-sigma to indicate a discovery, as the odds of being fooled being about one in two million.[3] This is similar to the chance of tossing a random coin heads up twenty-one times in succession, or betting successfully on the same number coming up on a roulette wheel four times in a row (in an honest casino).

The actuality was rather less dramatic than the build-up had promised, the value of "sigma" being around 2 to 3. The media presented headlines suggesting that the Higgs Boson had been found, meanwhile covering themselves by quoting an ATLAS scientist: "We cannot say anything today but it's clearly intriguing." A leader of CMS explained that more data were needed to decide if the correlations—the excess of sixes in our analogy—were due to a "statistical fluctuation or hints of a real signal."

A major conference in Mumbai in August 2011 was the next public outing of the results, based on more data than had been available at Grenoble. Hints of a signal were still present, but a spokesman said there was "no striking evidence of anything that could resemble a discovery" in the hunt for the Higgs Boson.

Many media interpreted this as a metaphorical U-turn. "Ripples of excitement" of July were replaced by "dashed hopes."[4] That hopes of immediate discovery were dashed was true, but in some quarters this was widely and falsely escalated to suggest that the Higgs Boson might not exist at all. Absence of evidence is not evidence of absence—certainly not with the limited amount of data available in mid-2011. This was not going to be a short-term business. Cool heads advised waiting until the end of the year, by which time there would be enough data to clarify some, though not all, of the uncertainties. Nonetheless, whenever I gave a talk about this, I found myself being asked: "What if there is no Higgs Boson?"

11.12.13 (2011, DECEMBER 13)

The LHC continued to work well, and by December excitement began to grow once more.

In the second week of December, members of CMS met behind closed doors at CERN to assess their results and decide what they were confident enough to announce. A similar private meeting took place in the ATLAS conference room. Attempts by television crews to be present in the closed discussions were politely rebuffed, and requests for advance news were met with smiles.

On December 13, representatives of the two experiments presented their results in an open session in the CERN auditorium. This was the first time that ATLAS learned what CMS had, and vice versa. For the rest of us, this was the first news at all. If discovery of the Higgs Boson was going to be announced, everyone wanted to be able to say: "I was there." Like passengers awaiting boarding of an overcrowded aircraft, fearing being bumped, scientists crowded at the doors of the auditorium, anxious for seats. More than an hour before the scheduled start, the hall was full, with people sitting on the steps, leaning against the walls, or standing like commuters on a rush-hour train. The Internet was saturated with attempts to access the webcast, and the event went viral on Twitter. The results were tantalizing.

To those unable to read the physics code, initial reaction was that nothing dramatic had occurred. The experiments were able to rule out the existence of a Higgs Boson over almost the entire range of masses accessible to the LHC, with the exception of a small domain around 125 GeV. The ATLAS and CMS data individually gave tantalizing hints that something unusual might be present in this region but were still far from proof.

Independent confirmation is an axiom of science, which in this occasion created an enigma: the masses that ATLAS and CMS favored for the putative Higgson were almost, but not quite, the same. Some feared that this discrepancy might be evidence that we were being tricked, reading more into some chance fluctuations and misinterpreting as signals what was, in reality, noise. Most, however, felt optimistic that the two experiments were broadly in agreement, and that 125 GeV was the place to look. Expectations were that, all being well, the answers should be clear by the end of 2012, possibly earlier.

THE DAYS BEFORE JULY 4

Twenty years from now the textbooks may say that December 13, 2011, was the day when the Higgs Boson first announced its presence, and that July 4, 2012, was when hypothesis matured to discovery. That CERN would make an announcement that day had been known for weeks. What was not known was whether they would announce a clear discovery or simply more enticing hints. Rumors on the Internet abounded, feeding one on the next such that it was impossible to tell if there was any signal or just noise.

On June 2, I was at a literary festival in Hay, Wales, with the CERN director- general, Rolf Heuer. We discussed the latest news. Heuer agreed with me that the ability of the Internet to spread misinformation around the globe at the speed of light is a real problem. It was also clear that even he did not know what ATLAS and CMS would say.

The LHC had taken as much data in the first half of 2012 as it had accumulated previously in total. As the amount of data had doubled since December 2011, the anticipated statistical significance of any genuine signal could be expected to increase by the square root of 2. So long as the hints from 2011 were indeed signs of a genuine Higgson, simple statistics led us to anticipate a sigma around 4 or above.

For ten days at the end of June, I was privileged to be lecturing at a summer school on particle physics in Erice, Sicily, together with Peter Higgs. It was like being at the eye of a hurricane. While rumors raged all around the globe as to the likely birth announcement of the Higgs Boson, the central protagonist was in a haven of peace, at the Ettore Majorana Centre, on a Sicilian hilltop. We were there together with several physicists from the CMS and ATLAS experiments—they either knew little more than we did or were remarkably leak-proof—as well as Gerard 't Hooft, whose breakthrough had brought the quest alive and who also knew nothing of what CERN had found. The general ignorance was not surprising: like guardians of a precious secret, the collaborations had made internal firewalls, so that only a handful had access to all the pieces.

Over dinner on June 26, I asked Higgs how he felt, just a few days before the hopes of nearly half a century could be confirmed or dashed. He was remarkably sanguine. He had no plans to go to CERN; he was cautious about being pressed by the media to comment on something that might still have been tenuous. Even Higgs himself would have had to admit, like Manuel in *Fawlty Towers*: "I know nothing." However, he thought, "Things could be coming to a head in the autumn."

That was the case until at least June 28. After that point, however, opinions were beginning to change.

There was no need for Higgs's neighbors to alert him to "watch out for CERN on July 4," but I contacted their son and cheekily suggested that if there was any message for Peter, he was in Erice with me. The reply was cautious: all that could be revealed, even to Higgs, let alone me, was that the ATLAS data were good enough to be able to say something "one way or the other." As one cannot prove a negative in science, this looked positive. Similar clues came to us from CMS also. But even with the most optimistic interpretation, until the results of ATLAS and CMS were compared, no one would know whether they agreed.

On June 29 my diary of the saga expressed cautious optimism, with the conclusion: "Place your bets. This is, after all, a matter of weighing chance." Over late-night drinks, a group of us managed to convince ourselves that the signs looked good and that a forty-eight-year wait might soon be over. The following morning, these arguments were shared with Peter Higgs. He seemed convinced, but dampened the enthusiasm somewhat by saying: "In any event, I haven't been invited to come to CERN yet."

Contrary to reports in some media that weekend, CERN never issued any formal invitations to the five surviving theorists. How could CERN have done so without being certain both that the experiments had a confirmed discovery, and moreover that it was the eponymous boson? Nevertheless, the rumor grew.

During the last week of June, it became known at CERN that the two Americans in the Gang of Six, Gerry Guralnik and Dick Hagen, had decided to miss Independence Day at home in order to be present for the announcement. This led CERN to alert the remaining trio of survivors, including Higgs, on June 30 that they would be welcome to attend. Although the source could release no details, suspicion was that "Peter [Higgs] will regret it if he is not there."[5] So the presence of four of the theorists (Tom Kibble remained in London) had less formal significance than many believed.

Following this, events moved quickly. At Erice we learned that ATLAS had finalized their presentation and that their signal for a discovery was near to the critical "5-sigma." Peter Higgs's response was understated: "That sounds good enough," with the afterthought, "provided that CMS are more or less in agreement."[6] And disagreement, of course, was one big possible hiccup remaining.

When the CMS leader, who was due to come to Erice the next day, July 1, canceled at the last minute, this was taken to be positive; the alternative possibility—that he had discovered a bug in some crucial program after two years—seemed unlikely. Higgs rebooked his tickets so as to fly from Palermo to Geneva via Rome, on July 2. I was due to return to the UK via Milan. Aware of the approaching media storm, and in the hope of protecting Higgs for some while further, Alan Walker, his aide from Edinburgh, took a photo of Higgs and me after check-in. This photo was tweeted to media, with reference to our separate travels but written in such a way that it would be easy to expect that Higgs would be arriving in Geneva from Milan, rather than Rome. This seems to have bought him one more day of privacy.

A CERN car whisked him from the Geneva airport to lodgings at the site. The next day, July 3, Higgs got a preview of life as a celebrity.

Arrangements had been made for him to have lunch in a private area. To reach it involved passing through the main restaurant, and although his aides cleared the way, en route he was surrounded by scientists wanting his autograph, to shake his hand, to take his photo. One image posted on the Web showed a bald head in the midst of a crowd of people, captioned "Looking for Higgs: It's hard to see the signal for the noise."

At last, July 4 arrived.

THE DISCOVERY: A HIGGS BOSON OR *THE* HIGGS BOSON?

The frenzy exceeded even that of December 2011. The auditorium was locked overnight to prevent opportunistic campers from taking early occupation. Expectations were high, and the hints could hardly have been stronger: Higgs, Englert, Guralnik, and Hagen would be there; directors-general, past and present, had reserved seats awaiting them. In the excitement, CERN had jumped the gun by releasing a video on the Web of a CMS spokesman describing the significance of the "discovery." This was removed as soon as the error was realized, but the genie had already escaped.

A Higgson has its greatest affinity for massive particles. At a mass of 125 GeV it is too light to decay into pairs of massive *W* or *Z* bosons, let alone into the even heavier top quark and antiquark, its dominant direct coupling being to a bottom quark and antiquark. This combination is hard to isolate at the LHC but has been seen at Fermilab.[7]

Although too light to decay directly into *W* or *Z* bosons, or top quarks and antiquarks, the Higgson can do so transiently, courtesy of quantum uncertainty, so long as these particles immediately self-destruct, leaving signals in the form of a pair of photons or four leptons. CMS and ATLAS each announced that they saw clear signals of a boson decaying into two photons and into two *Z* (revealed by the decays into four leptons).[8] First CMS made their presentation, and the sight of a clear signal around 125 GeV in the two-photon data was exciting, but it was when the speaker announced that their data on this together with four leptons reached the magic "5-sigma" that the audience burst into prolonged applause. The presentation of ATLAS followed, with a similar result, but this time the confirmation that the two experiments agreed, and with each at "5-sigma," led to cheering.

The cameras caught Peter Higgs wiping a tear from his eye, and he was not alone. That equations written on paper can know nature, and that forty-eight years later experiments can prove this, is awesome—an overworked adjective, but on this occasion justified. This was not just a discovery of interest to specialists, but one that touches on the nature of the material universe. If you have experienced a total solar eclipse, recall that moment when the last sliver of the sun disappears and a shimmering black hole appears in its place: there is a momentary intake of breath, cheering, and many adults are in tears. Such outpourings can occur when listening to your favorite music, when some phrase or sequence of chords sends tingles down your spine. In such moments, you may sense the profound sensations that many of us felt as the full import of the discovery began to sink in, as an epoch of hypothesis ended and a new era of certainty was born.

There was one wrinkle: the actual rate at which the Higgson produces pairs of photons seems slightly different to what the standard model would predict. However, it remains to be established whether this is a genuine phenomenon, and if so whether its cause is mundane—for example, due to uncertainties in the production mechanism—or a hint of novel physics.[9] The one-line summary, following the announcement, is that the collection of results is in line with the 125 GeV particle being a Higgs Boson.

It is not yet *the* Higgs Boson. Indeed, theorists have wondered for several years whether the boson is a singleton or just one member of an extended family, thanks to growing maturity of our understanding of particles and forces. Concepts such as supersymmetry have emerged, as we saw earlier. If "SUSY" is indeed part of reality, and not just some mathematical musing in theorists' notebooks, there could be a family of higgsons, which includes fermions as well as bosons. Thus the sighting of a new particle could be the entrée into an entire new realm of particles. Whether it is "the one and only," or the first sighting of an extended family, was not for July 4. That question is for the future.

GLITTERING PRIZES

During the year leading up to the discovery, the discussion sessions after my talks always featured two questions. One was: "What if there isn't a Higgs?" This we can now leave for philosophers to debate: the scientific method has established the nature of reality, and what was once speculation is now lore. The other question, which assumed that the Higgs Boson does exist and would be found, was most prominent in a gracious review of the original version of this book in *The Economist.* This review said that although the book "offers no unambiguous advice to the Nobel committee . . . the judges would be wise to give it a thorough read anyway."[10] In its bluntest form, the second question, as posed in many media articles is: How do six people who had an idea share a Nobel Prize that is limited to three? The answer is: they don't. To paraphrase George Orwell: all may appear equal, but some are more equal than others.

At a meeting at the Royal Irish Academy in Dublin, just ten days after the discovery, I sat on a panel that included Rolf Heuer, director-general of CERN, and Steve Myers, Director of Accelerators at CERN, and as such, responsible for the LHC. After ninety minutes of questions and answers, a prominent media star asked: "So who gets the Nobel Prize?" Heuer, to his credit, responded immediately: "I do not care. This is a triumph for humanity."

And, in a nutshell, that is the real legacy. But the question persists, and, as we have seen throughout this book, whether one approves or not, the reality is that a huge majority do care. For just one out of a multitude of such conjectures, see *Atlantic Wire*, "... Who Will Win the Nobel Prize?," not least so that if my following summary turns out to be wide of the mark, you can compare it with other assessments in the immediate aftermath of the discovery.[11]

First and foremost, in my opinion, the discovery was a triumph for engineering—the design, construction, and successful operation of the most extensive, complex precision instrument in history. If the new Queen Elizabeth Prize for Engineering, which has ambitions to rival the Nobel, were to recognize the pioneers who conceived, designed, and successfully commissioned the Large Hadron Collider, few would argue; although deciding whom to honor out of a cooperative involving hundreds—possibly thousands—of people, working together in an effort that spanned more than two decades, may be a challenge. Second, the discovery of the Higgs is not just a triumph for theorists. Indeed, the discovery would not have been possible without the skills of hordes of experimenters, who used this amazing machine to find the nature of reality.

In the discussion of who deserves the Nobel Prize in conjunction with the discovery of the Higgson, there are both candidates who seem obvious, and at least one whose work I insist not be overlooked. In my opinion, one of the greatest unrecognized achievements was Bjorken's. The discovery of quarks followed from his work in the 1960s, and in one form or another this breakthrough has underpinned decades of research. It led to Quantum Chromodynamics, and thereby the modern ability to design and interpret the results of experiments involving quarks and gluons. Nearly all experiments today rely on Bjorken's techniques and their derivatives. In particular, these enabled Rubbia to design the means to produce the *W* and *Z* in proton-antiproton collisions at the SPS and, critically, to identify those jewels by means of the signature that Bjorken himself had proposed. Having inspired QCD and having helped establish the players in Quantum Flavordynamics—and so having already led to several Nobel Prizes—Bjorken's ideas brought us to the threshold of finding the source of electroweak symmetry breaking.

Now that these experiments have sighted Higgs's boson, which is the first step toward exposing how electroweak symmetry is broken, who should take the credit for the idea? We have identified a long line of theorists—the Gang of Six (sadly now five), plus Anderson and others before them, and 't Hooft, for rediscovering the principle in his own magnum opus—who may each lay some claim.

Higgs's name is a household word, but as we have seen, Brout and Englert beat him into print, and some of their work was anticipated by Migdal and Polyakov. The latter duo, having received negative and incorrect criticisms from skeptical senior scientists, published nothing until after the above had done so. Guralnik, Hagen, and Kibble too, independently aware of the basic ideas around the same time, and ignorant of what Brout, Englert, and Higgs had done, were scooped.

Brout sadly died in 2011, unaware that in his intellectual dreams he had glimpsed reality. This leaves Englert undisputed as first to publish the mass mechanism in relativistic quantum field theory. Higgs, uniquely, mentioned the massive boson, which now carries his name. By 1966 he had noted that its decays can prove whether the mechanism is truly in nature's lexicon. Finally, as I explained earlier, there is the singular role of Tom Kibble.

In 1967, Kibble used the pieces of intellectual lego that he had earlier constructed with his collaborators, Guralnik and Hagen, to build the empirical model whose truth is now being revealed. He showed how the photon stays massless while other bosons, now realized to be the *W* and *Z*, gain mass. Kibble's work also inspired Weinberg and Salam to incorporate these ideas in the works that led to their Nobel awards in 1979. As one of the Guralnik, Hagen, and Kibble team, who were narrowly beaten to the tape in 1964, he is also special in having been involved throughout the entire construction. So, of this sextet, Englert, Higgs, and Kibble, in my judgment, have special roles in the saga.

In addition to these central players in the climax of this saga, we should not forget Bjorken. As I said earlier, our ability to plan and interpret several decades of experiments—not least those at the LHC—flows from Bjorken's work in the 1960s, which led to the proof of quarks and gluons, and became the inspiration for QCD. As for Goldstone, who inspired the work of the Gang of Six; the thousands of engineers who made the Large Hadron Collider the most remarkable scientific instrument in history; and the experimentalists who have used it so successfully, they all have experienced the profound mystery of knowing nature. The solution to the Infinity Puzzle is now becoming clear. Politics and sociology remain.

The events of July 4 occurred too late to influence the destination of the 2012 Nobel Prize. Nonetheless, during the weeks leading up to the announcement on October 9, there was much speculation in the media, and Peter Higgs was pestered for interviews. However, as most physicists expected, the prize went to another area of physics. Two days later I spoke at the University of Edinburgh, Higgs's home institution, about his life as a boson. I discovered that for Higgs, at least, every cloud has a silver lining, as after being inundated by media calls for many weeks, Higgs had once again been able to "enjoy a quiet life." For how long?

WHERE NEXT FOR THE LHC?

Powerful though the LHC is, it takes us less than halfway, in orders of magnitude from the heat energy of a summer's day to the Planck energy, where the as-yet-unknown quantum theory of gravity dominates. We can ignore quantum gravity in practice, but the very lack of any observable effects also leaves us clueless on how to proceed in constructing the theory that is necessary in principle.

In fact, it confronts us with a somewhat uncomfortable possibility: that whereas in the first half of the journey we have life, molecules, atoms, the atomic nucleus, protons, neutrons, and the Eightfold Way, quarks and now the Higgson—so many riches—there could be nothing but a desert from here to the Planck limit.

Of course, there may be much more in our journey's second half. There are theoretical arguments that favor the existence of a family of supersymmetric particles. As yet none has shown up at the LHC, unless the first hint of them is being revealed in the two-photon decays of the Higgs. Unknown, however, is whether the supersymmetric particles are within easy reach at the LHC, remote, or even inaccessible. If, by analogy, we compare finding the Higgson to discovering the continent of North America, and supersymmetry to reaching the gold fields of California, the current unknown is whether we have landed on the west or east coast. If the former, the discovery of supersymmetry could be soon; if the latter, it will be a long haul.

There are plenty of strictly empirical indications that we need to find more particles, as well. Although the Higgs Boson may indeed be the final piece in the cast of characters needed to describe our world, over 90 percent of the universe consists of "dark stuff," which does not shine but gives itself away by its gravitational pull on the galaxies of stars. There are no candidate particles known with the required dark properties, but supersymmetry theory contains such possibilities, which adds to its interest.

If supersymmetric particles exist, the question arises of how they get their mass. In such a world, it is anticipated that there is a whole family of "higgsons" —with a small *h*—awaiting discovery.[12]

Although I have mentioned some of the following elsewhere in this book, I repeat it here, as I was astonished at how much misunderstanding about what the so-called Higgs field does, or does not do, was propagated following the July 4 announcement.

First and foremost: it is not the source of all mass, only that of the most basic of particles. It is the atomic nuclei in your body that make up over 99.5 percent of your weight. This has nothing to do with the Higgs field, but is a consequence of quarks having large amounts of kinetic energy when confined within nucleons. This energy is manifested as the mass of the nucleon, due to the mass-energy equivalence of Einstein. What the Higgs field does is potentially give structure by acting on the fundamental particles, such as the electron found in the outer reaches of atoms, and the quarks, which are the ultimate seeds of the atomic nucleus. Your weight has little to do with the "Higgs mechanism," but your size does.

The size of the hydrogen atom is determined by the dimensionless quantity alpha, which is approximately 1/137, and the mass of the electron. Were the electron mass zero, the hydrogen atom would have infinite size—i.e., would not exist.

The mass of the proton is only slightly affected by whether or not quarks have mass. However, according to chiral symmetry, the masses of the up and down quarks are proportional to the square of the mass of a pion.[13] The pion transmits the powerful attraction between protons and neutrons, which form atomic nuclei. The range of this force is inversely proportional to the pion's mass, and hence also to the mass of the quarks. Were the quarks massless, and the pion also, the range of the nuclear force would be infinite. Thus the existence of compact complex nuclei, which seed the chemical elements, is linked to the quarks having mass.

If, as seems likely, we now know how the fundamental particles gain mass, this leaves open the question of why they have the particular masses that they do. If the electron were slightly heavier, essential examples of beta radioactivity would not occur, elements would not form, and we would not exist. Were it much lighter, these processes would change in other ways, once again unfavorable for life. Exactly what determines the strength of the Higgs's affinity for one particle or another is what experiment might reveal, but to do so will require some quirk in the data, some clue to guide us. At present, the pattern of particle masses, and the disparate forces, is an utter mystery.

The results announced in July 2012 imply that the Higgs Boson gives mass to the carriers of the weak nuclear force—the *W* and *Z* bosons—and probably the quarks too, but there is not yet unequivocal evidence that it gives mass to the electron and its siblings. The *W* and *Z* are the carriers of the weak force, which transmutes elements and keeps the sun shining; so we understand why the sun has lasted 5 billion years, long enough for sentient life to evolve. Proving that the Higgs Boson is responsible for the mass of the electron, and hence for the origins of chemistry, will be harder to establish, but should be settled one way or the other in a few years.

As to the nature of the field—its shape, structure, and dynamics—these may take much longer to decode. Contrary to many popular simplistic accounts, the discovery of a boson, with mass of 125 GeV, teaches us little about the nature of the mechanism.

Fields associated with forces, such as the gravitational or electromagnetic fields, have some source. The massive sun creates the gravitational field that entraps the planets; electrical currents within the earth create its magnetic field, which influences compass needles. The "field" associated with the mass mechanism is not like these, but is the vacuum itself. This has enormous implications for our concept of space, which is filled with some weird stuff with which particles can interact. A century after

Einstein did away with the ether, it has re-entered physics, though configured in such a way that it satisfies the constraints of relativity.

In 1964, the Gang of Six uniformly sidestepped the details of how the field performs this magic, and made the simplest generic assumptions about its properties. As we saw earlier, Brout and Englert recognized that there is a field but said nothing about its nature, while Guralnik, Hagen, and Kibble didn't even include a field, but assumed that spontaneous symmetry breaking nonetheless occurs. Goldstone's example of the wine bottle or "Mexican hat" potential, which Higgs adopted, is useful pedagogy to simulate the situation where an empty vacuum has more energy and is less favored than one containing this enigmatic stuff. However, it says nothing about the source of that field.

Discovery of the boson enables this fundamental issue to be pursued. To determine how higgsons condense into the ubiquitous field is trickier. A first step could be to produce two of them in a single collision and see how they interact with one another. While the LHC in principle is capable of producing such collisions, they are exceedingly rare. To observe and study the Higgs field may need a dedicated machine, such as a collider of electrons and positrons whose energies are tuned to produce pairs of higgsons, and that is likely to be a long-term hope at best.

What we have seen in this narrative is how, after half a century of effort, confronting the paradox of the Infinity Puzzle has brought us to the threshold of being able to explain the origins of structure, which underpin the existence of atoms, molecules, and life itself. This story began with the success of QED, which proved that what we call the vacuum, far from being empty, is actually seething—a foam of particles of matter and antimatter. Today, we have the first hints that in addition to gravity and electromagnetic fields, the vacuum is filled with another influence also—the Higgs field. Whereas the known phenomena described in QED are insensitive to the Higgs field, Gerard 't Hooft showed that the weak force feels its presence in a profound and pervasive way.[14] It was ignorance of this fact that had ruined the early attempts to build a description of the weak force. 't Hooft showed that the weak force can be explained, if what for centuries we have referred to as mass is actually a manifestation of fundamental particles interacting with the Higgs field.

The discovery of the Higgs Boson, with the implication that higgsons have condensed into an all-pervading field that is the source of the mass of the fundamental particles, is an essential stage toward understanding why the particles and forces have the characteristics that they do. This is an important step toward addressing what has been characterized as the most fundamental question of philosophy: "Why is there something rather than nothing?" Since Leibnitz first posed this conundrum in 1697, philosophers have argued the issue like seals tossing a ball. "Well, why not?" has satisfied some, but not physicists. With the Big Bang accepted as the origin of our universe, the question becomes: "Why this one?" The discovery of the Higgs Boson may be a first step toward understanding *why* things are as they are.

Whether the answers will be found with the proton blunderbuss that is the LHC, or whether they will require the forensic skills that only a future electron-positron machine can offer, is currently unknown.[15] Physicists are already thinking about what will follow twenty or more years from now, and a facility for colliding beams of electrons and positrons, customized to investigate higgsons, is already being discussed.

For particle physicists, and all who are curious about the origin of our material universe, July 4, 2012, marks the end of the beginning, not the beginning of the end.

POSTSCRIPT

There may be many who will insist upon reading this: "It did not happen like that." You may be right, and if some different version is established, I can update the history, but I believe this narrative to be consistent with the best information currently available. Documents, diaries, or records with dates trump the most adamant crystal-clear memories: That is the advice that I received from some respected historians of science and from psychologists.

I know this too from my own experience. In November 1974 a momentous discovery occurred in particle physics—my area of expertise and the stage for this book. To illustrate how singular this was, the memory of how we heard the news of the "November Revolution" is for many physicists as clear as when we learned about the assassination of John F. Kennedy or, in more recent times, of 9/11—events that one would expect to stay burned in one's memory forever. The drama of November 1974 was so important for science that I made a daily diary of the unfolding events on an old-fashioned pocket tape recorder, which I then mislaid. During a house move, some twenty years later, the tape resurfaced. When played, its message differed so much from my apparently perfect memories in several respects—such as the order of events, locations, and even of the people involved—that I imagined that someone must have interfered with it.

It is possible that my case is atypical, but experienced professionals would suggest not. In researching *The Infinity Puzzle*, I interviewed people who were present at some event, but whose memories differ—not always on trivial peripheral issues. I have attempted to tell the different versions in such cases. In other examples, where memories seriously conflict—such as the events surrounding the independent discoveries of what has become known as the "Higgs Mechanism" in 1964, or the emergence of the "Weinberg-Salam" model thereafter—I have left some questions open. Hopefully, others reading this account may be able to fill in details or provide further documents, which may someday help to clarify the history.

To those who will nonetheless insist, with certainty and integrity, that their own version is the one and only correct one, I have to admit that you may indeed be correct, and if subsequent events can confirm this, the record may be updated. However, in the interim, please bear in mind the experiences in this postscript, and the quotation at the very start, which Shakespeare in his wisdom expressed much more elegantly four hundred years ago.

FRANK CLOSE
Oxford

GLOSSARY

Alpha: $1/137$, the quantity that appears in Feynman diagrams when a photon is absorbed or emitted by an electrically charged particle.

Angular momentum: A property of rotary motion analogous to the more familiar concept of momentum in linear motion.

Antimatter: For every variety of particle there exists an antiparticle with opposite properties such as the sign of electrical charge. When particle and antiparticle meet, they can mutually annihilate and produce energy.

Anti(particle): Antimatter version of a particle: e.g., antiquark, antiproton.

Atom: System of electrons encircling a nucleus; smallest piece of an element that can still be identified as that element.

Baryon: Class of hadron; made of three quarks.

BCS Theory: Theory of superconductivity due to John Bardeen, Leon Cooper, and Robert Schreiffer. A key feature is the idea that electrons act cooperatively in pairs—"Cooper pairs"—while moving through a solid lattice of atoms. Individual electrons are fermions, but the Cooper pairs act like bosons. These bosons condense into the lowest energy level, which enables them to move through the solid without resistance. See page 138 et seq.

Beta decay (beta radioactivity): Nuclear or particle transmutation caused by the weak force, causing the emission of a neutrino and an electron or positron.

Boson: Generic name for particles with integer amount of spin, measured in units of Planck's quantum; examples include carrier of forces, such as photon, gluon, *W* and *Z* bosons, and the (predicted) spinless Higgs Boson.

Bottom quark: Most massive example of quark with electric charge of –⅓.

CERN: European Centre for Particle Physics, Geneva, Switzerland.

Charm quark: Quark with electric charge of +⅔; heavy version of the up quark but lighter than the top quark.

Collider: Particle accelerator where beams of particles moving in opposing directions meet head-on.

Color: Whimsical name given to property of quarks that is the source of the strong forces in the QCD theory.

Conservation: If the value of some property is unchanged throughout a reaction, the quantity is said to be conserved.

Cosmic rays: High-energy particles and atomic nuclei coming from outer space.

Down quark: Lightest quark with electrical charge of –⅓; constituent of proton and neutron.

$E=mc^2$ (energy and mass units): Technically, the unit of MeV is a measure of the rest energy, $E=mc^2$ of a particle, but it is often traditional to refer to this simply as mass and to express masses in MeV or GeV.

Electron: Lightweight electrically charged constituent of atoms.

Electroweak force: Theory uniting the electromagnetic and weak forces. See also *QFD*.

eV (electron volt): Unit of energy; the amount of energy that an electron gains when accelerated by one volt.

Fermion: Generic name for particle with half-integer amount of spin, measured in units of Planck's quantum. Examples are the quarks and leptons.

Flavor: Generic name for the qualities that distinguish the various quarks (up, down, charm, strange, bottom, top) and leptons (electron, muon, tau, neutrinos); thus, flavor includes electric charge and mass.

Gamma ray: Photon; very high-energy electromagnetic radiation.

Gauge invariance: A gauge is a measuring standard. Gauge invariance means that the result of a calculation does not depend on the scheme of measurement.

Generation: Quarks and leptons occur in three "generations." The first generation consists of the up and down quarks, the electron, and a neutrino. The second generation contains the charm and strange quark, the muon, and another neutrino, while the third, and most massive, generation contains the top and bottom quarks, the tau, and a third variety of neutrino. We believe that there are no further examples of such generations.

GeV: Unit of energy equivalent to a thousand million (10^9) eV (electron volts).

Gluon: Massless particles that grip quarks together, making hadrons; carrier of the QCD forces.

Hadron: Particle made of quarks or antiquarks or both, which feels the strong interaction.

Hidden symmetry (spontaneously broken symmetry): See Chapter 8.

Higgs Boson: Massive particle predicted to be the source of mass for particles such as the electron, quarks, and *W* and *Z* bosons.

ICTP: International Centre for Theoretical Physics, near Trieste.

Ion: Atom carrying electric charge as a result of being stripped of one or more electrons (positive ion), or having an excess of electrons (negative ion).

keV: A thousand eV.

Kinetic energy: The energy of a body in motion.

LEP: Large Electron Positron collider at CERN.

Lepton: Particles such as electron and neutrino that do not feel the strong force and have spin ½.

LHC: Large Hadron Collider; accelerator at CERN.

Magnetic moment: Quantity that describes the reaction of a particle to the presence of a magnetic field.

Mass: The inertia of a particle or body and a measure of resistance to acceleration; note that your "weight" is the force that gravity exerts on your mass so you have the same mass whether on Earth, on the moon, or in space, even though you may be "weightless" out there.

Meson: Class of hadron; made of a single quark and an antiquark.

MeV: A million eV.

meV: A millionth of an eV.

Microsecond: One-millionth of a second.

Molecule: A cluster of atoms.

Muon: Heavier version of the electron.

Nanosecond: One-billionth of a second.

Neutrino: Electrically neutral particle, member of the lepton family, feels only the weak and gravitational forces.

Neutron: Electrically neutral partner of proton in atomic nucleus, which helps stabilize the nucleus.

Nucleon: Collective name for proton or neutron—the constituents of an atomic nucleus.

Parity: The operation of studying a system or sequence of events reflected in a mirror.

Perturbation (expansion or theory): A way of simplifying sums that involves a small parameter, such as *alpha* in QED. As a first approximation, ignore contributions that include *alpha-squared*, *alpha-cubed*, or higher powers, relative to those with simply *alpha*. Every extra presence of *alpha* merely perturbs slightly the previous estimate.

Photon: Massless particle that carries the electromagnetic force.

Picosecond: One-millionth of a millionth of a second.

Pion: The lightest example of a meson; made of an up or down flavor (or both) of a quark and an antiquark.

Planck's constant: A very small quantity, *h*, that controls the workings of the universe at distances comparable to, or smaller than, the size of atoms. The fact that it is not zero is ultimately the reason the size of an atom is not zero, why we cannot simultaneously know the position and speed of an atomic particle with perfect precision, and why the quantum world is so bizarre compared to our experiences in the world at large. The rate of spin of a particle also is proportional to *h* (technically, to units or half-integer units of *h* divided by 2 pi).

Positron: Antiparticle of electron.

Proton: Electrically charged constituent of atomic nucleus.

QCD (Quantum Chromodynamics): Theory of the strong force that acts on quarks.

QED (Quantum Electrodynamics): Theory of the electromagnetic force.

QFD (Quantum Flavordynamics): Theory of the combined weak and electromagnetic forces; sometimes called quantum electroweak dynamics, or electroweak theory.

Quarks: Seeds of protons, neutrons, and hadrons (see Figure 12.2 on page 255).

Radioactivity: See *Beta decay*.

SLAC: Stanford Linear Accelerator Center, Stanford, California.

Spin: Measure of rotary motion, or intrinsic angular momentum, of a particle, measured in units of Planck's quantum.

Spontaneous symmetry breaking: Hidden symmetry. See Chapter 8.

Strangeness: Property possessed by all matter containing a strange quark or antiquark.

Strange particles: Particles containing one or more strange quarks or antiquarks.

Strange quark: Quark with electrical charge of –⅓, more massive than the down quarks but lighter than the bottom quark.

Strong force: Fundamental force, responsible for binding quarks and antiquarks to make hadrons, and gripping protons and neutrons in atomic nuclei; described by QCD theory.

SU2, SU3, and U1: SU2 is an example of the "special unitary group" of 2 × 2 unitary matrices (i.e., the sum of their diagonal members is zero—"traceless"—and their determinant is one). SU3 analogously involves 3 × 3 matrices. See "special unitary group" at http://en.wikipedia.org/wiki/Special_unitary_group. For U1, see http://en.wikipedia.org/wiki/Unitary_group. For simple examples of matrices, see pages 112 and 114 of *The New Cosmic Onion* by Frank Close.

SUSY (supersymmetry): Theory uniting fermions and bosons, where every known particle is partnered by a (yet to be discovered) particle whose spin differs from it by one-half.

Symmetry: If a theory or process does not change when certain operations are performed on it, then we say that it possesses a symmetry with respect to those operations. For example, a circle remains unchanged after rotation or reflection; it therefore has rotational and reflection symmetry.

Synchrotron: Modern circular accelerator.

Tau: Heavier version of muon and electron.

Top quark: The most massive quark; has a charge of +⅔.

Uncertainty principle: One cannot measure both the position and momentum of a particle, or its energy at a given time, with perfect accuracy. The disturbance is so small that it can be ignored in everyday affairs, but is dramatic for the basic particles. This is an intrinsic property of nature, not simply a failure in the measuring apparatus. One consequence is that the total energy of particles can fluctuate by some amount, *E*, for a short time, *t*, so long as the product of *E* times *t* does not exceed Planck's constant divided by 4 pi. See also note 32 in Chapter 1.

Unified theories: Attempts to unite the theories of the strong, electromagnetic and weak forces, and ultimately gravity.

Virtual particle: A particle whose energy and momentum are not conserved and exists briefly courtesy of the uncertainty principle.

***W* boson**: Electrically charged massive particle, carrier of a form of the weak force; sibling of the *Z* boson.

Weak force: Fundamental force, responsible inter alia for beta decay; transmitted by *W* or *Z* bosons.

Yang-Mills theory: Mathematical theory of forces based on gauge invariance. The theory predicts that the force is transmitted by a vector boson, which is massless. The latter can gain a mass due to spontaneous breaking of the gauge symmetry. QCD, QED, and QFD are examples of Yang-Mills theories.

***Z* boson**: Electrically neutral massive particle, carrier of a form of the weak force; sibling of *W* boson.

NOTES

PROLOGUE

1. See R. P. Crease and C. C. Mann, *The Second Creation*, 327. Veltman colorfully disagreed with 't Hooft's version of some events in an interview with the author, December 17, 2009.

2. M. Veltman, *Facts and Mysteries in Elementary Particle Physics*, 275.

3. The electrons are held in place by the electromagnetic force, following its basic rule: "Opposite charges attract." The corollary—that like charges repel—gave a paradox for the existence of the nucleus itself, where large amounts of positive electric charge can survive in this compact kernel, magically resisting the electrical disruption. This hinted at the existence of some "strong" force, which prevents the nucleus from blowing itself apart. And finally, the ability of elements to transmute, as in the sun and stars, is due to a third force, which is feeble and known as the "weak" force. Gravity acts on matter in bulk, but has no measurable effect on individual atoms or particles, and plays little part in this book.

4. This simile, and its implications, originates with J. C. Ward, whom we will meet in Chapter 6. He used it on the only occasion I saw him, in Canberra in the 1980s, and included it in his work *Memoirs of a Theoretical Physicist*, available at http://www.opticsjournal.com/ward.html. He was concerned with the "conflict between premature publication and the fear of being scooped," which, as we shall see, permeates our tale. It hints also at the bitterness he felt in his later years, having missed sharing a Nobel Prize.

5. Or even that a cosmic ray passed through your laptop and short-circuited the computation.

6. This is just a preview. The details are in Chapters 1–3.

7. Veltman, interview by the author, December 17, 2009.

8. Memories of different participants vary in details but have common features, which I have built on here. I have followed most closely the reminiscence

of my colleague Chris Korthals-Altes. 't Hooft himself was so focused on making his debut that he can remember little other than the room had no windows; see also page 223.

9. A. Tenner and M. Veltman, eds., *Proceedings of the Amsterdam International Conference on Elementary Particles, June 30–July 6, 1971* (North Holland, 1972), 415. The mention of Ward at this critical moment is ironic given how events subsequently develop. The omission of Ward from the Nobel Prizes that flowed, stimulated by 't Hooft's breakthrough, is one of the major controversies in the field. This will be examined in Chapter 15.

10. Actually, they were asking if he had indeed proved that a Yang-Mills theory of the electromagnetic and weak interactions is "renormalizable"—that is, viable. This is what I have subsumed by "Infinity Puzzle." Quite what this means and why it matters will be one of this book's main themes.

11. Weinberg anecdotal comment. In *Dreams of a Final Theory*, he writes, "At first I was not convinced by 't Hooft's paper. I had never heard of him, and the paper used a mathematical method . . . that I had previously distrusted" (120).

12. Glashow made this comment to David Politzer, reported in Crease and Mann, *The Second Creation*, 463n30.

CHAPTER 1

1. 380 This was probably November 1957 when he described the nature of space and time and fundamental particles. This led to his idea that a massless neutrino can be linked to an absolute distinction between left and right: "parity violation." A Nobel Prize was awarded to T.-D. Lee and C. N. Yang for their work on parity violation, but Salam felt that his work on this had been overlooked and later featured in several letters nominating Salam for the Nobel Prize (see Chapter 15). Parity violation is not a theme in this book; for more, see A. Pais, *Inward Bound*, 525; and F. Dyson, *From Eros to Gaia*, 108.

2. As we shall meet several research institutions in this narrative it may be worth giving some sense of their different natures. There are many universities, where research takes place alongside teaching, and there are also dedicated research centers. Most of the latter are experimental laboratories, such as CERN, where theoretical research also occurs. The ICTP is unusual in being an international center dedicated to theoretical physics. If scientific research is compared to a pyramid, ICTP through its visitor programs and advanced schools has broadened the base considerably and helped to increase its height. The peak of the pyramid has been sharpened primarily by work at other places in this narrative, including Princeton's Institute for Advanced Study, or Salam's British home for many years, Imperial College.

3. G. Fraser, *Cosmic Anger*, 73.

4. Obituary of P. T. Matthews by Abdus Salam; Royal Society, London, and Imperial College Archives.

5. My memory of the 1950s was that the establishment was more than happy for immigrants to take menial tasks at low wage, meanwhile cloaking more extreme prejudice with comments like, "Very nice people but not really one of us."

6. Quoted in Fraser, *Cosmic Anger*, 81.

7. Abdus Salam, introduction to Matthews's inaugural lecture at Imperial College, ICTP Archives.

8. Kellogg's Corn Flake packets advertised themselves as "the original and best."

9. Obituary of Matthews.

10. According to Einstein's theory of general relativity, an orbiting mass will lose energy by "gravitational radiation." Although no direct detection of this radiation has yet been made, the orbit of the binary pulsar PSR 1913+16 is decaying at the exact rate predicted by Einstein's theory of general relativity if it is emitting gravitational radiation. The planets in the solar system likewise should also be experiencing this phenomenon, but it is far too trifling to detect. The only direct measurement of an evolving orbit is that of the Moon around Earth. The tidal forces between Moon and Earth are causing Earth's rate of spin to slow—the length of a day is increasing—and the Moon is receding from us. Measurements using laser reflectors, which were left on the Moon by the Apollo astronauts, show that the average Earth–Moon distance is increasing by about four centimeters per year.

11. The quantum nature of radiation, encoded in "$E = h\,nu$"—where E represents energy, *nu* is frequency, and h is "Planck's constant"—will play a central role throughout our saga. The relation implies that high energy is allied with high-frequency oscillation, at least in the case of electromagnetic radiation. High frequency means "many per second"; thus, individual oscillations occur on very short time scales. Since the speed of all electromagnetic radiation is the same, and speed is distance traveled in an interval of time, the short time scale just alluded to is equivalent to the beam having traveled but a short distance. This association of high-energy beams as a means of resolving short distances underpins the science of high-energy physics. In quantum theory, not just electromagnetic waves appear as particles—photons; any particle, such as an electron, can take on wavelike character. In all cases, the higher the energy of the particle, the shorter the length of the associated wave. Hence, beams of electromagnetic radiation, or electrons, or indeed any particle, can resolve short distances if their energy is large.

12. For examples, see http://www.ipac.caltech.edu/outreach/Edu/Spectra/spec.html.

13. Pais, *Inward Bound*, 172.

14. In 1853, the Swedish physicist Anders Angstrom first observed the spectrum of hydrogen. He saw three lines: one in the red, one in the blue-green, and one in the violet regions of the spectrum. Together with a fourth line that was found soon after, also in the violet region, these form what became known as the "Balmer Series," after Johann Balmer. The four lines that had initially inspired him lay in the visible spectrum. The infrared and ultraviolet reaches of the electromagnetic spectrum revealed further lines. One set, discovered in the ultraviolet by American Theodore Lyman, fits $1 - 1/n^2$: hence, $m = 1$ and where n starts at 2. Lines in the infrared turned up, such as those found by Frederich Paschen, which corresponded to $1/9 - 1/n^2$: hence, $m = 3$ and where n starts at 4.

15. See note 11.

16. The energy latent within a hydrogen atom consists of the energy locked within an electron and proton—the sum of their respective mc^2—minus the energy used up in providing their electrical attraction. We now know that an electron in hydrogen has a speed that is some 1/137 of the speed of light. This is fast to our daily experience, but is slow enough relative to the speed of light that the theory of relativity could be ignored when calculating how its kinetic energy is related to its momentum. The latter depends on the number of oscillations in the circuit, and for nonrelativistic situations, such as here, the energy of an electron is proportional to the square of its momentum. In quantum theory this is the same as being inversely proportional to the square of the number of wavelengths in the orbit. Thus, the total energy of the hydrogen atom is proportional to the sum of (mc^2) of an electron and proton minus A/n^2, where A is a constant and n denotes the number of oscillations in the circuit. The change in energy when the number n changes is thus $A(1/m^2 - 1/n^2)$, which gives Balmer's empirical formula.

17. This property is called "spin."

18. The more famous implication of Dirac's equation is that Special Relativity does not permit the electron to exist alone. Instead, it implies that for every variety of charged particle, such as the electron, there exists a counterpart with the same mass but opposite sign of electric charge, known as its antiparticle. As particles build up matter, so could antiparticles combine to form antimatter. Thus, the negatively charged electron has a sibling—the positive "positron." Dirac had predicted this in 1931, and he was proven right when Carl Anderson discovered the positron in cosmic rays in 1932. The history is described in S. Weinberg, *The Quantum Theory of Fields*, 11–13, and G. Farmelo's biography of Dirac, *The Strangest Man*. Dirac's equation is cast in stone, at his memorial in Westminster Abbey.

19. F. Wilczek, "Paul Dirac Centenary Symposium," Tallahassee, Florida, December 2002.

20. Dirac's original conception of QED was in 1927, but at that stage he had no relativistic description of the electron. Having created his relativistic equation, in 1928, the modern relativistic QED immediately followed.

21. These transient or "virtual" particles do not manifest themselves directly in detectors, but their presence influences other phenomena. We know they are not simply figments of the imagination, as it is possible to design experiments where the virtual particles convert into directly observable ones. For example, two virtual photons may meet and turn into an electron and its antimatter counterpart, a positron.

22. In QED, *alpha* is expressed in terms of the speed of light, c, the magnitude of Planck's quantum h, and e, the magnitude of the electric charge. It thereby connects the dynamics of electrically charged particles in a profound way with the great theories of the twentieth century—Special Relativity (*via* c) and Quantum Theory (*via* h). The tantalizing feature is that the particular combination of these quantities, *alpha*, is dimensionless.

23. Arthur Eddington had built a set of sixteen equations, involving fundamental constants, with which he hoped to construct a theory of the universe. He then claimed that *alpha* followed from $(16 \times 16 - 16)/2 + 16$, which equals 136. When the data settled on its value nearer to 137, Eddington announced that he had forgotten to include *alpha* itself in his formula, so added 1 to 136. Today we know that "one divided by *alpha*" is not exactly 137 and that its magnitude has no mystical significance. But none of this was apparent in 1928 when Dirac first wrote his equation for the electron, and then extended it to include the interaction between an electron and the electromagnetic field. The London magazine *Punch* dubbed Eddington "Sir Arthur Adding-One."

24. Crease and Mann, *The Second Creation*, 112n4.

25. Schwinger's license plate is mentioned in ibid., 140.

26. P. Varlaki, L. Nadai, and J. Bokor, "Number Archetypes and Background Control Theory Concerning the Fine Structure Constant," *Acta Polytechnica Hungarica* 5, no. 2 (2009), available at http://bmf.hu/journal/Varlaki_Nadai_Bokor_14.pdf.

27. R. P. Feynman, *QED: The Strange Theory of Light and Matter*, 129.

28. See F. Close, *The Void*, or in paperback: *Nothing: A Very Short Introduction*.

29. Similar remarks can be made for a magnet, which is surrounded by a magnetic field. One implication of Einstein's theory of relativity is that electric and magnetic fields are intimately related. Where one observer may perceive an electric field, with no magnetic field at all, someone moving relative to the first person will find there is a magnetic field also. Relativity implies that electric and magnetic fields are united in an "electromagnetic" field, whose properties all observers agree on.

30. Whereas gravity is always attractive, the positive and negative electrical charges give rise to both attraction and repulsion. In bulk matter the counterbalancing effects of positive and negative tend to cancel out, leaving gravity as the dominant long-range force, though not always—for example, Earth's magnetic field extends over thousands of kilometers and can be sensed by a small compass.

31. Its mass is about 10^{-30} metric kilograms. You or I contain more than a thousand trillion trillion electrons, and, even if overweight, are certainly not infinite. This enigma had been known since the nineteenth century, even before the invention of quantum theory, and dismissed: So long as the electron has a size, the self-energy is finite. However, in Dirac's theory the electron has no size, and the paradox erupts. (Our mass is dominantly due to the nuclei of our atoms, which outweigh the electrons by several thousand.)

32. A fundamental principle of quantum mechanics, discovered by the German Werner Heisenberg, is the "uncertainty" or "unknowable" principle: It is impossible to know both the position and motion of anything with perfect precision. There has to be an uncertainty in one or the other, or both. The product of the uncertainty in position multiplied by the uncertainty in motion (or "momentum" in its modern terminology) cannot be smaller than the amount known as Planck's quantum, denoted by the symbol h, divided by 4 pi. The magnitude of h is trifling, effectively nothing on the scale of large distances familiar to our human senses. As a result, it places no practical limitation on our ability to be certain about motion in daily affairs. Thus, if I know where a billiard ball is, and strike it accordingly, I can with certainty guarantee that it will head toward the desired goal—or would if I were more skilled. For atomic billiards, by contrast, even a champion could not be sure of the outcome. The better the location is known, the less certain will you be of the motion. The reason that it is impossible to measure both the size and the momentum of an atom to unlimited precision is not because of any deficiency in our measuring implements, nor is it the result of the act of observation somehow disturbing the atom and hence introducing a post hoc error in the subsequent measurement of its state. Instead, it is an inherent property of nature, an axiom of how things are. In particular, it limits the precision with which we can know the magnitude of a field, such as an electromagnetic field, from point to point in space. All that we can ever know is some time average. As we try to measure more precisely the value at any point, the greater and more uncontrolled become the random fluctuations. One consequence is that energy conservation can be put on hold for exceedingly short time spans.

33. Energy becoming infinite corresponds in wave mechanics to a time interval becoming zero (footnote 32). The infinite sum in QED therefore is equivalent to supposing that the theory applies literally all the way to a point.

This is actually very radical, and indeed it would be remarkable if no new structures or phenomena lurked at short distances, beyond the limits of our present vision. However, QED was built with that very assumption: Fundamental particles such as the electron and positron, together with the necessary photon, were assumed to be the entire menu. An alternate philosophy to renormalization (page 41) is that new physics occurs at a short distance, which can be revealed by experiments at high energy; the summation all the way to infinity is then cut off at the threshold for the new physics. In that case, the troublesome infinite quantities become finite, though still very large, their value depending on the cutoff. In this language, instead of "removing infinities," the renormalization procedure removes the dependence of all calculated quantities on the cutoff. This makes *renormalizable* theories very attractive (though not necessarily empirically true!), since they are insensitive to unknown physics occurring at very short distances. I am indebted to Ian Aitchison for long discussions about this.

34. J. R. Oppenheimer, *Physical Review*, vol. 35, p. 461 (1930).

35. Quoted in Crease and Mann, *The Second Creation*, 96.

CHAPTER 2

1. Crease and Mann, *The Second Creation*, 125; J. Gleick, *Genius*, 233.

2. When an electron is in the lowest energy state—the "ground state"—its waves envelop the center of the atom with perfect spherical symmetry, appearing the same in all directions. In this spherical *S* state, the electron has energy but, on the average, no net rotary motion—no "angular momentum." Moving up the energy ladder, Dirac's equation predicted that at the next level two states occur with the same energy. One is also spherically symmetric (known as the 2*S* to distinguish it from the ground state, *1S*), while in the other, denoted *P*, the electron orbits the center with angular momentum. According to Dirac's theory, the energy in the $2S_{1/2}$ and $2P_{1/2}$ levels (the subscripts denoting the combined angular momentum and intrinsic spin of the electron) is identical.

3. F. Dyson, letter to his parents, March 8, 1948, quoted in Gleick, *Genius*, 3.

4. Bethe later won the Nobel Prize for his explanation of the nuclear processes in stars.

5. National Academy of Sciences (NAS), biographical notes by P. C. Martin and S. L. Glashow, http://www.nasonline.org/site/PageServer?pagename=MEMOIRS_S.

6. Crease and Mann, *The Second Creation*, 130.

7. Oppenheimer's early career was based in Zurich, Cal Tech, Berkeley, and Los Alamos. See A. Pais, *J. Robert Oppenheimer: A Life*, chap. 4.

8. NAS biographical notes, 336.

9. Action has the same dimensions as Planck's constant, *h*. When the action is smaller or of similar size than *h*, quantum mechanics rules. When the action is much larger than *h*, quantum democracy gives way to the more rigid phenomena that we term *classical mechanics*. Thus, classical mechanics emerges from the fundamental quantum mechanics when the action is large compared to Planck's h.

10. The actual path that nature chooses is that where the action stays the same when you try to change the path very slightly; we call it a "stationary value" of the action. This is often a minimum value of the action but not always (for example, when you see the reflection of an object in a mirror, the light rays reflected from the mirror take a longer route, with a larger action, than if they had gone from the object to your eye directly; however, the constraint of a stationary value gives the rule that angle of incidence equals the angle of reflection).

11. A simple example illustrates the idea of Lagrangians and Newton's standard method. Suppose we have a smooth block of ice able to slide freely on a flat ice sheet. As friction is negligible, the block will travel in a straight line at constant speed in accord with Newton's laws of motion. How does Lagrange explain this? The potential energy is the same always, so let's ignore it entirely—define the ice sheet to be "ground" level. Lagrange's rule then says that in going from one point to another in a given time, the kinetic energy must be the least. It turns out that this too implies that it must move at a constant speed. We can see this is so by imagining what would happen if it were to have been different: sometimes going faster and at others slower (in order to keep the total time the same). Kinetic energy varies as the square of the speed. The square of something that deviates around an average value is always greater than the square of the average. For example, suppose the average speed in some units is 2, made up of a speed of just 1 for half the time and 3 for the rest. At constant speed the kinetic energy would have been proportional to 2 squared = 4; in the latter case 1 squared gives 1, but the 3 squared gives an enormous value of 9, confirming that the minimum kinetic energy is equivalent to constant speed. No one would seriously determine the motion in this particular example using Lagrangians rather than Newton's laws, but in many realistic problems, such as gyroscopes spinning on moving machinery, Lagrange's techniques are far superior.

12. Farmelo, *The Strangest Man*, 216, tells how Dirac first developed some of these ideas and had been largely ignored.

13. If we denote the magnitude of the action for some path by S, the amplitude is proportional to $e^{iS/h}$ where i is the square root of minus 1 and h is Planck's quantum divided by 2 times pi.

14. There are phenomena that are easiest to understand with rays and others with waves. Diffraction is an example of the latter. Light exhibits properties that are familiar to anyone who has watched ripples on the surface of water. At the opening in a harbor wall, the waves will bend—diffract—at the edges and spread into the harbor. If there are two narrow openings in the wall, waves diffract from each. When these meet, their peaks and hollows mingle, producing alternating large splashes, where two peaks coincide, or troughs when two hollows meet, or level water when a hollow meets a peak. The resulting web of peaks and dips is called a diffraction pattern. A similar phenomenon occurs when light waves pass through narrow slits in a screen. On the far side of the screen a pattern of alternating brightness and dark lines occurs. That two beams of light can combine to make darkness would be bizarre were it not for the insight that light consists of waves.

15. Originally, Hero of Alexandria thought that the shortest path was the rule. However, this does not explain refraction, such as when light passes from air to water, making a partly submerged rod appear to be bent. The rule of shortest time, however, describes refraction perfectly. "Action" is the modern generalization that subsumes these.

16. The way that light decides on the "optical path" is by sampling nearby paths and comparing them. The feature not present in geometric optics is wavelength, and it is this that determines how far away light must do the sampling. If the light is restricted, such as by a narrow slit whose size is similar to the wavelength, the sampling is disrupted and the wave diffracts, leading to interesting interference effects, as when water passes through openings in a harbor wall. See, for example, Feynman's *Lectures on Physics*, 1:26–28, for an illustrated example, and also his excellent *QED*.

17. Pais, *Inward Bound*, 452.

18. In the 1930s, as the plague of infinities threatened to destroy faith in QED, Hendrik Kramers, a Dutch physicist, ruminated about the classical problem of how an electron behaves when it is near a magnet. At first sight this appears straightforward: The magnetic force sets the electron in motion, and the ratio of the force to the acceleration defines its inertia, or mass. However, the electron is itself the source of an electrostatic field throughout the surrounding space. As the electron begins to move, it drags its electrostatic field along with it. Consequently, the resistance to acceleration involves not just its material mass but also the inertia in its electric field. Kramers's idea was to make a pragmatic trade-off. Instead of trying to calculate the behavior of both a material electron and its electrostatic field, use the experimentally measured mass of the electron in your equations and forget about the associated electrical effects. His point was that the experimentally measured inertia subsumed this "self-energy" already. His idea was not quantum mechanical,

nor was it in accord with relativity, but it was nonetheless direct, intuitive, and, we now realize, the essence of modern quantum field theory's way of coping with infinities.

19. This sign of the beta slope will later—in Chapter 13—become central to the plot.

20. A reason that QED is a renormalizable theory is that electrons, positrons, and photons bubble in and out of existence, and the overall image is like a fractal—there is no means to tell at what resolution your microscope is operating. In quantum field theory, the range of a force is linked to the mass of its carrier. The electromagnetic force has infinite range, which in QED is because of the fact that the photon has zero mass. An infinite-range force appears the same—infinite range—in all microscope images, which is a reason for QED's being renormalizable. Contrast this with a finite-range force, such as the weak force in Chapters 7–11, which will not appear the same at all microscopic scales. At high resolution, where a small distance fills the entire image, the range will appear to fill more of the picture than in a lower-resolution example covering a greater span. The finite-range weak force is carried by a massive W boson. It is this mass, and its link to a finite range of distance, that spoils the invariance of the microscopic image and contributed to the difficulty of constructing a renormalizable theory of the weak force. This analogy is described in more detail in G. 't Hooft, *In Search of the Basic Building Blocks*, 67.

21. The transitory photon takes up energy, which means that the motion of the electron differs from what it was at the outset. A slow-moving electron feels a smaller magnetic force than a fast one. Consequently, the change in the electron's motion means that its magnetism also subtly alters. This is the source of its "anomalous" magnetism—anomalous relative to what the Dirac equation implied.

22. Schwinger's calculation gives *alpha*/2 *pi*, which = 0.00116. Today the precision on this number is known to one part in a trillion, and is 0.001159652181(1) where the experimental uncertainty is 0.000000000000(7). See http://physics.nist.gov/cgi-bin/cuu/Value?ae.

23. Quoted by P. Martin and S. Glashow in "Biographical Memoirs of the U.S. National Academy of Sciences," http://www.nap.edu/openbook.php?record_id=12562&page=333.

24. American Institute of Physics interviews, quoted in Gleick, *Genius*, 252.

25. The first particle discovered at Berkeley was the electrically neutral form of the pion in 1949. This was soon followed by the production of novel forms of strange particles and short lived "resonances" such as the *Delta*—see Chapter 12.

26. Pais, *Inward Bound*, 458, has Schwinger "on day 1"; Gleick, *Genius*, 256, has "the next morning."

27. J. Mehra and H. Rechenberg, *Historical Development of Quantum Theory*, 6:1055.

28. Pais, *Inward Bound*, 459; Gleick, *Genius*, 258.

29. His colleagues were Edward Teller and Gregor Wentzel.

CHAPTER 3

1. The occasion was the centenary of William Henry Bragg's appointment as professor at the age of twenty-four. He and his son William Lawrence Bragg shared the 1915 Nobel Prize in Physics for their work on X-ray crystallography.

2. Gleick, *Genius*, 262.

3. See F. Dyson, *Physical Review*, vol. 75, pp. 486, 1736 (1949).

4. F. Dyson, *From Eros to Gaia*, 116.

5. Quotation from Feynman's *Nobel Lecture*, http://nobelprize.org/nobel-prizes/physics/laureates/1965/feynman_/lecture.html. For versions of this episode, see also Gleick, *Genius*, 271; B. Feldman, *The Nobel Prize*, 121; and Tony Hey, http://www.youtube.com/watch?v=9miKIWIYi4w.

6. Feldman, *The Nobel Prize*, 121.

7. Slotnick's calculation had dealt with an electron scattering off a neutron with no deflection. Even more remarkable was that Feynman's calculation was for an electron scattering at any arbitrary angle, which included Slotnick's six-month quest as a special case.

8. Quotes from Feynman, *Nobel Lecture.*

9. That these two quantities are sufficient to make QED viable was due also to J. C. Ward, whom we shall meet in Chapter 6.

10. This is actually in probability. The quantum amplitude is proportional to the square-root of *alpha*-cubed.

11. Feynman, *Nobel Lecture*; Gleick, *Genius*, 378.

CHAPTER 4

1. The discovery of the electrically neutral pion in 1949 completed the cast of characters.

2. Crease and Mann, *The Second Creation*, 234. Dyson's paper is *Physical Review*, vol. 75, p. 1736 (1949). J. C. Ward proved that overlapping infinities are harmless in his paper, which contained the eponymous Ward identities: *Physical Review*, vol. 78, p. 182 (1950).

3. Abdus Salam, *Physical Review*, vol. 82, p. 217 (1951).

4. Salam had sent an advance copy of his paper on overlapping divergences (see ibid.) to Dyson, who replied on September 11, 1950: "I think it is quite correct and a useful piece of work, [but] you have not anywhere made any rigid

mathematical proof that your subtraction procedures will be successful. Everything is described, and nothing is proved." He urged him to go on and settle whether the pion-nucleon theory can be renormalized and makes the fateful introduction: "You must get in touch with [J. C.] Ward, Clarendon Laboratory, Oxford and learn what he has done." Ward had found quantities in QED, which were identical to one another; these "Ward identities" were pivotal in completing the proofs that QED is renormalizable: *Physical Review*, vol. 78, p. 182 (1950). We shall meet Ward in Chapter 6 and see how Salam's and Ward's careers intertwined for a quarter of a century. Salam clearly responds, as on September 28 Dyson is excited to hear that Salam has "essentially completed the renormalization [for pion-nucleon]," Dyson describing it as "a major event if you have really done it." (This was eventually published in 1952, with P. T. Matthews, in *Physical Review*, vol. 86, p. 715.) They exchanged further letters, and on November 8, 1950, Dyson invited Salam to Princeton. ICTP Archives.

5. Letter from Dyson to Salam, November 8, 1950, ICTP Archives.

6. Salam appears to have taken time to complete this, as two papers on scalar electrodynamics, the theme mentioned in Dyson's letter, eventually appeared in 1952. These are in *Proceedings of the Royal Society* (London), vol. A211, p. 276, and *Physical Review*, vol. 86, p. 731.

7. C. Isham article about Abdus Salam, November 2007; e-mail to the author, November 2, 2010, and interview November 4, 2010.

8. Date confirmed by N. Seiberg, e-mail to G. Farmelo, November 30, 2010.

9. This is a folklore story attributed to Feynman. Gleick, *Genius*, 378, reports this as a suggestion made to Feynman by a reporter from *Time*, as a means of avoiding questions.

10. In 1859 Charles Blondin was the first person to cross the Niagara Falls on a tightrope.

11. Suppose that x refers to a large number. The difference of x + 3 and x + 2 is 1, however large x may be. If x itself becomes infinite, each of the separate expressions is also infinite, but their difference and ratio remain finite: 1. Contrast this with x + 3 and 2x + 2. Here the individual terms and also their difference are infinite. In renormalizable theories, only combinations where (infinite) x cancels or becomes impotent, as in ratios, can occur, though they are usually much more complicated than these simple examples. Functions such as logarithm, exponential, and complex numbers all may enter. The fact that infinities mutually cancel is remarkably profound. Most field theories are not renormalizable.

CHAPTER 5

1. Quotes in this chapter come from e-mails exchanged between R. Shaw and the author, July 10, July 13, and August 8, 2010, and his Web site, http://www.hull.ac.uk/php/masrs/reminiscences.html.

2. Shaw e-mail to the author, August 3, 2010.

3. Shaw to Schwinger, December 8, 1995, 8/12/95; Shaw e-mail to the author, July 13, 2010.

4. This is "gauge invariance," which is described later; see page 81.

5. Schwinger's equations mathematically were abelian SO(2); Shaw restructured them with nonabelian SU2.

6. J. C. Polkinghorne to the author, July 8, 2010. Shaw's memory differs: "Very probably I presented the idea in a dismissive way, since the requisite zero mass particles did not exist in nature, and so I think we moved on to discuss other things." E-mail to the author, July 12, 2010.

7. The chronology of who had done what and when will be clearer later on.

8. Quoted in Crease and Mann, *The Second Creation*, 120.

9. A photon has no mass and travels through the vacuum at nature's speed limit. Yet the vacuum is not empty, according to QED, and the photon is immersed in a sea of virtual electrons and positrons, which can ensnare it, interrupting its flight. As an electron gains an infinite self-energy as a result of interacting with these virtual particles, how does a photon manage to avoid a similar fate? Schwinger demonstrated that gauge invariance in QED underpins this phenomenon.

10. A nice example in 't Hooft, *In Search of the Ultimate Building Blocks*, 60, is that his dishwasher would make the same amount of noise if he added 100,000 volts to both of the holes in the electrical outlet.

11. Imagine a counter moving around the face of a clock, such that twelve o'clock corresponds to a peak, six o'clock to a trough, and onward back to twelve. You don't have to make this choice. Instead, you could start at, say, three o'clock for the peak, leading to nine o'clock for the trough, completing the circuit at three. The angle, or hour, around the dial is known as the "phase" of the wave.

12. Most particles spin, at a rate proportional to Planck's quantum, *h* (more precisely, to *h* divided by 2 *pi*). In quantum mechanics the spin is either an integer or half-integer multiple of this quantity. Half-integer examples are called fermions; those with integer values are bosons. Some have no spin at all—these are called scalar bosons. Those with one unit are vector bosons.

13. Crease and Mann, *The Second Creation*, 193.

14. In the mathematical jargon, the proton and neutron are said to be two "isospin" states of the nucleon. Mathematically, it is isospin conservation that underpins the strong interaction of nucleons and also pions. The observed properties of nuclear forces show that isospin is a quantity that is conserved during the interactions between particles. Enrico Fermi confirmed this when, in 1952, his experiment proved that the strength of interaction between pions and nucleons cared naught for their electric charges, in accord with what isospin had predicted. Although up to that time the concept of isospin had

not made a big impact in the physics community, Fermi's work changed things. Yang had been a student of Fermi and so was well prepared to recognize the importance of isospin.

15. Ibid., 194.

16. Ibid., 195.

17. Abdus Salam to Shaw, October 1, 1988.

18. This seems to me to be overly generous of Shaw. It was not until February 1954 that Yang gave the public presentation, which drew Pauli's ire, by which time Shaw had already shared this experience with Salam. Given later debates about priorities, as in the Salam-Weinberg Nobel Prize (see Chapter 15) and the "Higgs Mechanism" (see Chapters 9–10), Shaw has a legitimate right to be the third name in the Yang-Mills-Shaw theory.

CHAPTER 6

1. G. Guralnik, interview by the author, March 1, 2010; H. Burkhardt, e-mail to the author, January 5, 2010.

2. Quotes come from Ward, *Memoirs of a Theoretical Physicist.*

3. At Oxford, a Ph.D. (known locally as D.Phil.) thesis is examined by two people. One, the primary expert, or "external examiner," comes from an independent institution. The other, the "internal examiner," is a member of Oxford University who has not supervised or collaborated with the candidate. Kemmer was thus chosen to be the "external examiner."

4. J. C. Ward, "Some Properties of the Elementary Particles." The nature of the thesis suggests a more prosaic explanation of what occurred. His original thesis, completed on February 1, 1949, had to be extended, the additions clarifying some points in the original text. The original could well have been as short as folklore claims, as the final version contains a "postscript—March 30" for the thesis that was eventually resubmitted in April. Even then it contains little more than three thousand words and various equations. It is not unusual for an examiner to find a thesis unacceptable but advise that it may pass if additions or modifications are made. If these are simply amplifications and explanations of what already has been done, as distinct from substantial corrections or modifications, it may be left to the Oxford authorities to decide if the revised version meets the specific requirements, without further call on the external examiner. Thus, at the examination in February, Peierls may well have objected—for example, at the overly concise argument or unclear proof, which led to his demanding that Ward "add some notes and paragraphs" (quoting from Ward's postscript dated March 30, 1949), the Oxford authorities subsequently being satisfied that Ward had responded to Peierls's require-

ments. Ward's claim that Peierls "retired hurt from the contest" is amusing anecdotally, but perhaps no more than that.

5. Ward's Identities rely on gauge invariance. Thus, renormalization and gauge invariance are intimately linked.

6. Ward, *Memoirs of a Theoretical Physicist*, 11.

7. Ward's role and importance are highly contentious. The version in the text follows Ward's own account and the investigation by Norman Dombey and Eric Grove in "Britain's Thermonuclear Bluff," *London Review of Books*, October 22, 1992, available at http://www.lrb.co.uk/v14/n20/norman-dombey/britains-thermonuclear-bluff. Although Ward's ideas were not put into practice initially, the official version is that Britain first exploded an H-bomb in June 1957. According to Dombey and Grove, this was no more than a powerful atomic bomb, and the word that it was actually an H-bomb was allowed to propagate for political advantage. Dombey and Grove concluded, "The evidence is very strong that it was not until 28 April 1958 at 1905 GMT that Britain first tested a real H-bomb incorporating the Ulam-Teller concept." It was Ward who had discovered this concept independently, and so it was Ward's insights that had eventually led to Britain's H-bomb. Dombey and Grove wrote, "Within four years of Churchill's decision to launch a British H-bomb programme, Britain's overriding political objective—a special nuclear relationship with the United States—had been achieved. That this goal was reached appears to have been due mainly to the contribution of the physicist John Ward. The closeness of the US-UK special relationship in nuclear weapons, which resulted from the combination of this bluff and Ward's genius, has also been a carefully kept secret." However, Peter Knight and others claim that Ward's role was that of "a" rather than "the" theorist involved, and that other features of his claims are inaccurate. Some of these issues remain classified, even fifty years later. It is unfortunate that we are unable to assess Ward's reliability by comparing his version of his role with a generally agreed-upon record.

8. On page 82 we met Scalar and Vector—quantities defined respectively by magnitude or by both magnitude and a direction. Tensors are quantities that depend on more than one direction: for example, shear forces in a sheet of rubber. The role of mirror symmetry—the behavior of quantities under spatial reflection—has special interest for the theory of the weak interaction. A scalar, such as height above sea level, appears the same in a mirror. In quantum theory there are cases where a magnitude changes under mirror reflection; such quantities are known as Pseudo-scalar, shorthand: *P*. There are cases where a quantity with sense of direction remains the same under mirror reflection; such quantities are known as Pseudo-vector, or more traditionally, Axial-vectors, shorthand: *A*.

9. Marshak's paper is coauthored with E. C. G. Sudarshan, *Physical Review*, vol. 109, p. 1860 (1958).

10. Ward's account, described in *Memoirs of a Theoretical Physicist*.

11. This is Ward's account. There is no copy of this letter in the Abdus Salam archive at ICTP.

12. Crease and Mann, *The Second Creation*, 214.

CHAPTER 7

1. The *W* is named for "Weak." The nomenclature in particle physics is rich and can even appear ridiculous. The names of some particles are obvious, others less so, and some appear to be whimsical for no apparent reason. In this chapter we meet particles that are the carriers of forces. The photon, the quantum carrier of the electromagnetic interaction, was introduced by Einstein, but the name appears to have originated with Gilbert Lewis in 1926. See http://www.nobeliefs.com/photon.htm. The *W* was named by Schwinger, in recognition of its role in the weak force. Electrically charged, either + or –, it is denoted W^+ or W^- respectively. The prediction of a massive sibling with zero charge led to the *Z*. This is traditionally written Z^0, which like New York, New York, is so good they named the zero twice. Later we will meet gluons—the carriers of the strong force, which "glue" together the constituents of protons and neutrons. These constituents are known as quarks—of which more later.

2. In QED the chance of some outcome is controlled by a number—*alpha*—$1/137$. In Fermi's model of the weak interaction, the analogous quantity is proportional to the inverse square of a mass. The magnitude is approximately $1/100{,}000$ if the proton or neutron mass is used to set the scale of mass. It is the small value of $1/100{,}000$ relative to $1/137$ that led to the force being named "weak." If we rewrite this as $R^2/100{,}000$, where *R* is the ratio of some mass to that of a proton, then if *R* is about 27 neutron or proton masses, $R^2/100{,}000 = 1/137$. In Schwinger's more explicit calculation, and Fermi's model, there is a factor of square-root of 2, which elevates this qualitative example to the 40 of the text. Modern theory takes account of the empirical fact that the weak interaction violates parity symmetry, unlike the case for QED, and of the Weinberg angle appearing in the SU2 × U1 model, introduced later in this chapter. The effect of these is to raise the required value of *R* to be around 90 proton masses. The *W* boson, carrier of his weak force, is now known to have such a mass—see Chapter 14.

3. Crease and Mann, *The Second Creation*, 218.

4. S. Glashow and B. Bova, *Interactions*, 15.

5. S. Glashow, "The Renormalizability of Vector Meson Interactions," *Nuclear Physics*, vol. 10, p. 107 (1959), received November 24, 1958.

6. "Weak and Electromagnetic Interactions," *Il Nuovo Cimento*, vol. 11 (February 16, 1959).

7. Crease and Mann, *The Second Creation*, 224.

8. S. Glashow, "Unification Then and Now," talk at CERN, December 4, 2009. The papers were later published: A. Salam, *Nuclear Physics*, vol. 18, p. 681 (1960); and S. Kamefuchi, *Nuclear Physics*, vol. 18, p. 691 (1960). A. Salam and I. A. Komar, *Nuclear Physics*, vol. 21, p. 624 (1960), was later cited in Salam's annually updated curriculum vitae for Nobel nominators as demonstrating that Glashow's claims for renormalization were wrong. ICTP Archives.

9. Crease and Mann, *The Second Creation*, 224.

10. S. Glashow, interview with the author, December 15, 2010; Glashow, *Interactions*, 144.

11. Glashow interview.

12. S. Bludman (*Il Nuovo Cimento*, vol. 9, p. 443 [1958]) too had a model with parity violation, a neutral current, the electron and neutrino as an SU2 doublet, and noted that the charged massive vector bosons had nonrenormalizable electromagnetic interactions.

13. For any other value of N, the number of gauge bosons is $N^2 - 1$, hence three in the case of SU2, as exemplified by W^+, W^-, and a neutral sibling. Nature uses SU3, as we shall see in Chapter 13. Quarks, the seeds of protons, neutrons, and other particles that feel the strong interactions—hadrons—interact by exchanging gluons in eight ($3 \times 3 - 1$) varieties.

14. More technically, the photon is a mixture of the SU2 and U1 neutral members; the new neutral Z is then the "orthogonal" combination of this pair. In quantum mechanics, if the photon is some amount *a* of one and *b* of the other, then the "orthogonal" combination is an amount *b* of the first, with an amount *a* of the second removed from it. Quantum mechanics implies that the sum of the squares of *a* and *b* is one, so we can picture the situation as a right-angled triangle, with sides length *a* and *b*. The size of the square along the base represents the probability that you have the first option (the U1 member), while that along the vertical represents the chance of the second option (the SU2). As the chance that you have one or the other must be 100 percent, the length of the hypotenuse will be defined as unity. This pictorial representation illustrates that the mathematics can be summarized in an angle, *theta*. From basic trigonometry, cosine of *theta* and sin of *theta* are thus equal to *a* and *b*, respectively (see Figure 7.2 on page 182). This angle *theta* summarizes how much the neutral piece of the SU2 group (the W_3) and the U1 piece are mixed up to make a photon and the Z. It is this angle that in the 1970s became known as the Weinberg angle. This essential parameter of the SU2 × U1 mathematics had actually been introduced by Glashow several years before and is more correctly called the "weak mixing angle."

FIGURE 7.2. THE PHASE ANGLE

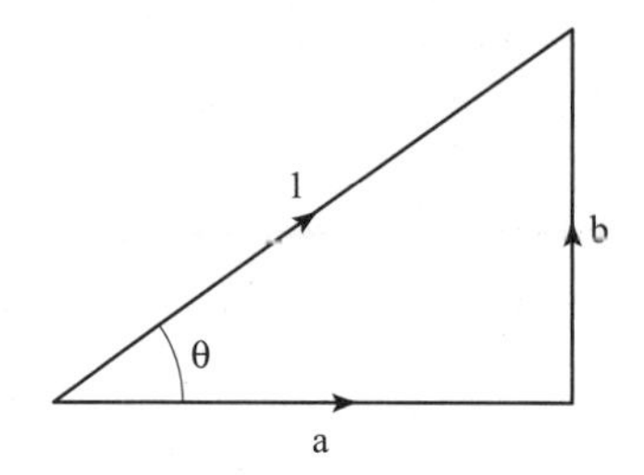

A right-angled triangle whose base and vertical are lengths a and b, respectively, where $a^2 + b^2 = 1$. The angle θ is the angle between the base and the hypotenuse.

15. S. Glashow, "Partial Symmetries of Weak Interactions," *Nuclear Physics,* vol. 22, p. 579 (1961).

16. If the photon itself were the only child in the U1 "family," while the Z^0 is the neutral member of the SU2 triplets, there would be no link between the weak, the SU2 part, and electromagnetic, the U1, at all. In such a case the neutral partner of the W^+ and W^- would imply that there has to be a new neutral weak force, utterly independent of electromagnetism.

17. Salam and Ward in 1959 considered three fermions: the negatively charged electron, its positively charged antiparticle—the positron—and the neutral neutrino. The interactions between the three "photons" and these fermions then become the standard electromagnetic ones for electron and positron, together with weak interactions linking the electron or positron with the neutrino, as in beta decay. This is in effect an SU3 model. There is thus no mixing angle. The remark about renormalization is from their paper "Weak and Electromagnetic Interactions," 572.

18. Glashow, e-mail to the author, December 10, 2010.

19. See note 1.

20. The standard description of strangeness-changing weak decays of hadrons was then, and remains today, encoded in Cabibbo theory. A popular and succinct description is in M. Veltman, *Facts and Mysteries in Elementary Particle Physics,* 98ff. Salam and Ward, in *Physics Letters,* vol. 13 (November 1964), in their paper combining weak and electromagnetic interactions, remark in the penultimate paragraph that "the [Cabibbo] 'rotation' view is incompatible with the present theory" (171).

21. Glashow showed in 1971 that the existence of particles with "charm" solve the enigma of the unwanted strangeness-changing neutral processes. See Chapter 13.

CHAPTER 8

1. K. Follett, *The Pillars of the Earth*, viii.

2. R. Laughlin, *A Different Universe.*

3. Buridan didn't propose the paradox, nor did he choose a donkey when he discussed it, yet this is widely known as the paradox of Buridan's donkey or Buridan's ass. Aristotle had considered the paradox of a starving man in this situation. Buridan, in discussing this, chose a dog; others, parodying Buridan, chose an ass.

4. I first encountered this example in a talk by Abdus Salam in 1980. See J. Mulvey, ed., *The Nature of Matter*, 111.

5. In this analogy the phenomenon occurs because the guests are near-sighted. In the jargon: They experience a short-range interaction. For a long-range, far-sighted, situation, the Goldstone Boson disappears; see Chapter 9.

6. W. Heisenberg, *Zeitschrift fur Physik*, vol. 49, p. 619 (1928).

7. Today, the exceptional electric and magnetic properties of superconductivity have vast implications in technology, being used in powerful electromagnets, such as are found in MRI scanners in hospitals, in magnetic levitation systems for high-speed transport, and in the world's largest cryogenic facility—the twenty-seven-kilometer ring of superconducting magnets of the Large Hadron Collider, the particle accelerator at CERN in Geneva.

8. See Chapter 5, note 12. The Cooper Pairs act as bosons, but they do not transmit a force.

9. Ibid.

10. This analogy is familiar to students and widely used. Like all analogies it has its limits. A version, together with a lengthy description of superconductivity and the subtleties of gauge invariance, may be found in D. Langenberg, R. J. Soulen Jr., and M. Osofsky, *McGraw-Hill Encyclopaedia of Science*, s.v. "Superconductivity."

11. Profile of Nambu in *Scientific American* (February 1995).

12. Nambu found that disturbing the distribution of the chiralities of the nucleons and antinucleons generates a wave, analogous to what occurs in a ferromagnet. The energy in this wave is zero, and in quantum theory corresponds to a particle with zero mass. Y. Nambu, *Physical Review*, vol. 117, p. 648 (1960).

13. See Chapter 6, note 8.

14. The reasons for its nonzero mass are because it is made of quarks (see Chapter 12), which themselves have a small mass.

15. As I am focusing on a particular thread, I am not giving a detailed history of the development of particle physics during this era, nor of research topics such as Partially Conserved Axial-vector Current (PCAC), Goldberger-Treiman relation, or Adler-Weisberger relation, which may be found by a

search of Google. A more complete history may be found in Crease and Mann, *The Second Creation.*

16. J. Goldstone, *Il Nuovo Cimento*, vol. 19, p. 154 (1961).

17. It appears that Goldstone published the strong conclusions of this seminal paper only after Glashow had badgered him. See Crease and Mann, *The Second Creation*, 240, where Goldstone admits that he had no proof in 1960, but strongly believed his theorem to be true. Glashow, a colleague at CERN where they were based, said, "Publish it anyway."

18. The pion is traditionally recognized as the "Nambu-Goldstone Boson," arising from the broken chiral symmetry of the strong interactions, the more general examples being named for Goldstone alone. However, in the literature sometimes one sees both names, at others just Goldstone's. There is no agreed-upon convention.

19. Attention subsequently focused on the enigma of the unwanted massless boson. The massive boson, which will reenter the history in Chapter 9, appears three lines below figure 7 on page 162 of his paper, note 16. See also note 57 in Chapter 10.

20. Steven Weinberg, e-mail to the author, September 24, 2010. Weinberg has no memory of Salam being at the conference, but Salam's account, in Crease and Mann, *The Second Creation*, has him present: "They hit it off immediately. . . . [T]hey were both excited and appalled by a series of conversations with Jeffrey Goldstone" (239). In any event, there is no doubt that Weinberg and Salam collaborated at Imperial College later that year.

21. This is based on the airmail letter from Weinberg to Salam, which is in the Salam archive at ICTP, dated received in Zurich, December 16, 1961. The telegram quote is copied from that in the Salam archive. The eventual paper by the trio is in *Physical Review*, vol. 127, p. 965 (1962).

22. Weinberg, interview with the author, May 6, 2010.

23. Letter from Weinberg to Goldstone, ICTP Archives.

24. Crease and Mann, *The Second Creation*, 240.

25. P. Anderson, *Physical Review*, vol. 130, p. 439 (1963), received by the editor in November 1962.

26. The nature of the vacuum, from ancient times to quantum physics, is described in Close, *The Void.*

27. A metal is not plasma, as its positive ions form a lattice, giving it solidity, but electrons, which conduct electricity, flow freely and affect incoming electromagnetic waves much as do those in a genuine plasma.

28. An object may have any amount of kinetic energy, the amount getting smaller as the body slows toward rest. At this point the particle has its minimum energy; the amount of which, *E*, would then correspond to a mass *m* given by Einstein's famous equation $E=mc^2$. For a massless body, the minimum energy in principle can be zero.

29. A longitudinal wave involves compression and thinning along its path (top example in Figure 8.4). A transverse wave has a constant density in that direction. The constraints of relativity for a wave traveling at the speed of light requires invariance along the trajectory—no one point is preferred to another—and longitudinal density changes are forbidden (bottom example in Figure 8.4). The pedagogical example of a wave hitting a plasma does not satisfy the constraints of relativity—the plasma at rest defines a particular frame of reference. Thus, it was not obvious whether the arguments immediately carried over to the relativistic situation of quantum field theory, which led to some particle physicists' reaction to Anderson's example, as we see later.

30. This phenomenon is well known today by those attempting to achieve commercially viable nuclear fusion. Their experiments involve the use of plasmas at very high temperatures. When an electromagnetic wave hits the plasma, it experiences what the fusion scientists refer to as an "OX conversion": the O (ordinary) electromagnetic wave turns into an X—"extraordinary"—wave, whose photons act as if they have mass. (With thanks to A. Kirk.)

31. P. Anderson, *Physical Review*, vol. 130, p. 439 (1963), received by the editor in November 1962.

32. Technically: relies on perturbation theory. As the strength of this interaction, encapsulated in the number *alpha*, is indeed very small, 1/137, this was in practice a reasonable assumption, widely used when calculating Feynman diagrams as a series of approximations. However, if *alpha* had been large, this approximation scheme, known as perturbation theory, would have failed, in which case Schwinger would have had no general argument linking gauge invariance and a massless photon.

33. P. Anderson, *Physical Review*, vol. 130, p. 439 (1963), received by the editor in November 1962.

34. The BCS Theory, and Anderson's example, were nonrelativistic. The particle physics community, focused on relativistic field theory, took little notice. Goldstone had studied theories that respected special relativity, which the BCS Theory does not. This was not a failing of their explanation of superconductivity, as the conditions there were not relativistic—the superconductor or plasma is a particular frame of reference. Consequently, the fact that superconductivity had no massless Goldstone Boson had no obvious relevance for the interests of high-energy particle physicists. From their perspective, the central challenge was to find a way around the theorem in a fully relativistic theory. Tom Kibble, interview with the author, March 17, 2010; Weinberg interview. See also the Scholarpedia article, http://www.scholarpedia.org/article/Englert-Brout-Higgs-Guralnik-Hagen-Kibble_mechanism, by Tom Kibble, 2009.

35. Y. Nambu and G. Jona-Lasinio, *Physical Review*, vol. 122, p. 345 (1961). The remark is on p. 346.

CHAPTER 9

1. *Guardian*, April 13, 2010. The "[5,5]" indicates that the answer consists of two words, each with five letters. "Gosh! Big's no" is the anagram of "Higgs Boson."

2. The CERN budget is in Swiss Francs; the amount of U.S. dollars is similar.

3. See, for example, the BBC's weeklong extravaganza: http://www.bbc.co.uk/radio4/bigbang/programmes.shtml.

4. Letter to *CERN Courier*. The offending article on September 8, 2008, is at http://cerncourier.com/cws/article/cern/35887. The letter from Guralnik, Hagen, and Kibble is on December 8, 2008, at http://cerncourier.com/cws/article/cern/36683.

5. Web claims of prejudice favoring Europeans for the Nobel Prize were rife at the time. They resurfaced in 2010. see *Nature* 4 (August 2010), available at http://www.nature.com/news/2010/100804/full/news.2010.390.html.

6. http://news.bbc.co.uk/1/hi/sci/tech/7567926.stm.

7. Brout and Englert did introduce what has become known as the Higgs Field. That a field will lead to particle manifestation(s) is a small step. The crucial point for particle physics is that the resulting massive boson has observable consequences as a test for the entire mechanism, in that its affinity for various particles is in proportion to their masses. This first appears in P. Higgs, *Physical Review*, vol. 145, p. 1156 (1966), section 3i, equation 17.

8. The title "Gang of Six" was used at the Sakurai award ceremony.

9. The full citation reads: "For elucidation of the properties of spontaneous symmetry breaking in 4-dimensional relativistic gauge theory and of the mechanism for the consistent generation of vector boson masses." Higgs, letter to the author, September 1, 2010; François Englert, e-mail to the author, September 24, 2010. The Wolf Prize citation reads: "For pioneering work that has led to the insight of mass generation, whenever local gauge symmetry is realized asymmetrically in the world of sub-atomic particles."

10. To recap, Goldstone's Boson has no spin: It is a "scalar" boson. The gauge bosons, which transmit forces, have one unit of spin: They are "vector" bosons.

11. These ideas are today associated with the *W* and *Z* bosons, which is why I refer to them in this narrative, using hindsight. However, the original investigations in 1964 were more general, and it was not until 1967 (see Chapter 10) that the specific model was born, which applied these ideas to the weak interaction.

12. Higgs referred to the "ABEGHHKtH mechanism" in a talk at the conference on fifty years of weak interactions, at Wingspread, Wisconsin, in June 1984. Higgs, letter to the author, September 1, 2010. 't Hooft independently rediscovered the mechanism in 1970. See Chapter 11.

13. P. W. Anderson, book review, *Nature*, vol. 405, p. 726 (2000). Anderson's comments were in his review of a book that I had written, *Lucifer's Legacy: The Meaning of Asymmetry* (Oxford: Oxford University Press, 2000), in which I had attributed to Higgs more than Anderson felt was justified.

14. See note 12.

15. Higgs, interview by the author, August 28, 2010, and letter, September 1, 2010.

16. P. Pullman, *Northern Lights*.

17. Being able instantly to get from one place to a remote location would create enigmas of other sorts, however. If we satisfy relativity, the cat would take at least a few microseconds to make the trip and would not be "globally conserved" for the duration.

18. J. Schwinger, *Physical Review*, vol. 125, p. 397 (1962).

19. A. Migdal, http://alexandermigdal.com/prose/paradise1.shtml.

20. A. Polyakov, e-mail to the author, September 17, 2010.

21. A. Migdal and A. Polyakov, November 1965, English version published in *Soviet Physics*, vol. 24, p. 91 (1967). Their paper makes no mention of Higgs or the others in this chapter.

22. G. Guralnik, e-mail to the author, November 16, 2010. Gilbert's paper is D. G. Boulware and W. Gilbert, *Physical Review*, vol. 126, p. 1563 (1962).

23. Guralnik's memoir is available at http://en.wikipedia.org/wiki/Higgs_mechanism and Gerald S. Guralnik, "The History of the Guralnik, Hagen, and Kibble Development of the Theory of Spontaneous Symmetry Breaking and Gauge Particles," *International Journal of Modern Physics*, vol. A24, p. 2601 (2009). Quotes are from this and the letter to the *CERN Courier*, December 8 2008, http://cerncourier.com/cws/article/cern/36683.

24. G. Guralnik, e-mail to the author, November 16, 2010.

25. Though widely referred to this way, Goldstone in 1961 had presented examples; formal proof, which could justify it as a theorem, actually came from the paper published with Salam and Weinberg, referred to in the previous chapter.

26. G. Guralnik, Sakurai ceremony, February 17, 2010, http://il.youtube.com/watch?v=cagfWJFtv84&feature=related.

27. Guralnik and Hagen, interview by the author, March 1, 2010.

28. J. Charap, interview by the author, September 24, 2010.

29. See note 23.

30. Guralnik and Hagen, interview by the author, March 1, 2010.

31. In particular, when Goldstone's boson gets "eaten," giving mass to the vector boson, there is the question of what constitutes the physical degrees of freedom. This is discussed explicitly by Guralnik, Hagen, and Kibble. Guralnik noted that the lack of such an explicit treatment of this issue in the papers by

Brout, Englert, and Higgs was a reason "we did not take these papers seriously when Kibble found them." E-mail to the author, November 16, 2010.

32. A bizarre coincidence is that on Monday, October 5, just a week before Guralnik, Hagen, and Kibble's paper was received by the editor of *Physical Review Letters* in New York, and hence around the time that it would have been submitted to that journal, Peter Higgs gave a seminar about his mechanism at Imperial College. Neither Guralnik nor Kibble has any memory of this, and extensive correspondence between us has failed to shed light on this. Higgs's diary records that his seminar at Imperial College was arranged for 2:30 p.m. on October 5. Higgs, letter to the author, September 23, 2010, with a copy of the diary. Higgs recalls that his air flight from Edinburgh was two hours late and that the seminar was delayed. He recalled also that he had arranged a dental appointment (although Higgs had moved from London to Edinburgh in 1959, his dentist was still in London near Swiss Cottage, and remained so until 1975)—the diary records two options, one for October 5 and one for the next day. These appear to have been options offered after he had accepted the original invitation, but before he had decided on his travel plans. He chose the Tuesday appointment and stayed overnight at the Crofton Hotel in South Kensington. These events in his diary appear to rule out one hypothesis: that as a result of Guralnik, Hagen, and Kibble's paper and their discovery of his work, Higgs was invited on October 5 for a seminar that was to be given at a later date and that he had made the note in his diary on the day when he received the invitation. Others recall Higgs having given a seminar but were unable to identify a date. I asked Kibble whether it could have been the seminar, rather than sight of Higgs's paper (as reported in Guralnik's memorandum), that had occurred as Guralnik, Hagen, and Kibble were about to submit their manuscript, but Kibble independently recalled the appearance of Higgs's paper "around that time." Kibble, letter to the author, December 10, 2010. After fifty years, memories can be unreliable. Higgs's diary is a rare piece of surviving contemporary documentation. Any further information about this event would be welcome to help complete the history of "The Mechanism."

33. The middle equation at the foot of p. 586 in Guralnik, Hagen, and Kibble, *Physical Review Letters*, vol. 13, p. 585 (1964), refers to such a boson, but in their work it would have zero mass. Recall that the presence of unwanted massless Goldstone Bosons is what started the research. Guralnik, Hagen, and Kibble eliminated one of these and in the process discovered the mass mechanism for gauge bosons. This is the general result, of major interest in 1964. There is no explicit potential field in their model—the dynamics of the symmetry breaking is not discussed—and a second boson remains massless. At first sight this appears to ruin everything—nature has no place empirically for *any* massless scalar boson—but Guralnik, Hagen, and Kibble point out that it

is merely a passive spectator, which decouples from the rest, "and has nothing to do with Goldstone's theorem." If they had included a spontaneous symmetry breaking potential in their model—for example, as in Goldstone's 1961 paper—this second boson would have become massive. This is in effect what is now called Higgs' massive boson, which appears at equation 2b in Higgs, *Physical Review Letters*, vol. 13, p. 508 (1964). However, Goldstone's theorem, and its evasion, places no restriction on the mass of this boson. Its mass, existence, and dynamics depend on further model-dependent assumptions.

34. Due to Heisenberg, in 1932.

35. Brout and Englert's assessment of Anderson's contribution was, "That was all done without including the constraints of relativity. With hindsight this was the mechanism giving mass to gauge vector bosons. Anderson had the ideas. He had a model of electromagnetic waves propagating in plasma. Anderson had it essentially right, but the transcription to relativistic [field theory] was not there." Brout and Englert, interview by the author, February 2, 2010. Englert's assessment today of the investigation is that it was "easy to see that vector boson could get mass from symmetry breaking but keeping gauge invariance was trickier."

36. Quotations in this section are from interviews with Higgs, April 11, 2000, and August 28, 2010, and a letter of September 1, 2010.

37. In 1958 a tenured job opened up at Edinburgh. However, Higgs applied for a temporary post at University College, London, in part because he had a girlfriend in London, and also because he expected that there were strong candidates for the Edinburgh post. Higgs asked Kemmer to write a reference on his behalf for the UCL post, even adding a suggestion as to whom Kemmer might appoint to the Edinburgh position. The person whom Higgs had proposed had not applied for the post. Kemmer duly offered Higgs's nominee the job, having previously had no suitable applicants. When in 1959 Tom Kibble arrived at Imperial College, the aim by then was to get security and a tenured position. Higgs applied for a position at King's College, London, but this was in condensed-matter physics rather than particle physics, and he was not sure that he really wanted to be forced into that route. Also, "it had a low salary." Then out of the blue, in 1960, another job in particle physics was advertised in Edinburgh. As Higgs recalled, "Kemmer wanted Kibble back and had engineered the position." Kibble, meanwhile, was settling in at Imperial College. Higgs and Kibble discussed how to end up where they wanted. Higgs had been given a deadline at King's College. He said to me, "I told Tom Kibble that I had to force Kemmer's hand, in the knowledge that if Kemmer didn't get either of us, he had no more strong candidates. So I did."

38. The school is where my own career began to take a fortunate turn, courtesy of the wine. I had met and befriended Israeli theorist Haim Harari, who

later became head of the Weizmann Institute. He was just five years older than I, already a rising star and a wonderful lecturer. His wife was a teetotaler, and as they also had two young children, who sat with us at the same dinner table, this meant that the bottles of wine on offer went even further. This encouraged a nervous young student—me—to shed inhibitions while talking physics and ensured that when I took up my first postdoctoral fellowship later that year, in Stanford, at which Harari was on sabbatical, I had a helping hand. (The atmosphere at Stanford at that time was hectic, as we shall see in Chapters 12 and 13, and Harari helped point me in the right direction.)

39. Higgs, letter to the author, September 1, 2010.

40. A. Klein and B. W. Lee, *Physical Review Letters*, vol. 12, p. 266 (1964).

41. In this hall of mirrors, Gilbert is the same one who became Guralnik's adviser.

42. Boulware and Gilbert, *Physical Review*, vol. 126, p. 1563 (1962).

43. W. Gilbert, *Physical Review Letters*, vol. 12, p. 713 (1964), received March 30, published June 22, 1964. His publications are listed at http://garfield.library.upenn.edu/histcomp/gilbert-w_auth/index-tl.html.

44. Date from Higgs' diary and Higgs, letter to the author, September 1, 2010.

45. Date from Higgs' diary. Received by the editor in CERN on July 27. Published September 15, 1964, in *Physics Letters*, vol. 12, p. 132 (1964).

46. The editor's comments and Higgs's experiences with *Il Nuovo Cimento* are on page 9 of http://www.lnf.infn.it/theory/delduca/higgsinterview.pdf.

47. P. W. Higgs, *Physical Review Letters*, vol. 13, p. 508 (1964). The dynamics of the "Higgs" Boson is described at equation 2b in this paper. It is essentially the same as that immediately below figure 7 in Goldstone's paper of 1961. Most of the content after "When one considers . . . " (ten lines after equation 4) was added after the initial rejection. Higgs, interview by the author, August 28, 2010.

48. Guralnik's recollection was that this was "mid-summer—probably July." E-mail to the author, September 9, 2010. See Guralnik memoir, cited in note 23.

49. Higgs, interview by the author, August 28, 2010, and entry in his university diary, 1964. For an account, see I. Sample, *Massive*, 62.

50. Guralnik, e-mail to the author, September 9, 2010.

51. Boulware and Gilbert, *Physical Review*, vol. 126, p. 1563 (1962).

52. Guralnik, e-mail to the author, September 9, 2010.

53. Guralnik had discussions during the late summer with Robert Lange, in Oxford. Lange later published "Goldstone Theorem in Non-relativistic Theories," *Physical Review Letters*, vol. 14, p. 3 (1965), received November 12, 1964.

54. Gilbert, letter to Higgs, from Harvard University, written August 27, 1964; copy sent to the author, 2000. Higgs discusses this on page 11–12 of http://www.lnf.infn.it/theory/delduca/higgsinterview.pdf.

55. F. Englert, interview by the author, February 2, 2010.

56. P. W. Higgs, *Physical Review*, vol. 145, p. 1156 (1966).

57. A "Higgs Boson" occurs even for the global symmetry case—for example, the fluctuation in ferromagnetic susceptibility. In Goldstone's original paper, page 162, three lines beneath figure 7, he gives the equation of motion, which is formally identical to Higgs's equation 2b, referred to in note 47. On page 163 Goldstone writes, "The massive particles phi$_1'$ correspond to oscillations in the [radial] direction." This appears identical to what has subsequently become named after Peter Higgs. Kibble (interview by the author, March 17, 2010) and Higgs (interview by the author, August 28, 2010) each agree with my assessment. Goldstone confirmed that the massive mode on page 163 corresponds to the Higgs Boson, but emphasized that he was "certainly most concerned with the massless mode." Goldstone, e-mail to the author, December 9, 2010. Higgs uniquely associated a massive scalar boson with the mechanism for generating a vector boson mass gauge invariantly and considered the decay of this boson into those vector bosons. While Goldstone had the massive "radial" excitation, his model had nothing to do with the combined theory of scalar and vector bosons in the context of spontaneous symmetry breaking; its focus was on the enigmatic appearance of the massless scalar boson.

58. Technically, the scalar field expectation value.

59. Technically, fluctuations of the order parameter.

60. Higgs, interview by the author, August 28, 2010.

61. Guralnik and Hagen, interview by the author, March 1, 2010.

62. Kibble, e-mail to the author, February 2010.

63. P. W. Higgs, *Physical Review*, vol. 145, p. 1156 (1966).

64. Or even quarks, the seeds of hadrons, which we will meet in chapter 12. There are various flavors of quark, from lightweight up and down to heavy top, by way of the middleweight charm and bottom.

65. Higgs Boson—or whatever more complicated family of particles may actually be involved. The decay rate depends on a quantum mechanical "amplitude" squared and the available "phase space." The latter tends to favor lighter decay products over heavy ones; it is the amplitude that favors heavy particles in the case of the Higgs Boson decays and gives this unusual aspect to its decay products.

66. Englert, interview by the author, February 2, 2010, and e-mail March 8, 2010.

67. Guralnik, e-mail to the author, November 16, 2010.
68. S. Weinberg, *Physical Review Letters*, vol. 19, p. 1264 (1967).

CHAPTER 10

1. Kibble, interview by the author, March 17, 2010, and e-mail to the author, March 23, 2010.
2. Salam's diary records that this was March 1967; see later.
3. Weinberg, interview by the author, May 6, 2010.
4. The SU2 × SU2 symmetry was broken when you looked at phenomena that were sensitive to the spinning corkscrews, but in cases where the spin didn't matter, the symmetry is excellent.
5. On December 9, 2010, Higgs discovered a letter that he had written to his wife on her birthday, August 29, in 1967 from Rochester. This gives an account of his journey and his visit to Brookhaven on the Friday and to the New York Museum of Modern Art on the Saturday. The meeting with Weinberg is not directly mentioned, but the fresh memories induced by this letter reinforced the certainty of Higgs's memory, as expressed to me earlier on February 25, 2010. Higgs, letter to the author, December 9, 2010, received December 14.
6. Higgs, letter to the author, February 25, 2010.
7. Weinberg, in an e-mail to the author, September 24, 2010, confirmed that he had no memory of this. Having spent the summer away from Cambridge, and returned home in August, he has "strong doubts" that he would have visited Brookhaven. Weinberg did visit Brookhaven in 1968, but Higgs did not. Higgs's diary (February 25, 2010) and the subsequent discovery of the letter to his wife (December 14, 2010) confirmed the 1967 date for his visit. His memory is that he asked Weinberg if he was going to Rochester, and Weinberg said he was not, which if correct would confirm this occasion and not a conflation from elsewhere. No other participants could recall the occasion.
8. Weinberg, e-mail to the author, April 13, 2010, and phone conversation, May 6, 2010. The date often given, wrongly, as October 2, is based on his essay "The Red Camaro," originally published in the magazine *George* and reprinted in *Facing Up*. The editor of the magazine had pushed Weinberg to give a definite date, so he looked at the date *Physical Review Letters* had received his paper and guessed how long it had taken to work out. In *Facing Up* he writes that it was October 2 "as near as I can remember" (185). However, on this date he was in Brussels (see page 192). Weinberg agreed with me that the true date must therefore have actually been mid-September, as the MIT term starts around Labor Day.

9. At least for leptons.

10. S. Weinberg, *Physical Review Letters,* vol. 19, p. 1264 (1967).

11. There is however an unintended consequence for history in the order that Weinberg cited these authors, which by citing Higgs in front of Brout and Englert inadvertently thrust Higgs's name to the forefront in popular credits. Brout and Englert's paper in *Physical Review Letters,* volume 13, is chronologically earlier than that of Higgs, who published in volume 12 of a different journal: *Physics Letters.* In Weinberg's 1971 paper (*Physical Review Letters,* vol. 27, p. 1688) this order is maintained, with Higgs incorrectly cited in note 2 as publishing in *Physical Review Letters,* vol. 12. The editors of *Physical Review Letters,* who are infamous among scientists for their care in querying references and chronology, seem to have overlooked something on this occasion. This typo has propagated through the literature for four decades, most recently appearing in a review by the Particle Data Group in 2010 that has Higgs referenced as publishing in volume 12 of *Physical Review Letters.* Another common typo has Higgs correctly in volume 12 of *Physics Letters,* with Brout-Englert in volume 13 of *Physics Letters* [*sic*]. These show a widespread misconception that Higgs's paper came first.

12. Weinberg, interview by the author, May 6, 2010.

13. *Fundamental Problems in Elementary Particle Physics: Proceedings of the Fourteenth Solvay Conference on Physics, October 1967.*

14. Englert, interview by the author, February 2, 2010, and e-mail to the author, September 24, 2010.

15. Weinberg, interview by the author, May 6, 2010.

16. They were convinced that gauge invariance was right, because in their equations the propagator looked similar to that in QED, which was already renormalizable. It contained an extra contribution, due to the third—longitudinal—mode associated with the massive boson. This extra term stayed roughly constant as the energy became large. Overall, the dependence on energy appeared to be "well behaved at infinity."

17. Brout, interview by the author, February 2, 2010.

18. Englert and Brout, e-mail to the author, February 5, 2010.

19. Technically, the "propagator" contained a term proportional to $q_i q_j / m^2$,where *q* has four possible values corresponding to the energy or the components of the three-dimensional momentum of the gauge boson.

20. I was sent it by Tini Veltman who had received a photocopy from H. Dürr in 1999, around the time of Veltman's Nobel Prize.

21. Kibble, *Physical Review,* vol. 155, p. 1554 (1967).

22. Quote from S. Coleman, *Science,* p. 1290 (December 14, 1979).

CHAPTER 11

1. Unless otherwise specified, quotes from 't Hooft are from interview with the author on April 11, 2000 and e-mails of September 10 and September 13, 2010.

2. This is a statement of the perception when the problem remained unsolved, exemplified by the startled reactions when the news of 't Hooft's work became known (see page 12). However, remarkable though 't Hooft and Veltman's work was, the technical achievement cannot really be compared to that of Wiles, who had to build whole new lines of mathematics, and alone. See Simon Singh's book *Fermat's Last Theorem* for the story of Wiles's triumph.

3. M. Veltman, *Nobel Lecture*, http://nobelprize.org/nobel_prizes/physics/laureates/1999/veltman-lecture.html, 384.

4. M. Veltman, *Facts and Mysteries in Elementary Particle Physics*, 299.

5. Veltman, *Nobel Lecture*, 384.

6. Quotes are from Veltman's *Nobel Lecture* in 1999. In 1971 he rewarded Lee with a warm-up role on the occasion of Veltman's triumphant launch of his student Gerard 't Hooft (see page 222).

7. Veltman, *Nobel Lecture*, 384.

8. As we saw in Chapter 5, Yang-Mills originally tried to apply their ideas to the strong interaction. However, in the weak interaction, the phenomenon of beta decay converts a neutron into a proton, which will be the key to what follows. Other examples of beta decay were known, such as the decay of a pion into a neutrino and an electron, or a muon—an electrically charged particle like the electron but heavier. The muon itself also decays into an electron and a pair of neutrinos. All of these processes showed that the basic interaction converting a neutron into a proton, and an electron or muon into neutrino, had a common strength. This "universality" played a central role in understanding the weak interaction that is responsible. It underpinned the basic idea that the material particles form pairs: neutron and proton, electron and neutrino, and muon and (another type of) neutrino. These pairs form the basic templates of an SU2 Yang-Mills theory; the electrically neutral, positive, and negative vector bosons that emerge from the Yang-Mills equations then transmit the weak forces.

9. The absence of massless electrically charged vector bosons being one example.

10. G. 't Hooft, *In Search of the Basic Building Blocks*, 67.

11. Ibid., 63.

12. See ibid., 65.

13. Ibid., 63.

14. R. P. Feynman, *Acta Physica Polonica*, vol. 24, p. 697 (1963).

15. S. Adler, *Physical Review*, vol. 177, p. 2426 (1969); J. Bell and R. Jackiw, *Il Nuovo Cimento*, vol. A60, p. 47 (1969).

16. 't Hooft, *In Search of the Basic Building Blocks*, 58.

17. Quotes are from interview in 2000 and e-mail, 't Hooft to the author, September 11, 2010.

18. Weinberg, interview by the author, May 6, 2010.

19. Quotes are from interview with 't Hooft on April 11, 2000.

20. The sigma model is not a gauge theory.

21. In quantum field theory, the behavior of fields also is subject to quantum rules. The standard approach is to start with a classical theory and reconfigure it to exhibit quantum effects. The way of doing so was developed in the 1920s; this "canonical" approach became a staple part of physicists' education in the subsequent four decades.

22. Weinberg, interview by the author, May 6, 2010.

23. Veltman quotes in these paragraphs are taken from his *Nobel Lecture*, 390.

24. Abdus Salam Archive, ICTP, Trieste, index University of Bath.

25. 't Hooft, *In Search of the Basic Building Blocks*, 78–79.

26. Ibid., 65.

27. Veltman, e-mail to the author, September 27, 2010. Veltman's first attempts had assumed that the massless theory was renormalizable, and he tried to write the massive theory (an SU2 model) as closely as possible to the massless case. They discussed 't Hooft's first paper—on massless Yang-Mills—at the turn of the year 1970–1971.

28. Quoted in Crease and Mann, *The Second Creation*, 324.

29. He was able to show that the first quantum corrections, involving diagrams where particles form just one closed loop, are anomaly free. This was proved by using a mathematical device—introducing a fifth dimension. That things would work at all loops could be proved only by using 4+epsilon dimensions, and then letting epsilon approach zero, to recover the "real world" of 3 space and one time dimension.

30. 't Hooft paper proving renormalizable massless Yang Mills: *Nuclear Physics*, vol. B33, p. 173 (1971).

31. Veltman, e-mail to the author, September 27, 2010.

32. As reported in Crease and Mann, *The Second Creation*, 325; and A. Pickering, *Constructing Quarks*, 178. The slight difference is based on Veltman's memory. Veltman, e-mail to the author, September 27, 2010.

33. Veltman and 't Hooft's memories differ in detail. I have included both, not least to show how perceptions can differ after many years—something to bear in mind for other examples throughout this record. As Veltman also

wrote, "Anyway, it is not that important, the gist of the story is right." E-mail to the author, September 11, 2010.

34. Veltman, e-mail to the author, September 27, 2010.

35. He used unitarity gauge, then one where infinity could be checked, and showed how to move between one and the other.

36. As recalled by 't Hooft in an e-mail to the author, September 11, 2010.

37. This version follows 't Hooft. However, see also the following note for Veltman's memory. 't Hooft comments also, "There were a few factors like that (actually ¼ or ½); they come about because the Higgs Field has isospin ½, while the ghost terms behave as having isospin 1, so Veltman thought I had made some mistakes there." E-mail to the author, September 11, 2010.

38. Veltman, e-mail to the author, September 27, 2010.

39. Ibid.

40. Veltman, *Nuclear Physics*, vol. B21, p. 286 (1970), and with J. Reiff, *Nuclear Physics*, vol. B13, p. 545 (1969).

41. Veltman, e-mail to the author, September 27, 2010.

42. 't Hooft then added a footnote on page 180, section 7, of *Nuclear Physics*, vol. B33 (1971). Veltman, e-mail to the author, September 11, 2010.

43. Veltman, e-mail to the author, September 27, 2010.

44. Ibid.

45. 't Hooft, interview by the author, April 11, 2000.

46. Veltman, e-mail to the author, September 27, 2010.

47. After I completed the manuscript for this book, sadly Robert Brout died in May 2011.

CHAPTER 12

1. J. D. Bjorken, e-mail to the author, October 1, 2010.

2. Note this involves lower-case "bj", which for any particle physicist equates with "bee-jay." For another personal view of "bj" see "Quantum Diaries," http://qd.typepad.com/17/2005/05/got_bjorken.html.

3. A colleague, Bob Johnson from Durham University, is six foot ten. He and "bj" once compared, standing back to back, determining that "Big-B Big-J minus little-b little-j is approximately ten centimeters" (Bob Johnson remark at CERN Theory Group Introductions, 1973).

4. G. Zweig, e-mail to the author, October 3, 2010, and *Memories of Murray and the Quark Model*: talk presented at the "Conference in Honor of Murray Gell-Mann's 80th Birthday," Nanyang Technical University, Singapore, February 24, 2010.

5. Bjorken, e-mail to the author, October 1, 2010.

6. Recall that in 1960 the weak interaction was still an enigma. The SU2 × U1 models emerged only during 1961 to '67, and their viability was not established until 1971.

7. H. J. Lipkin's team at the Weizmann Institute in Israel, G. Morpurgo in Genoa Italy, and a team in Moscow were the other leading players. Elsewhere in the world, as Weinberg recalled on page 256, the quark model was largely ignored. See A. Pickering, *Constructing Quarks*, for a history of this enterprise.

8. In SU3, family groups of eight and ten occur. Examples of these had been established by 1963. The importance of SU3 as an organizing scheme for hadrons was independently discovered by the Israeli theorist Yuval Ne'eman, working in Abdus Salam's group at Imperial College, London.

9. R. Serber, with R. P. Crease, *Peace and War*, 199. Gell-Mann's version is at http://www.webofstories.com/play/10658.

10. Quotes in this section are from George Zweig, e-mails to the author, September 22 and October 1, 2010.

11. Ibid.

12. Ibid., October 19, 2010.

13. The first preprint was shorter and less complete than the second. Zweig had to choose which one to ask his wife to retype, so he chose the second. Ibid.

14. G. Zweig, "An SU3 Model for Strong Interaction Symmetry and Its Breaking II," *CERN Report 8419/TH. 412* (February 21, 1964), published in D. B. Lichtenberg and S. P. Rosen, eds., *Developments in the Quark Theory of Hadrons: A Reprint Collection*, available at http://cdsweb.cern.ch/record/570209?ln=en.

15. See J. J. J. Kokkedee, *The Quark Model.*

16. This illustrates the metaphor that led to the name of "color." As three primary colors, Red, Yellow, and Blue, combine to white, so do three quarks each carrying a different "primary" color charge combine to make a proton or neutron (say), which has no net color. The rules of attraction and repulsion for color charge match those of electric charge: Like colors repel, and unlike colors attract (for readers interested in the finer print of quantum detail, they attract if their quantum state is antisymmetric under interchange of any pair of colors). Thus, attraction between three colors occurs if they are each different (and overall antisymmetric). Hence, RYB for example (this unique overall antisymmetric combination in SU3 corresponds to a "color singlet," or, in the jargon, colorless).

17. M. Gell-Mann, *The Quark and the Jaguar*, 182.

18. M. Gell-Mann, *Physica*, vol. 1, p. 63 (1964).

19. The basic idea of current algebra was to assume the existence of quarks, whose motion gives rise to currents, such as electric current. These currents determine how the quarks absorb or emit quanta such as a photon, W^+, W^-, Z^0, and so on. In current algebra, the responses of a quark current to these various quanta are related, the actual interrelationships forming a set of rules or "algebra." Under certain circumstances it was possible to elevate these relationships

from the basic quarks to the hadrons made of those quarks, without making any assumptions about the details of how those quarks actually interact among themselves when forming those hadrons. This led to very powerful predictions, which were amenable to experimental test. Their success then raised a question: Did these results imply the existence of the actual quarks, or did they follow more generally from the current algebra, which though "discovered" by assuming the existence of quarks did not necessarily require them to exist—in Gell-Mann's analogy, discarding the veal. Fifty years later, there is a vast amount of data testing the role of the quarks themselves, and revealing their dynamics, which goes far beyond the more general results of current algebra.

20. S. Adler, *Physical Review Letters*, vol. 14, p. 1051 (1965); S. Adler and F. J. Gilman, *Physical Review*, vol. 156, p. 1598 (1967).

21. An electron scatters through some angle and gives up some of its energy to the target. These two quantities—effectively the amounts of energy and momentum transferred to the target —can be varied independently in an experiment. Bjorken predicted that the data do not depend on the two independently, but in a correlated manner that depended only on the ratio of the energy to a combination of the energy and momentum. This ratio is traditionally denoted x_{bj}, and the dependence of the data on this single variable is known as Bjorken-scaling.

22. An elastic collision is where kinetic energy is conserved; inelastic is where it is converted into some other form of energy. Examples of inelastic collisions are when a lump of mud hits a wall, or a car runs into a heavy barrier. In particle physics, an example of an elastic collision is where an electron bounces off a proton without creation of other particles; inelastic is when some kinetic energy is converted into one or more new particles by the mass-energy relation $E=mc^2$. The "deep" inelastic refers to an extreme situation, where the electron probes deep into the target proton, with creation of a maelstrom of particles and consequent large loss of kinetic energy and momentum of the electron. Rather than attempt to detect all the debris, the idea of the experiment was to record solely the energy and momentum of the electron after it had scattered from the target.

23. Bjorken recalled (in an e-mail to the author, October 1, 2010) that the proposal included several measurements: elastic scattering, production of proton resonances, deep inelastic, and comparison of scattering electrons and positrons, pretty much in that order of priority. "I of course pushed for taking the deep inelastic very seriously, and it is that strong push that I am most proud of. In conversations with Dick Taylor, I asked whether the deep inelastic would have been done without that push. He said 'of course—when you invest so much in setting up such a big facility, you measure the hell out of everything

you possibly can.' I believe that to be so, but do feel that my pushing on them did make a difference."

24. The logbook of Elliott Bloom (viewed July 31, 2009), one of the experimentalists involved, records one committee member remarking: "This [is] probably enough to make the inelastic electron experiment impossible."

25. Bjorken, e-mail to the author, October 1, 2010: "I looked at the problem from many viewpoints; pointlike constituents was only one way (but a not unattractive way, albeit rather naïve even to me)."

26. R. Taylor, interview by the author, July 10, 2009.

27. The confirmation that the pointlike constituents are quarks comes from combining results of the SLAC experiments, which used electron beams, and analogous experiments at CERN using neutrinos.

28. Bjorken, e-mail to the author, October 1, 2010.

29. See note 21.

30. H. Kendall in his Nobel Prize speech recalls his astonishment; Bjorken quote from an interview by the author, August 4, 2009.

31. Manny Paschos, e-mails to the author, October 2010.

32. Elliott Bloom, interview by the author, July 31, 2009.

33. J. Friedman retained a copy of the page from Feynman's notes. The original is in the Cal Tech Archives.

34. J. Friedman, e-mail to the author, October 14, 2010.

35. Manny Paschos, e-mails to the author, September and October 2010.

36. M. Riordan, *The Hunting of the Quark*, 150; e-mail correspondence with the author, September and October 2010.

37. Paschos, e-mails to the author, September and October 2010.

38. R. P. Feynman, recorded in 1985: Crease and Mann, *The Second Creation*, 305.

39. Cal Tech archives; see also Riordan, *The Hunting of the Quark*, 149.

40. S. J. Brodsky, interview by the author, October 3, 2009.

41. Paschos, e-mails to the author, September and October 2010.

42. Bjorken, interview by the author, August 4, 2009.

43. Ibid.

44. J. D. Bjorken and E. A. Paschos, *Physical Review*, vol. 185, p. 1975 (1969).

45. Bjorken, interview by the author, August 4, 2009.

46. Ibid. and Feynman memorabilia at Cal Tech Archives.

47. R. Taylor, interview by the author, July 10, 2009.

48. Crease and Mann, *The Second Creation*, 306.

49. J. Friedman, e-mail to the author, October 14, 2010.

50. W. Panofsky, *Proceedings of the XIV International Conference on High Energy Physics* (Vienna, 1968), 37.

51. Taylor, interview by the author, July 10, 2009.

52. Bjorken, *Physical Review*, vol. 148, p. 1467 (1966).

53. Weinberg, interview by the author, May 6, 2010.

54. The strange feature was that these hadrons survived the ravages of the strong interaction, eventually decaying due to the influence of the weak interaction, even though it was the strong interaction that had led to their production. This strange property led to their eponymous descriptor. The explanation of this behavior is outside the scope of this book. See my *The New Cosmic Onion* or *Particle Physics: A Very Short Introduction* for more about strange particles and strange quarks.

55. Today we know of a third pair carrying electric charges of $-\frac{1}{3}$ and $+\frac{2}{3}$. They are in effect much heavier versions of the strange and charm, or down and up. Their names once again reflect this complementarity: "bottom and top."

56. Charm was discovered in 1974. Bjorken's insight into the behavior of electron-positron annihilation played a seminal role in this. See Pickering, *Constructing Quarks*, and Close, *The New Cosmic Onion*, chap. 9.

57. Weinberg, interview by the author, May 6, 2010.

CHAPTER 13

1. The theory has become known as Quantum Flavordynamics, although not everyone is happy, some preferring "Quantum Electroweak Dynamics." The rationale here is that as Feynman, Schwinger, Tomonaga, and Dyson created the viable full theory of Quantum Electrodynamics, so 't Hooft and Veltman did the same for quantum electroweak dynamics. The citation for 't Hooft and Veltman's Nobel Prize says: "for elucidating the quantum structure of electroweak interactions." The crucial word here is *quantum* (namely, with virtual processes and radiative corrections included). (I am indebted to Ian Aitchison for discussions about this.)

2. Quoted by D. Gross in his *Nobel Lecture*, http://nobelprize.org/nobel_prizes/physics/laureates/2004/gross-lecture.html; *Reviews of Modern Physics*, vol. 77 (July 2005).

3. Sidney Coleman was at Cal Tech before winning his position at Harvard, and references were obtained from Gell-Mann and Feynman. A story, possibly apocryphal, is that Gell-Mann asserted that "Coleman is smarter than Feynman," Feynman attesting that "Coleman is smarter than Gell-Mann." Even if this is not true, it illustrates the esteem in which Coleman was held.

4. T. Goldman, e-mail to the author, August 10, 2010, and interview by the author, August 9, 2010.

5. Ibid.

6. Tony Zee, interview by the author, October 28, 2010.

7. H. Fritzsch and M. Gell-Mann, *Proceedings of the XVI Conference on High Energy Physics*, vol. 2, p. 135 (1972), available at arXiv:hep-ph/0208010.

8. Quark color is explicitly assumed to follow the mathematics of SU3, in which case hadrons are "color singlets." See also Chapter 12, note 16. Gluons are assumed to be octets of SU3, though Gell-Mann says somewhat cautiously: "They could form a color octet of neutral vector fields obeying the Yang-Mills equations."

9. Ref. 5 in Fritzsch and Gell-Mann, *Proceedings of the XVI Conference on High Energy Physics*, cites "J Wess, private communication to B Zumino."

10. Harald Fritzsch, e-mail to the author, October 7, 2010.

11. Having raised the idea, Gell-Mann's talk then ignores this possibility.

12. For example, Politzer's *Nobel Lecture*, http://nobelprize.org/nobel_prizes/physics/laureates/2004/politzer-lecture.html, 89, mentions "folk-lore" and "variant versions"; and Wilczek, 123, comments on the reliability of accounts drawn from memory long after the event. In the final paragraph of his references, Politzer refers to "mutually contradictory perceptions of what transpired, [which] may shift as time passes."

13. This is as described in his book, *In Search of the Basic Building Blocks*, 86. In interview with me on April 11, 2000, he said that he was unaware of the phenomenon.

14. 't Hooft "moved on." 't Hooft, *In Search of the Basic Building Blocks*, 86–88; and interview by the author, April 11, 2000.

15. 't Hooft, interview by the author, April 11, 2000.

16. Ibid.

17. 't Hooft, *In Search of the Basic Building Blocks*, 88; and interview by the author, April 11, 2000.

18. This is as told to me on April 11, 2000. In his book he writes more dramatically: "Symanzik practically forced me to stand up. . . . And I wrote the result on the blackboard." 't Hooft, *In Search of the Basic Building Blocks*, 88.

19. C. Korthals-Altes, interview by the author, October 8, 2010.

20. Ibid. 't Hooft had done the calculation in Landau gauge and thought that this might be a peculiarity of that choice.

21. Ibid.

22. Wilczek was interested in the foundations of quantum field theory and the consistency of the new electroweak theory. He discussed several possible ideas with Gross, and the sign of the beta function "got an enthusiastic reception because it fit in with [Gross's] program to show that Bjorken-scaling could not be achieved in quantum field theory." Frank Wilczek, e-mail to the author, October 8, 2010.

23. Tom Appelquist, e-mail to the author, October 26, 2010.

24. K. Symanzik, *On Theories with Massless Particles*, DESY Report 72/73.

25. Appelquist, e-mail to the author, October 26, 2010.

26. This is two years before Politzer, Gross, or Wilczek.

27. 't Hooft, interview by the author, April 11, 2000. He was also concerned that the quarks do not appear as free particles, and in consequence some of the basic tenets of field theory seemed questionable to him: "You have in and out states that you don't really understand. I thought this was not very clean physics, and didn't understand that it is clean if you make a certain number of assumptions. At the time it looked rather ugly to me."

28. G. Parisi, interview by the author, August 4, 2010.

29. "Half an hour" quote from Parisi interview with the author, August 4, 2010.

30. Parisi, interview by the author, August 4, 2010.

31. Noted by Bryan Webber, "Asymptotic Freedom," November 3, 2004, available at http://www.hep.phy.cam.ac.uk/theory/webber/asymf.pdf, 21. See also I. B. Khriplovich, "Green's Functions in Theories with a Non-Abelian Gauge Group," *Soviet Journal of Nuclear Physics*, vol. 10, p. 235 (February 1970), submitted December 21, 1968.

32. D. Politzer, *Nobel Lectures*, 89.

33. One piece of physics that intrigued him was the mystery of the "Landau Ghost." He was aware that Landau had shown the logical structure of QED to be flawed—the "smile of the Cheshire cat" phenomenon that we met on page 73. This arose because in QED the charge concentrates into the smallest pixels as resolution improves—in the jargon, "beta" is *positive*. The discovery that in QCD beta is negative, and the possibility that QED and QCD merge into a grand unified force at distances far smaller than currently accessible, can ameliorate this problem.

34. The techniques involved the renormalization group. The upsurge in interest followed papers by C. Callan, *Physical Review*, vol. D2, p. 1541 (1970); K. Symanzik, *Communications in Mathematical Physics*, vol. 18, p. 227 (1970); and two papers by K. Wilson, *Physical Review*, vol. B4, pp. 3174, 3184 (1971).

35. Frank Wilczek, interview by the author, November 1, 2010.

36. Anthony Zee, interview by the author, October 28, 2010.

37. Quoted from Politzer, *Nobel Lecture*, 88.

38. Ibid., 89.

39. D. Gross, *Nobel Lecture*, 70.

40. Ibid., 71.

41. D. Politzer, e-mails to the author, September 24, 2010, and November 11, 2010.

42. E. Weinberg, interview by the author, October 28, 2010.

43. Wilczek, interview by the author, November 1, 2010. Wilczek checked that the theory satisfied renormalization by looking at the diagram where only gluons are present—the "gluon self-coupling"—and compared the renormal-

ization of the effective charge ("coupling") as revealed in those diagrams with the cases of a gluon coupling to other fields, in particular fermions and ghosts and scalars, all of which should be the same effective coupling and renormalize the same way. Ibid.

44. Wilczek, interview by the author, November 29, 2010. He tells this in detail in the book *Longing for the Harmonies,* where he writes, "I had wasted many hours in consternation and confusion simply because I was too trusting, a little lazy, and hadn't given 5 minutes to check this thing." F. Wilczek and B. Devine, *Longing for the Harmonies,* 212.

45. Wilczek, interview by the author, November 1, 2010.

46. Ibid.

47. Gross, *Nobel Lecture.*

48. E. Weinberg, e-mail from the author, August 4, 2010.

49. At equation 6.68, he demonstrated that the sign of beta is negative for an SU2 Yang-Mills theory. Politzer was studying the SU3 case. E. Weinberg believes his calculation of the beta function was "a relatively straightforward matter" and does not regard that alone as worthy of huge credit. "The important thing, which David [Politzer] had, but which I didn't, when we began our calculations, was the understanding of the possible significance of a minus sign for the strong interactions." Interview by the author, October 28, 2010.

50. Politzer, *Nobel Lecture.*

51. Wilczek, interview by the author, November 1, 2010.

52. Politzer, *Nobel Lecture.*

53. E. Weinberg, interview by the author, October 28, 2010.

54. Politzer, e-mail to the author, December 3, 2010.

55. In 2010 to 2012 it includes March 16; in 2009–2007 it was the final week of March, which is Weinberg's memory of the break timing historically.

56. This is most likely to have occurred during the middle or last week of April. The timing of Weinberg's thesis—examined May 11, submitted mid-April—implies that the sign of the beta function was established, at least in Weinberg's mind, at the latest by mid-April. Politzer's paper was received by the editor of *Physical Review Letters* on May 3, 1973; the Gross-Wilczek paper was received on April 27, 1973.

57. Politzer, e-mail to the author, November 14, 2010.

58. Ibid.

59. Roman Jackiw, interview by the author and e-mail, December 1, 2010.

60. Gross recorded in his *Nobel Lecture,* "At almost the same time, Politzer finished his calculation and we compared our results. The agreement was satisfying." Politzer has no memory of ever checking any of his results with Gross. Politzer, e-mail to the author, December 2010.

61. Sadly, Sidney Coleman died in 2007.

62. See note 49.

63. E. Weinberg, e-mail to the author, August 4, 2010, and interview by the author, October 28, 2010.

64. E. Weinberg, interview by the author, October 28, 2010.

65. Wilczek, interviews by the author, November 1 and November 29, 2010.

66. Wilczek, interview by the author, November 1, 2010, confirmed by Politzer, e-mail to the author, November 12, 2010. This may also explain the apparent difference between the account in Gross's by *Nobel Lecture* and Politzer's denial of having checked with Gross (see note 60). The "*we* compared" refers to the collective, at that meeting, not to Gross specifically.

67. Wilczek, interviews by the author, November 1 and November 29, 2010.

68. Although this seems to me a likely chronology, there remains an unresolved possibility that the meeting took place later. With stakes high, there can be an urgency to get one's paper to a journal, where it is stamped with a submission date and has its priority established. There can be plenty of time to retract or correct a submitted paper, even one sent to *Physical Review Letters,* and the relative penalty for doing so may be minor compared to the benefit of doing something interesting and winning the plaudits. The analogy to the Klondike gold rush, made by John Ward to describe some attitudes in the 1960s, is not uncommon. Whether Politzer made his visit to Princeton and the comparisons were made after papers were in the mail to the journal, or whether this happened before any manuscripts had been submitted, is known only to those who were involved and, as we have seen in many examples of false memory after so long, possibly not even to them. The question of whether Politzer initially sent a draft to another journal is, for example, unresolved.

69. The papers are D. Gross and F. Wilczek, *Physical Review Letters,* vol. 30, p. 1343 (1973); and D. Politzer, *Physical Review Letters,* vol. 30, p. 1346 (1973).

70. Such particles are produced also when cosmic rays hit the atmosphere. However, few have actually been detected other than in customized experiments at accelerators on Earth.

71. In the research literature, the standard abbreviation for "Bjorken-scaling" is "bj-scaling," recognizing Bjorken's moniker.

72. The full story is more complicated; see Pickering, *Constructing Quarks.* The first hints of a threat, which turned into an opportunity, came from data at CEA, the Cambridge Electron Accelerator, an electron-positron collider at Harvard. This machine was overtaken by the new SPEAR—"Stanford Positron Electron Accelerating Ring"—which initially confirmed the anomaly and then resolved it in 1974 as due to the presence of a hitherto unrecognized particle, the J-psi, and followed this in 1976 with the discovery of charmed mesons, together with a "heavy lepton," the tau particle, a heavier sibling of the electron and muon, for which Martin Perl won the Nobel Prize in 1995.

73. The discovery of the J-psi is probably the most singular event in experimental particle physics of the latter half of the twentieth century. It came out

of the blue, heralded the discovery of charm, and proved seminal in establishing both the electroweak theory and QCD. Known as the November Revolution, this period has been the subject of articles and highly profiled in books. For full justice, see Riordan, *Hunting of the Quark*; Pickering, *Constructing Quarks*; Close, *The New Cosmic Onion*; and *CERN Courier* (December 2004): 25. See also my remarks about it in the postscript to this book.

74. In QED, an electron and a positron form an unstable atom known as positronium. By analogy, in QCD the strong force makes an analogous "atom" of charm and anticharm, known as charmonium. As the electrons in atoms can occupy various levels of the energy ladder (see Chapter 1), so can the constituents of charmonium. This leads to a collection of charmonium states, with different energies (masses), leading to a characteristic bar code of gamma rays, analogous to those found for conventional atoms (see page 23).

CHAPTER 14

1. B. W. Lee, *Physical Review*, vol. D5, p. 823 (1972); vol. D6, p. 1188 (1972).

2. Higgs's colleague was Ken Peach, who is today a colleague of mine at Oxford University. This story is told on page 13 of http://www.lnf.infn.it/theory/delduca/higgsinterview.pdf and confirmed to me by Peach.

3. F. Close, "The Light at the End of the Tunnel," *Guardian*, July 21, 1983, 19.

4. A neutrino's energy enables it to feel the pull of gravity, but this is so feeble as to be unmeasurable.

5. For more details and references to papers, see Crease and Mann, *The Second Creation*, chap. 17.

6. S. Bludman predicted a neutral weak interaction in 1958. *Il Nuovo Cimento*, ser. 10, vol. 9, p. 433 (1958). Bludman wrote to Salam about this on October 12, 1979, days before Salam was awarded the Nobel Prize. Bludman letter, ICTP Archives.

7. This parameter determined the relative probability that a neutrino, upon hitting a proton, might turn into an electron (the conventional form of weak interaction involving the exchange of an electrically charged W boson) or remain a neutrino (exchanging a Z^0). For a definition of the Weinberg angle, see Chapter 7, note 14.

8. F. Close, "Iliopoulos Wins His Bet," *Nature*, vol. 262, p. 537 (August 12, 1976).

9. A. Salam, *Proceedings of the 8th Nobel Symposium*.

10. For a discussion of this series of experiments and how science came to a consensus, see A. Pickering in *PSA: Proceedings of the Biennial Meeting of the Philosophy of Science Association*, 459–469.

11. F. Close, "Parity Violation in Atoms," *Nature*, vol. 264, p. 505 (1976).

12. Prior to this experiment, and essential to its eventual interpretation, the Weinberg angle had been measured. Having determined the value, it was in turn possible to predict that in electron scattering, there would be a preference for left-handed electrons, at a rate of about one extra interaction in every 10,000 collisions.

13. C. Prescott et al., *Physics Letters*, vol. 77B, p. 347 (1978), received July 14.

14. F. Close, "New Source of Parity Violation," *Nature*, vol. 274, p. 11 (1978).

15. F. Close, "A Massive Particle Conference," *Nature*, vol. 275, p. 267 (1978).

16. Correspondence of S. Weinberg, in response to Abdus Salam, received October 19, 1978, Salam Archives, ICTP, Trieste. Weinberg agreed that some (unspecified) aspects of my report had been "disturbing."

17. Glashow and Bova, *Interactions*, 269.

18. Ibid., 267, 270.

19. Ibid., 270.

20. Ibid.; Glashow, interview by the author, December 15, 2010.

21. Letters between Abdus Salam and I. Waller, ICTP Archives, include copies or references to exchanges on January 1 and February 15, 1965; November 14, 1966; October 1, 1969; November 4, 16, and 27, 1969; and September 23, 1970. Areas covered include reports on Salam's work; requests for Waller to visit ICTP, Imperial College or Pakistan; and reciprocal invitations for Salam to visit Sweden.

22. Glashow and Bova, *Interactions*, 270.

23. 't Hooft, interview by the author, April 11, 2000.

24. Shared by Arnold Penzias and Robert Wilson for their discovery of the microwave background radiation and Pyotr Kapitsa for his work in low-temperature physics.

CHAPTER 15

1. A. Salam, *Il Nuovo Cimento*, vol. 5, p. 299 (1957).

2. Salam's work on the neutrino was mentioned in nominations from Dirac, Kibble, and Matthews. ICTP Archives.

3. Quote from nomination letter to Nobel Committee for Physics by Kibble, January 12, 1979.

4. Peierls wrote to Salam on January 3, 1982: "I do not remember asking about the neutrino mass in your Ph D viva, though the question may have come up. However, we came back to the subject much later." At this latter meeting, which dealt with other issues—such as how to define a state of a single massless photon, which led to the analogous question concerning a massless neutrino—"we concluded that if the neutrino mass is zero . . . there must be some conservation law [that could define a single neutrino from a collection of them]."

Salam and Peierls then met again, and Salam suggested that "gamma-5 invariance" might give the required conservation quantity. Peierls then pointed out, "It was immediately clear that this could involve a violation of parity."

5. Abdus Salam, letter to I. Waller, October 1, 1969, ICTP Archives.

6. Erik Hulthen was chair of the Nobel Physics Committee from 1958 to 1962.

7. The letter in the ICTP Archives is undated, but includes the statement that Salam was "awarded the Oppenheimer Prize and also elected to a Fellowship of the USSR Academy of Sciences this year." Both of these honors were bestowed in 1971.

8. S. Weinberg, *Physical Review Letters*, vol. 19, p. 1264 (1967).

9. S. Coleman, *Science*, p. 1290 (December 14, 1979).

10. Salam, *Proceedings of the 8th Nobel Symposium*.

11. There were questions from three physicists: Pais, Sudarshan, and Stech. Pais and Sudarshan focused primarily on the technical nature of a propagator (in a rather confusing way, given modern understanding), and Stech on the mass for the *W* boson, which was a somewhat peripheral issue. The spontaneous symmetry breaking, which is what subsequently became the raison d'être, is not debated at all.

12. Graduate students and postdoctoral researchers of the time, several of whom became established as leading figures in the field, were soaking up the new ideas that were flowering in the early 1960s, but when I asked for any memory of Salam's lectures about weak and electromagnetic unification, the typical response was, "I don't recall anything like that."

13. Kibble's paper does not mention SU2 × U1 explicitly, but it is a trivial step to extract this specific example from Kibble's general discussion. The March date comes from Salam's personal diary; see Chapter 10, note 2.

14. Kibble, e-mail to the author, December 18, 2009.

15. C. Isham, interview by the author, November 10, 2010. Isham was asked by Salam, years later, if he had taken any notes of the lectures, as Salam wanted to find if there was any written record of them. He had none.

16. R. Delbourgo, e-mails to the author, December 1, 2009, and October 23, 2010, and interview by the author, April 19, 2010.

17. That year nine papers shared John Strathdee's name as coauthor; a similar work rate had taken place in 1966 and continued into 1968. All of this work focused on the strong interaction and gravity, not on attempts to unify the electromagnetic and weak interactions.

18. I found no record of the 1967 lectures in the 350 boxes of his papers, which include the drafts of many of his published articles, along with notes of talks that he had either given or heard. The draft manuscript of his Göteborg talk survives and appears to be identical to the published version.

19. A point made by Ward himself—see note 49 below.

20. This is in the written version of his own talk, though it is unclear whether he made these comments in the discussion session at the time. Veltman interview by the author, December 17, 2009.

21. University of Bath archival index contains the entry "Paper A98: Letter from P. T. Matthews confirming that he heard Salam describe the unified gauge theory in a lecture in 1967, written in support of Salam's candidacy: 26 July 1976." The quotes are taken from the copy of this letter in the ICTP Archives, July 27, 2010.

22. Anne Barrett, e-mail to the author, October 22, 2010.

23. Information from Abdus Salam's diaries was kindly provided by his wife, Louise Johnson, in e-mails to Close, December 5 and December 11, 2010. They make no mention of him giving any lectures, but this was not abnormal. Salam's Nobel lecture referred simply to autumn 1967.

24. R. Delbourgo, in an e-mail to the author, October 23, 2010, confirmed that he is certain that they were completed before he saw Weinberg's paper in the library, in November.

25. Isham, e-mail to the author, December 12, 2010.

26. Louise Johnson, e-mail to the author, December 12, 2010.

27. Memories of people at this conference are in http://www.mth.kcl.ac.uk/~streater/salam.html.

28. J. C. Polkinghorne, interview by the author, July 8, 2010.

29. Memoranda written about Salam's notable research, and forwarded to the Nobel committee, did not mention this work until after 1971. See also note 7 and Salam's letter to Matthews.

30. There is no record in Salam's curriculum vitae of his publishing any significant research with these regular visitors. The quantity of letters exchanged with Swedish academicians on the physics committee is matched only by correspondence with some of his research collaborators, or those with whom he had conflicts over citation.

31. S. Weinberg, *Physical Review Letters*, vol. 27, p. 1688 (1971). This erroneously cites Higgs, first, as having published in *Physical Review Letters*, volume 12 (rather than *Physics Letters*)—with Brout and Englert in volume 13 of that journal. This incorrect citation of Higgs's 1964 paper has propagated through the literature for four decades, like some Darwinian mutation, giving Higgs apparent chronological priority over Brout and Englert.

32. S. Weinberg, *Physical Review*, vol. D5, p. 1412 (March 15, 1972).

33. A. Klein and B. W. Lee, *Physical Review Letters*, vol. 12, p. 713 (1964), published March 9, 1964.

34. B. Lee, *Proceedings of XIII International Conference on High Energy Physics*.

35. Correspondence between Abdus Salam, June 26, 1972, and Ben Lee, n.d., ICTP Archives.

36. Ben Lee died in 1977. Mary Gaillard, his collaborator, found no copy of the correspondence with Salam. E-mails with Gaillard, J. D. Jackson, and C. Quigg, January 21 and February 5, 2010.

37. His argument could today have renewed relevance in view of the independent discovery of the mass mechanism by the "Gang of Six"—Guralnik, Hagen, and Kibble completing their independent work only after those of Brout, Englert, and Higgs had appeared. In turn some of these ideas were anticipated by Migdal and Polyakov.

38. Correspondence between Abdus Salam, June 27, 1972, and Ben Lee (undated), ICTP archives.

39. D. Bailin and N. Dombey, "SU2 × U1: A Gauge Theory of Weak Interactions," *Nature*, vol. 271, p. 20 (January 5, 1978).

40. This prediction relies on the quantity widely referred to as the Weinberg angle.

41. N. Dombey, interview by the author, December 11, 2009.

42. Dombey and D. Bailin, e-mail to the author, November 2, 2010.

43. Veltman, interview, December 17, 2009.

44. When Veltman eventually won the Nobel Prize in 1999, he even challenged Weinberg about his perceived role in this bias (ibid. and Veltman, *Facts and Mysteries in Elementary Particle Physics*, 274: "Some explanations that I found hard to swallow"). Weinberg was present at that ceremony due to having been an assessor for that year's prize, thus having himself evaluated and approved the award for both 't Hooft and Veltman. He made clear to me that he has never had any animus toward Veltman. S. Weinberg, interview by the author, May 6, 2010. As for citing 't Hooft, some of the seminal papers carried his name alone; hence, there was some reason for his prominence.

45. Nomination letter for Salam, Nobel Prize, 1979, ICTP Archives.

46. G. Fraser, interview by the author, April 13, 2010.

47. Gell-Mann's role and opinions about Salam and Ward are preserved at http://www.webofstories.com/play/52253.

48. Ibid.

49. Quotes from J. C. Ward, letter to the editor of *Scientific American*, August 14, 1974, copied to Salam, ICTP Archives.

50. S. Weinberg, letter to Salam, November 24, 1971, ICTP Archives.

51. This is the same point made by Dombey on April 3, 1978, in response to Salam's objections at Bailin and Dombey's assignment of credits. Dombey, interview by the author, December 11, 2009.

52. Others claim that Ward's version has exaggerated his role. For a version supporting Ward's claims, see Dombey and Grove, "Britain's Thermonuclear Bluff" (see Chapter 6, note 7). For a contrary version, see L. Arnold, *Britain and the H-Bomb*. That Ward was involved in the UK atomic research program in the 1950s is not in doubt; whether his version of this event is accurate, I

cannot say, which is unfortunate as this is not the only place where we have had to rely on Ward's version of events. His record of his Ph.D. viva with Peierls seems, as we saw earlier, to have egged the pudding somewhat; his claim to have inspired Salam to take up unification seems likely to remain unverifiable, as although the ICTP Salam archive contains correspondence between Ward and Salam, with references to Ward in a letter from Dyson to Salam as early as 1950, there was no sign of the letter that Ward claims to have written in 1957, referred to on page 105. The drafting of the manuscript of their 1964 paper seems to have been driven by Salam. Original manuscript with Louise Johnson, seen by the author, December 12, 2010. See also Figure 10.2.

53. At least, until he became a religious outcast. See Fraser, *Cosmic Anger*, 272.

54. Ward died on May 6, 2000, in Canada.

55. Ward, *Memoirs of a Theoretical Physicist*, 18.

CHAPTER 16

1. Colliding beams were not new—electron positron annihilation at low energy, and the ISR (Intersecting Storage Rings) at CERN collided protons during the 1970s, but not at energies capable of producing the *W* or *Z*.

2. Intersecting Storage Accelerator + "belle."

3. Whereas electron positron annihilation can produce a single Z^0, the conservation of electric charge requires that the W^+ and W^- emerge as a pair, hence there has to be enough energy to produce the two, rather than just one.

4. John Adams's life and work is described in M. C. Crowley-Milling, *John Bertram Adams, Engineer Extraordinary.*

5. Adams, biographical notes by G. Stafford, Royal Society, London.

6. Llewellyn Smith, interview by the author, March 11, 2010.

7. *UA* stood for "Underground Area." Rubbia was the leader of UA1. Frenchman Pierre Darriulat led UA2. For an account of this period, see G. Taubes, *Nobel Dreams.*

8. Technically correct is that W^- produces an electron and *anti*neutrino.

9. Weinberg's paper made this explicit; Glashow, Salam, and Ward gave no explicit measure of both the *Z* and *W* masses.

10. The student was Gunnar Ingelman.

11. Llewellyn Smith, *Nature*, vol. 448, p. 281 (2007). Quotes in this section come from this article and interviews by the author, March 11 and October 28, 2010.

12. 1 TeV is 1,000 GeV, where the energy locked into the mc^2 of a single proton at rest is almost 1 GeV. 1 GeV is roughly 40 billion times the energy of body heat.

13. C. Llewellyn Smith, *Nature*, vol. 448, p. 281 (2007).

14. F. Gilman, interview by the author, October 10, 2010, and e-mails October 17 and November 10, 2010.

15. On January 8, 1992, George Bush Sr., was in Japan, prepared to agree concessions with the import of Japanese auto parts to America in return for Japanese contribution to the SSC. As the Collider was sited in his home state, he certainly was warm to the possibility. As to what happened next, the *Houston Chronicle* on October 1, 2008, lists "Top 10 Travel Blunders." At the head of its list is "vomiting on your host." Bush was taken ill after a tennis game, or food poisoning, depending on which politically spun version you prefer. In any event, he was ill and slid to the floor of the state banquet, giving rise to the Japanese slang "bushusuru" for such public mishaps. The day after the banquet there was supposed to be a meeting, but it was canceled because President Bush was still unwell. The SSC had been on the agenda, which U.S. scientists had seen as a hopeful sign. The Japanese, however, had not forgotten rejection of their earlier wish to be in on the ground floor. Later, when Llewellyn Smith himself was negotiating with the Japanese to join the LHC at CERN, he learned a lot about what had taken place with the SSC negotiations. He told me that this experience led him to believe that the Japanese were stalling, and that they were never going to come on board at the level of $2 billion. First, their high-energy physics community was relatively small, and although one hundred names appeared on a draft proposal, not all of these were real physicists. Wataru Mori, the vice minister for science and technology during the SSC era, told Llewellyn Smith during LHC negotiations, years later, that the prime minister had said, "I'm getting this request from the Americans for $2 billion for this [SSC] project in Texas. Please go around the entire scientific community—biologists, chemists, other physicists—asking them do they think that from the Japanese science budget we should put $2 billion into this project in Texas." And according to Llewellyn Smith, Mori replied, "I don't have to go 'round the community asking that question, I can tell you the answer right now—it will be no!" To which the prime minister gave a cunning answer: "In that case, do it very slowly." So the Japanese never said no to the Americans, but the outcome was clear. Llewellyn Smith, interviews by the author, March 11 and October 28, 2010.

16. Energy from the Big Bang has been locked into matter for billions of years, and the challenge for our industrial society today is to liberate that energy for use. Chemical and nuclear reactions are able to deliver about 1 percent of the total; that limitation is an unavoidable consequence of nature's laws and is not because of any inefficiency on our part. However, were we able to annihilate even a few grams of matter and antimatter, this would unshackle all of their latent energy—enough to power a city like New York for a day. That is what excites authors of science fiction, who imagine antimatter as the fuel for

Star Trek, or inspired the antimatter bomb in Dan Brown's fanciful *Angels and Demons.* In practice this is impossible, as there is no antimatter at large to light the explosive spark. However, the annihilation of beams of electrons and their antiparticle counterparts—positrons—has proved to be a marvelous tool for science, the resulting concentrated release of their energy simulating the aftermath of the Big Bang itself. For facts and fiction, see Frank Close, *Antimatter.*

17. The World Wide Web was invented by CERN scientist Tim Berners-Lee in 1989. It original purpose was to share information among scientists working in different universities and institutes all over the world.

18. After four varieties of quarks had been established—down and up, strange and charm—in 1977 a fifth variety was discovered. With electric charge of -1/3 this is a yet heavier version of down and strange. Known as bottom, it was confidently predicted to have a partner, charge +2/3, named top.

19. Quotes from Crease and Mann, *The Second Creation,* 326, 327.

20. Llewellyn Smith, interview by the author, March 11, 2010.

21. Llewellyn Smith, *Nature,* vol. 448, p. 281 (2007).

22. Llewellyn Smith, interview by the author, March 11, 2010.

23. Llewellyn Smith, *Nature,* vol. 448, p. 281 (2007).

24. Llewellyn Smith, interview by the author, March 11, 2010.

CHAPTER 17

1. Technicolor theory blossomed in the mid-1970s. For a description and list of references, see http://en.wikipedia.org/wiki/Technicolor_(physics).

2. I am using Higgs Boson as shorthand for whatever manifestation(s) of the field nature may present to us.

3. As an electron accelerator. In the 1970s SLAC provided beams for SPEAR, which broke new ground in the field of electron-positron annihilation, notably with the discovery of charm and of the tau lepton. In the 1990s its very high-energy electron-positron collisions enabled study of the Z^0.

4. For example, see Weinberg, *Dreams of a Final Theory.*

5. B. Greene, *Elegant Universe,* describes the ideas of string theory.

6. *CERN Bulletin* (September 2009).

EPILOGUE

1. In *Physics World,* August 2012, Gordon Fraser and Michael Riordan argued that, following discovery of the Higgs Boson, "higgson" (small *h*) should be used to denote the generic family of particles associated with this all-pervading field. It is possible that this family may consist of only one member—the Higgs

Boson as found in July 2012—or may contain many examples. This is one of the questions that the LHC may eventually reveal. In honor of the discovery, in this final chapter I shall endeavor to refer to Higgs Boson or Higgson (with a capital *H*) when referring to the particle at 125 GeV, retaining "higgson," with small *h*, for the generic concept, following Fraser and Riordan.

2. Peter Higgs remark to F. Close, June 26, 2012.

3. The psychological impact of "5-sigma" was illustrated when the discovery was eventually announced at CERN on July 4, 2012. The audience burst into applause when the speaker said "5 sigma," not when the actual data with signals were displayed.

4. Reported in *The Guardian*, August 22, 2011.

5. The source is utterly reliable, but anonymous. It is not Peter Higgs, nor John Ellis.

6. Peter Higgs remark to F. Close at 21.55 CET on June 30, 2012.

7. Fermilab reported evidence for this decay at a special meeting on July 2. The evidence was marginal and not enough in isolation to claim discovery of a new particle at 125 GeV, let alone the Higgs Boson. However, when combined with the data from the LHC, announced two days later, this becomes very significant in establishing that the 125 GeV particle has affinity for massive fundamental particles, in line with expectations for the Higgs Boson. Thus, although July 4 will generally be accepted as the date when the Higgs Boson was first definitively presented to the world, with hindsight, July 2 at Fermilab was its first public appearance—or with even more hindsight, December 13, 2011, may also lay some claim.

8. The observation of the boson in two photons almost certainly implies that it has spin 0, in accord with the theory. It is not possible for a boson with spin 1 to decay into two real photons, whereas spin 0 or 2 can do so. Eventually angular distributions of the decay products will show which of these is correct, but as there is no other example in particle physics where a spin 2 arises lighter than a spin 0 sibling (apart perhaps from the massless graviton), I expect it to be spin 0. Data in 2013 suggest that this is indeed the case.

9. The possibility that this decay may be a path to further discovery is because of what we noted above: the Higgson decay to two photons is due to the intermediate role of virtual particles, courtesy of quantum field theory. The production of the Higgson is predicted to be dominantly due to the intermediate role of top quarks—see F. Wilczek, *Physical Review Letters*, vol. 39, p. 1304 (1977). Its decay rate is a delicate balance between the two contributions: virtual top quarks and *W* bosons. It is possible that the relative role of the *W* and the quark channels is slightly different than thought, and that their combination in quantum mechanics modifies the production rate radically. An exciting possibility is that there may be a third significant contribution,

additional to the *W* and top quarks, due to as-yet-unknown massive particles. Any such contributions would enter the mathematical accounting as an equal partner to the top and *W* contributions, and need not be mere "perturbative corrections" to what is already included.

There remains a tantalizing possibility that all is far from over with the Higgs Boson. The accumulation of data over the coming months and years will provide precision measurements of branching ratios, and may reveal the presence of novel virtual particles.

10. See www.economist.com/node/21548911.

11. See http://www.theatlanticwire.com/global/2012/07/now-higgs-has-been-found-who-will-win-nobel-prize/54392/#.

12. See note 1.

13. See, for example, J. Gasser and H. Leutwyler, *Nuclear Physics*, vol. B250, p. 465 (1985).

14. The Higgson plays a role even in QED, albeit in circumstances that lie beyond the reach of current experiments. In the absence of a Higgson, the annihilation of electron and positron to produce a W^+ and W^- at very high energies would have a chance exceeding 100 percent, which shows that such a theory is incomplete. A spin-0, electrically neutral Higgson can also contribute to the process, where the Higgson is a virtual intermediate state: $e^+ e^- \rightarrow H^0 \rightarrow W^+ W^-$. As the peak and trough of two waves can cancel, so can the wavelike nature of particles allow contributions among different processes to cancel. When the Higgson is included, chance no longer exceeds 100 percent and the theory is sensible.

15. Or perhaps even with a novel collider of muons; see G. Fraser and M. Riordan, *Physics World*, August 2012.

BIBLIOGRAPHY

Aitchison, I. J. R. *An Informal Introduction to Gauge Field Theories.* Cambridge: Cambridge University Press, 1982.

Arnold, L. *Britain and the H-Bomb.* London: Palgrave Macmillan, 2001.

Bjorken, J. D. *In Conclusion.* River Edge, NJ: World Scientific, 2003.

Close, F. *Antimatter.* Oxford: Oxford University Press, 2009.

———. *The New Cosmic Onion.* London: Taylor and Francis, 2007.

———. *Nothing: A Very Short Introduction.* Oxford: Oxford University Press, 2009.

———. *The Void.* Oxford: Oxford University Press, 2008.

Crease, R. P., and C. C. Mann. *The Second Creation.* New Brunswick: Rutgers University Press, 1996.

Crowley-Milling, M. C. *John Bertram Adams, Engineer Extraordinary.* Yverdon, Switzerland: Gordon and Breach, 1993.

Dyson, F. *From Eros to Gaia.* New York: Penguin, 1992.

Farmelo, G. *The Strangest Man.* London: Faber and Faber, 2009.

Feldman, B. *The Nobel Prize.* New York: Arcade, 2001.

Feynman, R. P. *Lectures on Physics.* New York: Addison-Wesley, 1964.

———. *QED: The Strange Theory of Light and Matter.* Princeton: Princeton University Press, 1985.

Follett, K. *The Pillars of the Earth.* London: Pan Books, 2007.

Fraser, G. *Cosmic Anger.* Oxford: Oxford University Press, 2008.

Fundamental Problems in Elementary Particle Physics: Proceedings of the Fourteenth Solvay Conference on Physics, October 1967. London: Interscience Publishers, 1968.

Gell-Mann, M. *The Quark and the Jaguar.* Boston: Little, Brown, 1994.

Glashow, S., and B. Bova. *Interactions.* New York: Warner Books, 1988.

Gleick, J. *Genius.* Boston: Little, Brown, 1992.

Greene, B. *Elegant Universe.* New York: W. W. Norton, 1999.

Kokkedee, J. J. J. *The Quark Model.* New York: Benjamin, 1969.

Langenberg, D., R. J. Soulen Jr., and M. Osofsky. *McGraw-Hill Encyclopaedia of Science.* 10th ed. London: McGraw-Hill, 2007.

Laughlin, R. *A Different Universe*. New York: Basic Books, 2005.
Lee, B. *Proceedings of XIII International Conference on High Energy Physics*. Berkeley and Los Angeles: University of California Press, 1967.
Lichtenberg, D. B., and S. P. Rosen, eds. *Developments in the Quark Theory of Hadrons: A Reprint Collection*. Vol. 1, *1964–1978*. Nonamtum, MA: Hadronic Press, 1980.
Mehra, J., and H. Rechenberg. *Historical Development of Quantum Theory*. New York: Springer, 2001.
Mulvey, J., ed. *The Nature of Matter*. Oxford: Oxford University Press, 1981.
Nambu, Y. *Quarks*. Philadelphia: World Scientific, 1985.
Pais, A. *Inward Bound*. Oxford: Oxford University Press, 1986.
———. *J. Robert Oppenheimer: A Life*. Oxford: Oxford University Press, 2006.
———. *Subtle Is the Lord*. Oxford: Oxford University Press, 1982.
Pickering, A. *Constructing Quarks*. Chicago: University of Chicago Press, 1984.
PSA: Proceedings of the Biennial Meeting of the Philosophy of Science Association. Vol. 2, *Symposia and Invited Papers*. Chicago: University of Chicago Press on behalf of the Philosophy of Science Association, 1990.
Pullman, P. *Northern Lights*. London: Scholastic, 1995.
Riordan, M. *The Hunting of the Quark*. New York: Simon and Schuster, 1987.
Salam, A. *Proceedings of the 8th Nobel Symposium*. Edited by N. Svartholm. Stockholm: Almqvist and Wiksell, 1968.
Sample, I. *Massive*. London: Virgin, 2010.
Serber, R., with R. P. Crease. *Peace and War*. New York: Columbia University Press, 1998.
Singh, S. *Fermat's Last Theorem*. London: Fourth Estate, 1997.
Symanzik, K. *On Theories with Massless Particles*. DESY Report 72/73. Hamburg: Deutsches Elektronen-Synchrotron, December 1972.
Taubes, G. *Nobel Dreams*. New York: Random House, 1986.
't Hooft, G. *50 Years of Yang-Mills Theory*. Hackensack, NJ: World Scientific, 2005.
———. *In Search of the Basic Building Blocks*. Cambridge: Cambridge University Press, 1997.
Veltman, M. *Facts and Mysteries in Elementary Particle Physics*. River Edge, NJ: World Scientific, 2003.
Ward, J. C. *Memoirs of a Theoretical Physicist*. Rochester, NY: Optics Journal, 2004. Available at http://www.opticsjournal.com/ward.html.
———. "Some Properties of the Elementary Particles." Ph.D. diss., Oxford University, February 1, 1949, revised March 30, 1949.
Weinberg, S. *Dreams of a Final Theory*. New York: Vintage, 1992.
———. *Facing Up*. Cambridge: Harvard University Press, 2001.
———. *The Quantum Theory of Fields*. Vol. 1. Cambridge: Cambridge University Press, 2010.
Wilczek, F. *The Lightness of Being*. New York: Basic Books, 2008.
Wilczek, F., and B. Devine. *Longing for the Harmonies*. New York: W. W. Norton, 1989.

INDEX

Adams, John
background, 316, 319
CERN accelerators, 316, 317, 319, 327
wife (Renie), 319
Adler, Stephen, 241
Alpha
1920s QED/phenomena explained, 27
equations predicting infinity, 27–28
Feynman diagrams, 61
gauge invariance, 84
Schwinger's work, 49, 109
significance of, 26
size mystery, 27
value of, 26
Anderson, Philip
gauge invariance, 142, 146–147
Goldstone Bosons, 148, 149–150, 157, 171
Higgs Mechanism/Boson credit and, 155, 364
"missing" longitudinal wave, 146, 149–150
plasma/photon gaining mass, 142, 145, 146–147, 149, 158
Anomalies in quantum theory, 211
Antimatter
about, 5, 29, 34, 73–74, 140, 276, 316–317
Big Bang and, 325–326, 338
Anti-Semitism, 36, 37
Appelquist, Tom, 259, 264–265
Atiyah, Michael, 78
Atlantic Wire, 363
Atomic particles interactions
mathematics of waves, 82–84, 84 (fig.)
See also specific interactions
Atomic particles knowledge
100 years ago, 338–339
over time, 339
Atoms
solar system analogy and, 22
See also specific components

Bailin, David, 307
Balmer, Johann
background, 23, 24
hydrogen's spectrum/formula, 24, 25, 28, 30
Bardeen, John/Nobel Prize, 137–138
See also BCS Theory
BCS Theory
hidden symmetry and, 138–139
paper (1957) and response, 138
superconductivity theory and, 137, 149
See also Nambu, Yoichiro/BCS Theory
Bell, John, 207

Beta decay
description/properties, 107, 108, 109–110, 116, 117
Fermi and, 101–102, 103 (fig.), 318–319
particles emerging from, 103, 103 (fig.)
SPTVA classification and, 103
Beta function
resolution scale and, 44, 74, 259, 262
sign in QED, 259
strong interaction sign and, 259–261, 262–263, 264–267, 268, 269–273
Bethe, Hans, 36, 37, 100
Big Bang
about/description, 326, 336, 338, 340
experiments and, 342–343
symmetry/unity and, 336
Bjorken, James "B.J."
book of, 206
Close and, 234
description/personality, 233–234, 235, 247
electron-positron annihilation, 276–278
electrons use, 234–235, 236
nickname, 233–234
Nobel Prize omission, 252–253
presentations of work/reception, 244–245, 246–247
quarks, 233, 234–235, 236, 244–245, 257, 260, 261, 316, 317, 339, 352
quarks and SLAC, 206, 233, 235, 241, 244–245, 246, 260
Standard Model and, 253
unified theory and, 225
work summary/significance, 252–253, 276–277, 279, 363, 364
"Bjorken-scaling," 244, 248–249, 248 (fig.), 258–261, 265, 268, 269, 276–278, 279
See also QCD (quantum chromodynamics)
Bloom, Elliott, 247
Bohr, Niels
Balmer's formula and, 25, 30
electrons movement/energy, 23, 24–25
Pocono Mountains meetings (1948), 50, 52
Bose condensate, 136, 213–214
Bose-Einstein condensation, 136
Bosons
known bosons, 255 (fig.)
penguin analogy, 136, 213
See also specific bosons
Boulware, D.G., 174
Brodsky, Stan, 250
Brookhaven
Alternating Gradient Synchrotron (AGS), 315
Isabelle machine, 314–315, 321
Brout, Robert
background, 162–163, 364
broken symmetry work/significance, 163, 165–166, 364, 368
Englert relationship/work collaboration, 162–163, 165–166, 175, 176, 179, 193–195, 197
Higgs Mechanism work/credit and, 152–154, 155, 161–162, 175, 176, 179, 337
Weinberg's model/renormalization, 193–195, 197
Buridan, 132
Bush, George H.W., 323

Cabibbo, Nicola, 122
Cargese summer school, 213–215, 219, 224, 305
Carrazone, Jim, 264–265
Cartan, Elie, 118
Case, Kenneth/Case's theorem, 59
CERN
ISR (Intersecting Storage Rings), 314
World Wide Web and, 326, 350

See also LEP (Large Electron Positron collider); LHC (Large Hadron Collider); SPS (Super Proton Synchrotron)
CERN Council, 320, 328
CERN Courier, 152, 308
Charap, John, 160–161
Cherwell, Lord (Frederick Lindemann), 99
Children's comics (1950s) formats/analogy, 40–41
Chiral symmetry, 139–140, 188, 212, 366
Churchill, Winston, 99
Classical mechanics, 22–23, 37–38
Clinton, Bill, 323
Close, Frank
  1984 Nobel Prize ceremony/dance, 319–320
  before CERN announcement, 357–359
  CERN house journal article, 351
  Fermat's Last Theorem and, 204–205
  Feynman/Stanford seminar (1972), 62–63
  Feynman's diagrams and, 46
  Kendrew's committee and, 322
  magnitude of electron's magnetism, 4–5
  meeting 't Hooft, 320
  Nature articles, 286, 287, 289, 291
  quarks model, 235, 237, 240–241, 242, 251–252, 286, 287
  renormalization of weak interactions, 224
  SLAC, 252, 268–269
  Tokyo conference, 290–291
  University of Adelaide, 56
  Waldegrave's competition, 330 (fig.)
  weak force, 289, 290
  Z boson, 284
Coleman, Sidney
  Amsterdam conference (1971), 223
  description/personality, 259
  Politzer/strong interaction work, 268, 269–270, 271, 272, 273, 274–276
  Weinberg and, 297
Collected Papers with Commentary (Yang), 88
Computing in early 1970s, 220
Conjecture defined, 70
Cooper, Leon/Nobel Prize, 136, 137–138
  See also BCS Theory
Cooper Pairs
  energy/"energy gap" and, 139
  fermions and, 136
  spontaneous symmetry breaking and, 137
  superconductivity, 136, 137, 139, 148
Cosmic Onion, The (Close), 322
Crick, Francis, 167, 172
Current algebra, 115–116, 236, 241, 244, 245

Dalitz, Dick
  Close and, 204, 240, 242
  description/personality, 235
  parity violation, 296
  quarks, 204, 235–236, 239–240
Dannie Heineman Prize, 94
Dark Materials trilogy (Pullman), 155
Davies, Paul, 56
De Wit, Bernard, 195
Delbourgo, Bob, 299, 301, 302, 303–304
"Delta Resonance," 72
Dirac, Paul
  background, 20, 21, 95, 166
  Cambridge, 20
  electron's self-energy, 29
  equation/QED, 25, 28, 29, 30, 34, 41–42
  influence on others, 20, 36, 38
  Pocono Mountains meetings (1948), 50, 52
  QED/renormalization, 72–73, 215
  stature, 99
Dombey, Norman, 306–307

Drell, Sid, 206
Dürr, Hans-Peter, 192–193, 195
Dyson, Freeman
  Close and, 56
  Feynman and, 56, 57
  Feynman's, Schwinger's, Tomonaga's work comparison, 55, 57, 58, 63
  Lamb shift and, 58
  Nobel Prize omission, 55, 63, 258
  overlapping infinities, 69–70, 97
  QED renormalization, 68, 73
  Salam and, 68, 70, 71, 97
  Schwinger and, 56, 57, 58–59
  stature, 99
  strong interaction, 258
  Trinity College and, 78
  University of Adelaide presentation, 56–57

Economist, The, 362
Eddington, Sir Arthur, 26, 110
Edinburgh Scottish Universities Summer Schools, 168–170
Einstein, Albert
  gravity, 85, 86, 347, 348
  photons, 23, 145
  special relativity, 3, 29
Electromagnetic field
  description, 176
  electrons and, 29, 34, 41–42, 46–47, 47 (fig.), 49, 61–62, 62 (fig.), 79
  Feynman's diagrams and, 46–47, 47 (fig.), 51, 61–62, 62 (fig.)
  gauge invariance and, 85
  infinity and, 29
  light and, 23, 145
  renormalization and, 41–42
  superconductors and, 149
  as vector field, 102, 107
Electromagnetic force
  description, 7 (fig.), 288 (fig.)
  Fermi's weak force model and, 102
  parity conservation and, 116, 287
  vector fields and, 102, 107
Electromagnetic waves
  description, 143, 144
  longitudinal waves and, 146, 146 (fig.)
  material types and, 148–149
  Maxwell's equations and, 145–146, 146 (fig.)
  in plasma, 143–145, 146–147
  transverse waves, 146, 146 (fig.)
Electrons
  energy changes/ladder rungs analogy, 23–24, 25, 30, 34
  magnetism and spinning electrons, 163, 164 (fig.), 165
  resolution and, 42–45, 48
  self-energy, 29, 47 (fig.), 48, 50
  See also Leptons; QED
Electroweak theory/work
  charm and, 287
  gauge invariance and, 108
  left-/right-handed electrons and, 290
  Nobel Prize maneuvering and, 286
  Schwinger's ideas/work, 108–111
  steps in uniting, 308–309
  SU2 x U1 model, 121–122, 179, 194, 224–225, 226, 286, 290, 291, 294, 298, 300, 307, 308–310
  vector bosons and, 107–108
  W boson/mass and, 108, 109 (fig.), 112–114
  See also QFD (quantum flavordynamics); *specific physicists*; Spontaneous symmetry breaking
Englert, François
  broken symmetry work/significance, 163, 165–166, 364, 368
  Brout relationship/work collaboration, 162–163, 165–166, 175, 176, 179, 193–195, 197
  Higgs Boson discovery and, 360
  Higgs Mechanism work/credit and, 152–154, 155, 161–162, 175, 179, 337

Weinberg's model/renormalization, 193–195, 197, 217
Evans, Lyn, 328
Everest analogy, 209, 210, 226

Fadeev, Ludwig, 218
Fermat, Pierre de/Last Theorem, 40, 204–205, 223, 226
Fermi, Enrico
beta decay theory/weak force model, 101–102, 103 (fig.), 318–319
Feynman and, 53
Schwinger and, 36, 52–53
thermonuclear programs (U.S.), 100
Fermi Lab Sixteenth International Conference on Particle Physics, 260
Fermions
description/properties, 136, 179, 189, 240, 255 (fig.), 338, 348
Weinberg and, 189, 345
See also Leptons; Quarks
Ferromagnetism
Brout/Englert and, 163
Heisenberg's model, 135
Feynman, Richard
approach to science, 37, 38–41, 46
background, 33, 35, 36
Bjorken and, 250, 251, 257
description/personality, 35, 37, 46, 239, 250, 251
finite quantum theory of gravity, 209
Los Alamos work, 37
notebooks/diary, 247, 248 (fig.), 251
"parton model," 233, 247, 249–250, 251, 258, 316, 317, 352
path integrals, 215–216
SLAC visits/notes (1968), 247–250, 248 (fig.)
stature, 36, 37, 40–41, 46, 94, 99, 249
weak force SPTVA classification, 104–105
Wheeler and, 95, 215, 216
Yang-Mills theories, 218
Feynman, Richard QED/Infinity Puzzle
about, 37, 40–41, 45
American Physical Society meeting (1949), 59
competition with Schwinger, 33, 40, 46, 48, 49, 51–52, 53, 58–59
Lamb shift, 45, 51, 55
Nobel Prize, 63, 73
number 137 and, 26
Pocono Mountains meeting presentation (1948), 50–52, 53
reconstruction of quantum mechanics, 38–40
Schwinger's presentation of theory (1948), 49
Shelter Island conference (1947), 37, 41
Slotnick and, 59
Stanford seminar (1972), 62–63
understanding by others (1940s), 41, 48, 51–52
work significance, 59
See also Feynman diagrams
Feynman, Schwinger, Tomonaga QED work comparison
Dirac's basic theory and, 58
Dyson and, 55, 57, 58
results and, 57, 58
Feynman diagrams
about, 40–41
alpha and, 61
electrons/photons, 46–48, 47 (fig.)
need for Higgs Boson and, 345–346, 347 (fig.)
"perturbation" theory, 61, 63
rules for calculating, 60–62, 60 (fig.), 62 (fig.)
significance/use, 40–41, 46, 97, 220, 221, 268, 270
single vertex, 60 (fig.), 61
successive emissions/absorptions and overlapping infinities, 68–69, 69 (fig.)

Feynman diagrams *(continued)*
three vertices, 61, 62, 62 (fig.)
two vertices, 60 (fig.), 61
Franklin, Rosalind, 167, 172
Fraser, Gordon, 308
Friedman, Jerome
Feynman and, 247, 249
MIT/Stanford and, 245–246
quarks/Nobel Prize, 251, 252
Fritzsch, Harald, 260–261

Galileo, 158
Gauge invariance
alpha and, 84
conservation and, 86
Einstein's theory of gravity, 85
electromagnetic force existence and, 81, 84, 85
electromagnetic gauge invariance, 83
"four vector" potentials/quantum phases, 82–83
overview, 81–85, 84 (fig.)
photon mass and, 81, 85
scalar type, 82
Schwinger and, 80–81
strong force and, 85
use as checking calculations, 82, 83–84
vectors and, 82–83
Ward's Identities and, 113
weak force and, 85
Gauge invariance examples
air pressure/wind speed, 82
electric current flow, 82
GMT/EST and journey time, 81
gravitational potential and, 82
Gell-Mann, Murray
current algebra method, 115–116, 241
electroweak force, 106
Glashow and, 115, 116, 117–118, 306
group theory mathematics, 118, 187, 209
hadrons/particles classification, 115, 118, 119, 237–238
pi sigma model, 213
QCD and, 260, 261
quarks and hadrons, 260
quarks/changing opinion, 115, 233, 234, 235, 237, 238, 239, 241, 242, 251, 253, 258
on Salam/Ward, 303, 309–310
"strangeness" particles, 115–116, 119
weak force SPTVA classification, 104–105, 116
Gilbert, Walter
background/Nobel Prize, 171–172
Goldstone theorem, 171, 172, 173–174
Guralnik and, 159–160, 173–174
Kibble's work, 174
GIM mechanism/paper, 254, 256, 277–278, 286, 305
Ginzburg, 100
Glashow, Sheldon
1960 Scottish Universities Summer School/Higgs and, 169–170
background, 111–112
Goldstone and, 142
quarks, 254, 277
Glashow, Sheldon/electroweak theory
credit for, 191, 291
criticism of Salam/Ward's work, 121
errors, 114–115
failure to unify forces, 119–120, 226
gauge invariance, 114
Gell-Mann and, 115, 116, 117–118, 306
group theory classification, 118, 209
Nobel Prize, 12, 106, 111, 112, 116, 121, 225, 227, 283, 286, 294, 309, 313, 319, 333
"Nobelitis," 291, 292–293, 295
parity problem, 116–117, 119–120
publishing papers on, 114, 119
renormalization, 113, 114–115
Schwinger's suggestion, 111, 112, 116

status of (1961), 119–120
W bosons/mass, 112–113, 114
Yang-Mills theories, 112–113, 117
Z bosons/implications, 117–118, 119–120
Gluons, 257, 258, 342, 349
Goldman, Terry, 259–260
Goldstone, Jeffrey
description/personality, 140
Gang of Six and, 364
Glashow and, 142
Higgs Mechanism, 152, 153, 337
rotary/radial oscillations, 154 (fig.)
spontaneous symmetry breaking/Nambu's work, 140–142
See also Goldstone Boson
Goldstone Bosons
about, 142, 176
Anderson and, 148, 149–150, 157, 171
as Nambu-Goldstone boson, 180
napkin pickups analogy, 132–133
as pion, 189, 212
vanishing, 142, 150, 156–157, 159, 161, 165, 171, 173, 175, 179
wine bottle/Mexican hat example, 133–135, 134 (fig.), 368
Goldstone theorem, 140–142, 149, 150, 156, 171, 172, 173–174, 183
Gravitational force
description, 7 (fig.)
Einstein and, 85, 86, 347, 348
of massive bodies, 28
Gribov, Vladimir, 158
Gross, David
beta function sign, 258–259
Nobel Prize, 258, 265, 267, 273, 333
QCD/strong interaction, 258–259, 261, 265, 267, 269, 270–272, 273, 274–276
Ground state of energy, 134–135
Group theory, 118
Guardian crossword puzzle and physics, 21, 151
Guralnik, Gerry
hidden symmetry, 122–123
Higgs Boson discovery and, 359, 360, 364
Higgs Mechanism work/credit, 152, 153, 155, 162, 180, 298, 337
Marshak and, 178
missed opportunities on Higgs Mechanism, 159–162, 174, 177, 178, 364, 368

Hadrons
classification, 115, 118, 119, 237–238
electroweak unification, 277–278
Hagen, Dick
hidden symmetry, 122–123
Higgs Boson discovery and, 359, 360, 364
Higgs Mechanism work/credit, 152, 153, 155, 162, 179, 180, 298, 337
missed opportunities on Higgs Mechanism, 161–162, 174, 364, 368
Haggar, Reginald, 77
Heisenberg, Werner, 135, 177, 178
Heuer, Rolf, 357–358, 362
Hey, Tony, 242
Hidden symmetry
about, 127
crystals on window, 131 (fig.)
response to concept, 158, 177
significance of concept, 123
snowflake, 130, 131 (fig.), 132
spiral galaxies, 128 (fig.), 129
spontaneous symmetry breaking, 130, 132, 133–135, 134 (fig.), 158–159
universe and, 336, 340, 341
water/phases, 129–130, 131 (fig.), 132
wine bottle/Mexican hat example, 133–135, 134
See also Goldstone Boson
"Hidden Symmetry" (Guralnik, Hagen, Kibble), 123

Higgs, Peter
  1960 Scottish Universities Summer School, 168–170
  background, 166–168, 329
  Close and, 168
  hidden symmetry and, 123, 337
  Higgs Boson discovery and, 355, 358, 359–360, 361, 364, 365
  Higgs Mechanism/Boson credit and, 152–155
  publishing paper on Higgs Mechanism, 160
  rotary/radial oscillations, 154 (fig.)
  stature, 351
  strong interaction theory, 189–190
  Weinberg's work and, 190
  work known for, 170–171, 172–173, 174–175, 176–177
Higgs Boson
  credit for, 151–155
  LHC and, 151, 152, 153, 155, 179, 329, 333, 345, 346, 352
  naming, 180, 183, 283–284
  significance of, 340
  theory on, 179
  wine bottle/Mexican hat example, 133–135, 134 (fig.), 176
Higgs Boson discovery/LHC
  ATLAS team, 354, 355–356, 357, 358–359, 361
  CMS team, 354, 356, 357, 358, 359, 360, 361
  credit/prizes (Nobel) and, 362–365
  days before July 4, 2012 and, 357–360
  December 2011 and, 356–357
  future research and, 365–367, 368–369
  Grenoble conference (July 2011), 355–356
  "higgsons" family and, 354, 362, 365, 366
  July 4, 2012 announcement/findings and, 360–362
  Mumbai conference (August 2011), 356
  rolling dice analogy, 353–354
  sigma and, 355, 359, 361
  signal vs. noise, 354, 355, 356, 357
Higgs Field, 153, 176, 329, 366–368, 369
"Higgs-Kibble" mechanism, 152, 153, 168, 185, 302
Higgs Mechanism
  attitude towards (late 1960s), 211–212
  British postal strike and, 161–162, 177
  chronology of published papers on, 170–176
  credit for, 151–155, 212
  Gang of Six, 153, 175, 178, 179, 212, 337
  missed opportunities with, 157–163, 165–166, 170–171, 173–175, 176, 177
  naming, 283–284
Higgsino, 349
Hiroshima/Nagasaki bombs, 2, 33–34, 166
't Hooft, Gerard
  Amsterdam conference (1971), 1, 2, 10, 11, 12, 222–225
  anomalies work, 211
  Cargese summer school/inspiration, 213–215, 219, 224, 305
  description/personality, 2, 203–204
  Higgs Boson discovery and, 358
  learning of his Nobel Prize, 293–294
  Marseille conference (1972), 261, 262–263, 264
  Nobel Prize, 12, 225, 227, 293–294, 307, 327, 333
  path integrals, 215, 216, 219
  physics environment and, 211–212
  pi-sigma model, 213–214, 219, 221, 224
  significance of work, 5, 12, 225, 227, 259, 364, 368–369
  spontaneous symmetry breaking/introducing mass, 213–214, 215

strong interaction, 158, 225, 227, 229, 257, 265
strong interaction/beta sign, 252, 260, 261, 262, 263–264, 265–266
Veltman and, 1, 2, 204, 205, 211, 216, 218–219, 327
work of others and, 226–227
Yang-Mills theories/weak force renormalization with massive particles, 219–221, 222–225, 258, 261, 283, 286, 340
Yang-Mills theories/weak force renormalization with massless particles, 203–204, 208, 211, 215, 217–219, 261–262
Hughes Medal of the Royal Society, 94
Human Genome Project, 172
Hydrogen atom
Dirac's "self-energy," 29
electric field and, 29
spectral lines/Balmer formula, 24, 25, 28
structure/simplicity, 24, 28

Iliopoulos, John, 254, 277, 286
Infinity Puzzle
description, 3–6, 7–10
See also Feynman, Richard QED/Infinity Puzzle; Overlapping infinities; QED (quantum electrodynamics) infinity problems; Schwinger, Julian QED/Infinity Puzzle; *specific physicists*
Institute for Advanced Study (IAS), Princeton, 71, 72, 85, 98, 138
International Centre for Theoretical Physics, UNESCO, 18, 304–305, 306, 307
Invisible Man, The (Wells), 285
Isabelle particle accelerator, 314–315, 321
Isham, Chris, 299, 301

Jackiw, Roman, 273, 274
Jackson, Dave, 169
J. J. Sakurai Prize (2010), 153, 179
Jona-Lasinio, Giovanni, 150
Jung, Carl, 26

Karl, Gabriel, 236–237
Kemmer, Nick
background, 20
Cambridge and, 21
Close and, 20–21
description/personality, 20–21, 322–323
Edinburgh and, 74, 79, 172
Higgs and, 167–168
neutron/proton conservation and, 86
pions, 79–80
Shaw and, 74, 79
strong force, 67–68
working with Salam, 21, 31
Kendall, Henry, 245–246, 252
Kendrew, Sir John/committee, 322, 327
Kennedy, John F., 180, 181, 329
Khriplovich, I.B., 267
Kibble, Tom
hidden symmetry, 122–123, 298, 308
Higgs and, 167
Higgs Boson discovery and, 359, 364
Higgs Mechanism work/credit, 152, 153, 155, 162, 180, 187, 298, 337
mass-generating mechanism work, 186, 191, 196
missed opportunities on Higgs Mechanism, 159, 161–162, 174, 179, 364, 368
Salam and, 185–187, 197, 298, 299, 301, 302, 308, 309
Kinetic energy, 38, 130
Klein, Abe, 171, 172, 174
Kokkedee, J.J.J., 239
Korthals-Altes, Chris, 222–223, 225, 263–264, 265

Lagrange, Joseph-Louis, 38
Lagrangian, 38
Lamb, Willis
  measuring energy shift in electrons, 34, 97
  microwaves for radar work, 34
  QED infinity problem, 34–35
  Shelter Island conference (1947), 33, 41
Lamb shift, 34, 45, 46, 50, 51, 55, 57, 58, 97
Landau, Lev/Landau's paradox, 73–74
Language invariance, 81
Large European Project. *See* LEP (Large Electron Positron collider)
Larkin, Anatoly, 158
Lee, Benjamin (Ben)
  Amsterdam conference, 223, 224
  background, 171
  Gilbert's criticism of work, 171, 172, 174
  "Higgs Boson"/"Higgs Mechanism" names, 180
  pi sigma model, 213, 214, 215, 262, 305
  Salam promoting himself and, 305, 306, 307
  weak force theory, 283–284, 305–306
Lee, T.-D.
  Amsterdam conference, 10, 11, 222, 223
  Nobel Prize/work, 10, 91
Leibnitz, 369
LEP (Large Electron Positron collider)
  Big Bang and, 325–326
  building, 320, 321, 322
  description/goals, 320, 323–327, 343, 344, 345, 346
  development history, 315, 316, 317, 318
  Higgs Boson and, 343, 345
  LHC plans/location and, 316, 321, 327, 343, 344, 350
  overview, 324–327
  't Hooft/Veltman predictions and, 326–327, 344–345
  top quark and, 326–327, 333
  W/*Z* and, 337
Leptons
  description/properties, 90, 189, 253, 255 (fig.), 337, 339, 340, 345
  Weinberg and, 253, 256, 277–278, 345
Les Houches summer school, 212–213
Levy, Michel, 213
LHC (Large Hadron Collider)
  accident/repair, 351–352
  Big Bang and, 12, 151–152, 351
  description, 321, 347, 350–351
  future research, 365–368
  hidden symmetry, 123
  Higgs Boson, 151, 152, 153, 155, 179, 329, 333, 345, 346, 352
  Higgs Mechanism and, 339
  history of, 151–152
  LEP and, 316, 321, 327, 343, 344, 350
  planning/funding, 322, 327–331, 330 (fig.), 344
  scare stories about, 351
  supersymmetry, 346, 348, 349, 362, 365–366
  't Hooft's model, 12, 13, 227
  W/*Z* mass, 142
  Waldegrave's competition and, 329–331, 330 (fig.), 341
  See also Higgs Boson discovery/LHC
Llewellyn Smith, Chris
  CERN accelerators, 318, 322, 328, 331
  Kendrew committee, 322, 323
Lorentz invariant, 171

Magnetism and spinning electrons, 163, 164 (fig.), 165
Maiani, Luciano, 254, 277
Mandelstam, Stanley, 218
Manhattan Project, 29, 37, 99

Marshak, Robert/getting scooped, 104, 105, 178
Matthews, Paul "P.T."
background, 20
on physics textbooks, 21–22, 30
relationship with Salam, 19, 20
as Salam's mentor/promoter, 21–22, 31, 67, 68, 69–70, 71, 301, 302
Matthews, Paul "P.T."/strong force
Dyson/overlapping infinities, 69–70
renormalization, 67, 68, 71
theory, 68, 71
Maxwell's equations, 107, 145–146
Maxwell's theory of electromagnetism, 3
Meissner, Walther, 135
Meissner effect, 148–149
Migdal, Sasha, 158–159, 177, 178, 364
Miller, David, 329–330, 341
Mills, Robert
meeting Yang, 86
See also Yang-Mills theories
Momentum conservation, 86
Mott, Neville, 166
Muon, 72, 179, 255 (fig.), 340
Myers, Steve, 362

Nagasaki/Hiroshima bombs, 2, 33–34, 166
Nambu, Yoichiro
background, 138
breaking chiral symmetry, 139–140, 188, 212
Close and, 138
Goldstone theorem, 140–142, 150
Higgs and, 175
pseudoscalar boson/pion, 140
significance of work, 138–139, 140, 150
Nambu, Yoichiro/BCS Theory
cold effects, 138–139
Cooper Pairs/"energy gap," 139
hidden symmetry, 138–139, 170–171
universe as superconductor idea/research, 139–140
Nature, 102, 286, 287, 289, 291, 307
Ne'eman, Yuval, 187
Neutrinos
description/properties, 90–91, 101–102, 103, 103 (fig.), 109 (fig.), 119, 241, 252, 284–285, 287, 288 (fig.), 289, 296, 339, 340, 345, 347 (fig.), 349
particle accelerators, 253, 285, 318
prediction of, 90, 101
Salam and, 296, 297
Weinberg and, 188, 189
See also Leptons
Neutrons
discovery, 101
See also Nucleons; Quarks
New York Times, 49, 321
Newton, Isaac, 22–23, 37–38, 78, 166, 346–347
Nobel Prize
Bjorken's omission, 252–253
Dyson's omission, 55, 63, 258
Higgs Boson discovery and, 362–363, 365
maneuvering following 't Hooft's presentation, 227, 304, 307
"Nobelitis"/anxiety over, 291–293, 294
Ward's omission, 12, 94, 105–106, 160, 197, 227, 294, 308, 309–311
See also Salam, Abdus/Nobel Prize; *specific Nobel Laureates*
Nobel Symposium/*Proceedings of the Nobel Symposium*, 224, 297–298, 299, 306
Nucleons
chiral symmetry/spontaneously breaking, 139–140, 212
description/properties, 86–87, 213, 241, 251, 339, 341
strong force and, 31, 68, 72, 254
Weinberg and, 188–189
See also Neutrons; Protons
Number 137 significance, 26–27
Nuovo Cimento, Il, 120, 173

Okun, Lev, 158
Oldstone on the Hudson QED meeting (1949), 59
Onnes, Heike Kamerlingh, 135
Oppenheimer, J. Robert
  American Physical Society meeting/Slotnick and, 59
  description/personality, 29
  Dyson and, 71
  Pocono Mountains meeting (1948), 49
  QED Infinity Puzzle, 29–30, 35, 56
  QED meetings purpose, 63
  Salam/Matthews strong force work and, 71
  Schwinger and, 110
  Shelter Island meeting (1947), 35
  Tomonaga and, 56
  Yang/Pauli's interruption and, 88
Organization
  painting masterpiece example, 129, 135
  water/phases examples, 129–130, 131 (fig.), 132
Orwell, George, 362
Overlapping infinities
  Dyson and, 69–70, 97
  Feynman diagrams, 68–69, 69 (fig.)
  Ward's Identities and, 97–98

Pais, Abraham, 41
Panofsky, Wolfgang "Pief," 242, 246, 251–252
Parisi, George, 265, 266, 267
Parity defined, 116
Particle accelerators
  "atom smashers," 313–314
  Big Bang and, 12, 151–152, 325–326, 343, 351
  Bjorken's predictions and, 318
  cosmic rays and, 72
  finding *W*/*Z* boson, 314, 315, 317–318, 319, 320, 321, 322, 324, 326, 333, 337, 345
  history, 313–331
  International Committee for Future Accelerators (ICFA), 321–322
  method description, 314
  proton-antiproton collisions, 317–318, 321, 323–324, 327, 344
  Weinberg's predictions and, 318
  See also specific machines
"Particle" physics, 26
"Parton model," Feynman, 233, 247, 249–250, 251, 258, 316, 317, 352
Paschos, Manny, 247–249, 250
Pauli, Wolfgang
  description/personality, 21, 30, 36, 88
  Kemmer and, 21
  neutrino, 90, 101
  number 137 and, 26
  Schwinger and, 36
  work similar to Yang-Mills, 87–88
  Yang and, 87, 88, 112, 157, 189
Peierls, Rudolf, 21, 96, 296, 297
Perkins, Don, 242, 252
"Perturbation" theory, Feynman diagrams, 61, 63
Peterborough Cathedral and asymmetry, 127, 128 (fig.)
Phillips, Roger, 78
Photons
  electrons changing energy and, 23
  frequency/energy relationship, 23
  group theory classification, 118
  properties, 23, 103
  as "vector boson"/"gauge boson," 85, 107
Physical Review, 21, 70, 142, 238, 239, 305
Physical Review Letters, 161–162, 173, 175, 194, 272–273, 274, 299, 302, 305
Physics Letters, 170, 173, 175, 273
Physics status summaries
  1950 QED status, 65
  1960 QED status, 125
  1960s (mid-1960s), 183
  1970s (early), 229
  1975, 281

change over 100 years, 342–344
end of twentieth century (1980/1990), 333
Physics textbooks
Matthews to Salam on, 21–22, 30
usefulness, 22
Pi sigma model, 213–214
Pion particle
Nambu's work on chiral symmetry, 140
properties/description, 212, 366–367
strong force and, 31, 67–68, 72
Plasma
description/properties, 143
electromagnetic radiation transmission, 144, 145
photon gaining mass, 145
Plasma frequency, 144–145
Pocono Mountains QED meeting (1948)
about, 49–52
effects, 52–53
Politzer, David
description/personality, 268–269
Nobel Prize, 257, 265, 267, 273, 333
QCD/strong interaction, 257, 265, 267, 269–270, 271, 272–273, 274–276, 283
Tokyo conference, 290
Wilczek and, 268
Polkinghorne, John, 78, 79, 80, 303
Polyakov, Sacha, 158–159, 177, 178, 364
Popov, Victor, 218
Potato Eaters (van Gogh), 299
Potential energy defined, 38
Potteries, 77
Powell, Cecil, 166–167
Prentki, Jacques, 173
Principle of least actions, 38
Probability amplitude, 39
Protons
antiproton, 316, 323
See also Nucleons; *specific components*
Pryce, Maurice, 95, 96–97, 99
Pullman, Philip, 155

QCD (quantum chromodynamics)
Bjorken and, 269, 363, 364
controversy over discovery, 269–276
description, 240, 257, 258
physicists coming close to describing, 259–267
physicists explaining/publishing, 267–273
QED (quantum electrodynamics)
Dirac's equation electrons vs. "real" electrons, 41–42
electric charge conservation, 86
gravity/nuclear forces and, 58
group theory classification, 118
Landau's paradox, 73–74
significance of, 58, 59–60
tracking four-vector potentials/ quantum phases, 83–84
uniting nature of light and matter, 25–26, 59–60
See also Alpha
QED (quantum electrodynamics) infinity problems
as "banished" (1940s), 30, 31
beta function and, 44, 74
description, 3–6, 8
Dirac's equation/theory, 25, 28, 29, 30, 34
electron charge/resolution using and, 43–45, 48
Feynman's, Schwinger's, Tomonaga's solution, 57–58
illusionary quest of, 42
Lamb's calculations, 34–35
Oppenheimer's calculations, 29–30
Pauli's criticism, 29, 30
railroad track "convergence" analogy, 42
travel agent trips/bill analogy, 44, 45
See also Renormalization (QED); *specific components*; *specific physicists*

QFD (quantum flavordynamics)
  description, 255 (fig.), 258, 308–309, 335, 336
  LEP and, 326, 343
  particle accelerators and, 320, 326, 343
  See also Electroweak theory/work; *specific physicists*
Quantum mechanics
  about, 3
  action paths, 38–40
  See also specific physicists
Quarks
  "almost free" paradox, 256, 258–259, 261, 278
  antiquarks and, 316–317
  "bottom," 340
  CERN experiments/results, 253–254
  color property, 240, 257–258, 260, 266–267, 339
  "down," 236–237, 253, 276, 278, 339, 340
  GIM mechanism/paper, 254, 256, 277–278, 286, 305
  hadrons and, 237–238, 239–240
  idea beginnings/name, 237
  known quarks, 255 (fig.)
  model presentation (Vienna conference 1968), 234, 242, 246–247, 251, 252
  naming quark types, 236–237, 254
  research status (late 1960s), 234–237
  "strange," 236–237, 253, 276, 278, 340
  "top," 326–327, 333, 338, 340, 345
  "up," 236–237, 253, 276, 278, 339, 340
  Yang-Mills-Shaw's work and, 89
  See also QCD (quantum chromodynamics); *specific physicists*
Quarks/"charm"
  about, 236–237, 252, 254, 256, 278, 340
  discovery, 252, 257, 277, 315
  J-psi and, 277, 278, 315
  QCD and, 277, 278, 281, 287
  significance of, 278, 281
Queen Elizabeth Prize for Engineering, 363

Rabi, Isador, 72
Radio hams, 143
Radio waves, 143–144
Reagan, Ronald, 321, 323
Renoir, Pierre-Auguste, 129, 135
Renormalization (QED)
  description, 42–45, 207, 208
  dissatisfaction with, 72–73
  electron/charge and resolution, 42–45
  Feynman diagrams, 60, 63
  gauge invariance importance, 113–114
  massless photon importance, 113–114, 236
  as solved, 59
  status today, 73
  Ward's Identities, 97–98, 113
Rho-meson, 189, 221–222
Richter, Burt/Nobel Prize, 315
Rochester Conferences, 283–284, 286
Rubbia, Carlo
  description/personality, 317
  Nobel Prize, 319, 333
  particle accelerators, 316, 317–318, 320, 327–328, 363
Rutherford, Ernest, 242, 335, 338, 341–342

Sakharov, Andrei, 93, 99, 100, 286
Salam, Abdus
  Amsterdam conference, 10, 11–12, 222, 223, 224, 297
  atomic bomb knowledge, 94
  background, 18–19
  BBC broadcast, 17
  as Cambridge professor, 74
  Cambridge undergraduate studies, 18–20
  Close and, 17–18
  description/personality, 18, 302, 303

getting scooped, 90, 185
Glashow and, 114–115, 120–121
gravity work, 11–12, 122, 297, 300
group theory, 187
Guralnik and, 159
hidden symmetry/Goldstone theorem, 140–142
Higgs and, 167
Kibble and, 185–187, 197, 298, 299, 301, 302, 308, 309
missing significance of Shaw's work, 74–75, 80, 89–90, 106, 187
parity violation, 100–101
Ph.D. examination/Peierls question, 296
Shaw and, 74–75, 79, 80, 89–90, 106
't Hooft's work and, 11, 12, 224, 227, 297, 300, 304
UNESCO's International Centre for Theoretical Physics, 18, 304–305, 306, 307
Ward relationship, 94, 98, 100–101
weak force/hidden symmetry, 142
writing skills, 71, 272
on Yang/Lee, 296
Salam, Abdus/Nobel Prize
Göteborg Nobel Symposium and, 299, 300, 304, 306, 309
Kibble's nomination, 308
making Weinberg link, 298–300, 304–305, 307
reviews on importance of his work, 308, 309–310
self-promotion, 10–11, 227, 295, 296–297, 298–307
separation from Ward, 200, 298, 307
"Weinberg-Salam" model, 197, 227, 283, 286, 287, 289, 290, 291, 297, 298, 304–305, 307, 308
winning, 12, 105–106, 160, 187–188, 197, 225, 227, 283, 293–294, 313, 319, 333, 364
Salam, Abdus/strong force
criticism of writing/effects, 71
Dyson and, 68, 70, 71, 97
overlapping infinities and, 68–69, 69 (fig.), 70
Physical Review paper, 70
Princeton and, 71
renormalization, 68, 215
unitary property and, 217
Salam and Ward/electroweak theory
1964 paper/results, 11, 121, 122, 123, 197, 200–201 (fig.), 224, 307, 309, 310
Cabibbo's model and, 122
errors, 121
final model, 121–122, 160, 226, 286
gauge invariance, 105–106
Glashow's work and, 114–115, 120–122
Kibble's work and, 185–187, 229
parity violation, 100–101, 104, 121, 122
published paper (1959), 120
renormalization, 114–115, 120–121
reviewing work, 308, 309–311
secrecy on, 160–161, 184, 186
SPTVA classification, 104, 105
symmetry breaking, 303
Saxon, David, 322–323
Schoonschip computer program, Veltman, 206–207, 210
Schreiffer, Robert, 137–138
See also BCS Theory
Schwinger, Julian
approach to science, 37, 40, 41, 46
background, 33, 35–36
competition with Feynman, 33, 40, 46, 48, 49, 51–52, 53, 58–59
description/personality, 35, 37
electroweak force, 108–111, 114
Glashow and, 106, 112
Guralnik and, 177–178
microwave radar work, 37
personal life, 45, 48
stature, 36, 37, 94
W boson/beta decay and, 110–111, 112, 157–158, 313

Schwinger, Julian QED/Infinity Puzzle
about, 40, 41, 45, 48, 58–59
American Physical Society meeting (1948) presentation, 48–49
equations/symbols, 49
gauge invariance, 80–81, 84, 147
Lamb shift, 46, 50, 57
Nobel Prize, 63
number 137 and, 26, 109
plasma/photon mass, 147
Pocono Mountains meeting presentation (1948), 49–50, 51, 52–53
Shelter Island conference (1947), 37, 41
use of relativity, 50
See also Feynman, Schwinger, Tomonaga QED work comparison
Science Citation Index, 297
Scientists Against Nuclear Weapons, 246
Serber, Robert, 237, 238
Shaw, Robert
background, 77–79
bridge/"fairy chess" and, 78–79
as Cambridge research student, 74–75, 79–80
description/personality, 89, 90
discovery/getting scooped, 79–80, 87, 89, 90, 106, 187, 229
Landau's paradox, 74–75
parents, 77–78
the Potteries and, 77
quarks, 89
Salam and, 74–75, 79, 80, 89–90, 106, 187
Schwinger paper and, 79, 80
Trinity College, 78–79
Shelter Island conference (1947), 33–35, 37, 41, 56, 80, 81, 97, 178
Sigma described, 213
SLAC (Stanford Linear Accelerator Center)
about/building, 206, 243–244, 253–354
Bjorken/quarks, 206, 233, 235, 241, 244–245, 246, 260
electroweak theory, 290
Feynman's visits/notes (1968), 247–250, 248 (fig.)
SPEAR, 277, 315
strong force and, 225, 258–259
Slotnick, Murray
American Physical Society meeting (1949), 59
Feynman's techniques and, 59
Smith, Chris Llewellyn
CERN accelerators, 318, 322, 328, 331
Kendrew committee, 322, 323
Sorensen, Ted, 180
Spontaneous symmetry breaking
about, 180–181
Cooper Pairs, 137
discovery, 130, 132, 133–135, 134 (fig.), 158–159
Nambu/Goldstone and, 140–142
research review/question unanswered, 337
Soviet Union physicists/politics and, 158–159
supersymmetry ("SUSY"), 349
't Hooft's work, 213–214, 215
wine bottle/Mexican hat example, 133–135, 134 (fig.)
SPS (Super Proton Synchrotron)
converting, 317–318, 320
description, 317, 318, 321
finding *W/Z*, 324, 344
SPTVA classification, 103, 104–105
SSC (Superconducting Supercollider)
demise, 323, 328, 344
plans for, 321–322, 323, 327
Standard Model/change, 253, 293, 307, 346–348, 347 (fig.), 349
"Strange" particles discovery, 72
Strong force
conservation of nucleons, 86
description, 6, 7 (fig.)

Matthews-Salam (temporary) solution, 71, 72
pions and, 31, 67–68
previously unknown particles and, 72
See also Matthews, Paul "P.T."/strong force; QCD (quantum chromodynamics); Salam, Abdus/strong force
Stuller, Larry, 217
SU2 x U1 model, 121–122, 179, 194, 224–225, 226, 286, 290, 291, 294, 298, 300, 307, 308–310
Sunday Times, 303
Superconductivity
Bose condensate and, 136
Cooper Pairs and, 136, 137, 139, 148
dancing analogy, 136–137
description, 135
discovery, 135
electric/magnetic fields and, 135, 149
"massive" photons and, 148
regular conductivity vs., 136–137
Supersymmetry ("SUSY")
description/properties, 348–350
LHC and, 346, 348, 349, 362, 365–366
spontaneous symmetry breaking and, 349
Symanzik, Kurt
beta function sign, 262–263, 264, 265–266
pi sigma model/'t Hooft and, 213, 214, 215
Symmetry
Buridan's donkey's dilemma, 132
chiral symmetry, 139–140, 188, 212
global vs. local symmetry, 155–157
tension with asymmetry, 132
ways of breaking/examples, 127, 128 (fig.), 129
See also Hidden symmetry

Tau, 255 (fig.), 340
Taylor, Dick, 245–246, 247, 251–252
Taylor, J. C., 225
"Technicolor" theory, 338
Teller, Edward/wife, 100, 177–178
Test-ban treaty talks, 100
Thatcher, Margaret, 94, 323, 329
Thermonuclear programs (U.S.), 100
Ting, Sam/Nobel Prize, 315
Tokyo conference (1978), 290–291
Tomonaga, Sin-Itiro
background, 55
description/personality, 55
Nobel Prize, 63
Oppenheimer and, 56
QED infinity problem/solution, 35, 55, 56
stature, 94
See also Feynman, Schwinger, Tomonaga QED work comparison
Tsai, Paul, 247, 248–249

Ubbink, Jon, 211, 212
UK Atomic Weapons Research Establishment (AWRE), 99–100
Ulam-Teller invention, 100
Unified theories of forces, 336–337
See also specific forces
Unitary gauge, 217
Universe
hidden symmetry and, 336, 340, 341
unity of, 335–336
without asymmetry, 341
U.S. Atomic Energy (McMahon) Act (1946), 99–100

Van der Meer, Simon/Nobel Prize, 318, 319, 333
Van Gogh, 299
Van Hove, Leon, 238–239, 318
Vectors defined, 82
Veltman, Tini (Martinus)
1960 Scottish Universities Summer School, 169
Amsterdam conference, 1, 2, 10, 11, 12, 222–224

Veltman, Tini *(continued)*
CERN policy committee, 318
description/personality, 1–2, 205
Nobel Prize, 1, 12, 225, 227, 327, 333
path integrals, 216
Schoonschip computer program/results, 206–207, 210
T.-D. Lee and, 206
't Hoof's model and, 220–221, 222, 226, 258, 286, 307, 340
verifying/computing 't Hooft's work, 220–221
weak force/Yang-Mills theories, 205–206, 207–209, 210, 211
W's quadrupole moment, 206–207
Von Neumann, 100

W boson
about/properties, 108, 284
beta decay/weak force and, 103 (fig.), 108, 109 (fig.), 110, 111
electroweak force and, 108, 109 (fig.), 112–114, 236
mass and, 142, 144
mass/electroweak theory, 108, 109 (fig.), 112–114
photon role similarity, 103 (fig.), 108, 109 (fig.), 110, 111, 236
quadrupole moment, 206–207
renormalization and, 207, 208
weak force and, 288 (fig.), 289, 337
Waldegrave, William, 328–329, 330, 341
Walker, Alan, 360
Waller, Ivar, 293, 297, 301
Ward, John Clive "J.C."
after unifying forces work, 122
Australia and, 95, 99
awards, 94
background, 93, 94–95
British H-bomb and, 93, 94, 99–100, 311
colleges/work moves, 95–97, 98–99, 100
congratulatory telegram to Weinberg, 311
description/personality, 93–95, 99, 100, 160, 161, 186, 311
doctoral thesis/examination, 96–97
first published paper, 95
Kemmer and, 96
Nobel Prize omission, 12, 94, 105–106, 160, 197, 227, 294, 308, 309–311
parity violation, 100–101
Pryce and, 95, 96–97, 99
renormalization of weak interactions, 224
stature, 93
Ward's Identities, 97–98, 113
See also Salam and Ward/electroweak theory
Water phases
conditions for, 129–130
ice/change to liquid, 130
movement of molecules, 130
snowflake and symmetry/hidden symmetry, 130, 131 (fig.), 132
temperature/pressure effects, 129–130
Watson, Jim, 167, 172
Wave, The (Renoir), 129
Weak force
description, 6, 7 (fig.), 91
Fermi's model, 101–102, 103 (fig.)
massive *W* boson and, 113–114
parity violation and, 91, 100–101, 104, 289, 290
radioactivity and, 91, 101
solar fuel example and, 101
SPTVA classification and, 103, 104–105, 106
vector bosons possibility, 107–108
See also Beta decay; 't Hooft, Gerard
Weinberg, Erick, 268, 270, 271, 272, 274
Weinberg, Steven
1967 paper/citings, 297
background, 112

credit given to others, 191, 192, 196
electron acquiring mass, 179
electroweak theory, 180, 191–192
Goldstone Bosons, 191–192
hadronic physics theory, 189, 190–192
hidden symmetry/Goldstone theorem, 140–142, 212
Kibble's work and, 191, 196
mass-generating mechanism, 191
Nobel Prize, 12, 112, 185, 188, 190, 191, 225, 227, 283, 286, 294, 309, 313, 319, 333, 364
"Nobelitis," 292–293
original manuscript, 196, 198–199 (fig.)
path integrals, 215–216
pions, 188–189
quark model, 253, 256
renormalization of electroweak model, 192, 193–195, 196, 215, 222, 224
response to model, 193, 196–197, 203–204, 226
Solvay Congress presentation/reaction (1967), 192–197, 302
strong force theory work (1967), 188–189, 190–191
Tokyo conference, 291
weak force/hidden symmetry, 142
"Weinberg angle," 118, 286, 289, 291, 293, 308
"Weinberg-Salam" model, 197, 227, 283, 286, 287, 289, 290, 291, 297, 298, 304–305, 307, 308
Wells, H.G., 285
Wenninger, Horst, 328
Wess, Julius, 260
Wheeler, John, 95–96, 215–216
Wick, Gian Carlo, 237
Wilczek, Frank
description/personality, 267–268
Gross and, 264, 268, 333
Nobel Prize, 257, 265, 273
QCD/strong interaction, 257, 265, 267, 269, 270–272, 273–276, 283
Wiles, Andrew, 205, 226
Wilkins, Maurice, 167, 172
Witten, Ed, 73
Wolf Prize (2004), 152–153
World Wide Web beginnings, 326, 350

Yang, Chen-Ning "Frank"
after Pauli's condemnation, 88–89
background, 80, 85
gauge invariance, 81, 85
Nobel Prize/work, 10, 91
Schwinger's talk at Pocono Mountains meeting, 53, 80–81
seminar/Pauli's interruption, 87, 88, 112, 157, 189
Yang-Mills theories
beta function sign, 259–261
electromagnetic, weak, strong forces, 206
"gauge bosons," 87
gauge invariance, 86, 114, 147, 186
gauge theory of strong force, 89
group theory classification, 118
mass of vector boson, 148
publishing their work, 89
QED modifications, 86–87
quarks, 89
renormalization, 114, 216–218
Shaw's work and, 79–80, 87, 89, 90, 106, 187, 229
significance of, 77, 106
unitary property and, 216–217
weak force, 91
See also QCD (quantum chromodynamics)
Yukawa, Hideki
pions, 31, 67
strong force, 31, 67

Z boson
CERN experiments, 285–286
electroweak theory and, 284–286

group theory classification, 118
mass generation and, 142, 144
"neutral current," 284
weak force and, 288 (fig.), 289, 337
Zee, Tony, 259, 260, 269
Zichichi, Antonino, 293–294
Z^0 boson. *See Z* boson
Zumino, Bruno, 138, 221
Zweig, George
Feynman and, 238
publishing quark paper, 238–239
quarks, 233, 234, 235, 237–239, 253